Volterra Integral
and Differential Equations
Second Edition

This is volume 202 in
MATHEMATICS IN SCIENCE AND ENGINEERING
Edited by C.K. Chui, *Stanford University*

A list of recent titles in this series appears at the end of this volume.

Volterra Integral and Differential Equations

Second Edition

T.A. Burton
DEPARTMENT OF MATHEMATICS
SOUTHERN ILLINOIS UNIVERSITY
CARBONDALE, ILLINOIS
USA

2005
ELSEVIER
Amsterdam – Boston – Heidelberg – London – New York – Oxford
Paris – San Diego – San Francisco – Singapore – Sydney – Tokyo

ELSEVIER B.V.   ELSEVIER Inc.   ELSEVIER Ltd   ELSEVIER Ltd
Radarweg 29   525 B Street, Suite 1900   The Boulevard, Langford Lane 84 Theobalds Road
P.O. Box 211, 1000 AE Amsterdam San Diego, CA 92101-4495 Kidlington, Oxford OX5 1GB London WC1X 8
The Netherlands   USA   UK   UK

First edition 1983
Second Edition 2005

Library of Congress Cataloging in Publication Data
A catalog record is available from the Library of Congress.

British Library Cataloguing in Publication Data
A catalogue record is available from the British Library.

ISBN: 0-444-51786-3
ISSN (Series): 0076-5392
ISBN First Edition (Volume 167): 0-12-147380-5

♾ The paper used in this publication meets the requirements of ANSI/NISO Z39.48-1992 (Permanence of Paper).

Transferred to digital printing 2007.

Dedicated to
Professor Nikolai N. Krasovskii

Preface

This book provides an introduction to the structure and stability of solutions of Volterra integral and integro-differential equations. It is primarily an exposition of Liapunov's direct method. Chapter 0 gives a detailed account of the subjects treated.

To most seasoned investigators in the theory of Volterra equations, the study centers in large measure on operator theory, measure and integration, and general functional analysis. This book, however, is aimed at a different audience. There are today hundreds of mathematicians, physicists, engineers, and other scientists who are well versed in stability theory of ordinary differential equations on the real line using elementary differentiation and Riemann integration. The purpose of this book is to enable such investigators to parlay their existing expertise into a knowledge of theory and application of Volterra equations and to introduce them to the great range of physical applications of the subject.

Stability theory of Volterra equations is an area in which there is great activity among a moderate number of investigators. Basic knowledge is advancing rapidly, and it appears that this area will be an excellent field of research for some time to come. There are elementary theorems on Liapunov's direct method waiting to be proved; really usable results concerning the resolvent in nonconvolution cases are scarce; much remains to be done concerning the existence of periodic solutions; good Liapunov functionals have abounded for 10 years and await development of general theory to permit really effective applications; and there is a great need for careful analysis of specific simple Volterra equations as a guide to the development of the general theory.

I am indebted to many for assistance with the book: to the editors at Academic Press for their interest; to Professor Ronald Grimmer for reading Chapters 1 and 2; to the graduate students who took formal courses from Chapters 1–6 and offered suggestions and corrections; to Professor John Haddock for reading Chapters 3–8; to Professor L. Hatvani for reading Chapters 5 and 6; to Mr. M. Islam for carefully working through Chapters

3 and 5; to Professor Wadi Mahfoud for reading Chapters 1–6; to my wife Freddä, for drawing the figures; and to Shelley Castellano for typing the manuscript. A special thanks is due Professor Qichang Huang for reading and discussing the entire manuscript.

Preface to the second edition

In the twenty-one years since the book was published the subject has experienced significant growth and remains central in many applications. This edition corrects misprints, adds a great many references, and contains several advances in the theory. Over the life of the book many readers have reacted positively to the introductory material and so it was mainly left alone. It remains a gentle transition from ordinary differential equations to integral equations, integro-differential equations, and functional differential equations.

The resolvent is prominent in the new edition. Section 2.8 of the first edition is shifted to Chapter 7 and becomes the introduction. The rest of Chapter 7 is entirely new. It centers on Becker's resolvent and several important consequences by Becker and Krisztin, Hino and Murakami, and Zhang. It is also noted that there is now extensive theory of stability by fixed point methods and its focus is on a great many examples. That theory is too large to be included here, but references are given.

There is a new section at the end of Chapter 5 which introduces recent work by Appleby and Reynolds on sub-exponential decay of solutions.

Stability theory of functional differential equations has advanced in several ways, yielding solutions of many challenging problems which were introduced in the first edition. Section 8.3 has been expanded by thirty pages describing recent stability work of Hatvani, Kato, Makay, Wang, Zhang, and others. The theory of periodic solutions has also increased enormously. Section 8.6 now contains references to such work and a result by Zhang and the author showing periodicity as a consequence of uniform ultimate boundedness in the supremum norm. All of the aforementioned investigators helped in the presentation of the new material.

There is a very brief new Section 8.8 mentioning that there is now material concerning Liapunov theory for integral equations with a focus on classical examples. References are given.

This second edition contains about sixty-five pages of new material. All of Chapter 7 of the first edition was removed to make space for some of that new material. Some of Sections 8.3 , 8.4, and 8.7 was also removed. New work by at least thirty investigators has been mentioned with varying degrees of depth. About fifty new references have been added.

I wish to thank the editors of Elsevier, especially Charles Chui and Keith Jones, for their interest in publishing a second edition. Thanks also to Andy Deelen and the production staff at Elsevier. A very special thanks goes to Charles Gibson for typing the Latex copy. Finally, I want to thank the many who helped proof read the material including Leigh Becker, Geza Makay, Laszlo Hatvani, Nell Kravchenko, and Bo Zhang.

In looking over the work on stability theory of the last sixty years, the contributions of N. N. Krasovskii stand out as being as modern today as they were when he made them in the 1950's. This volume is dedicated to him.

T. A. Burton
Northwest Research Institute
Port Angeles, Washington
November 2004
taburton @olypen.com

Contents

Chapter 0

Introduction and Overview

0.1 Statement of Purpose

Although the theory of Volterra integral and integro-differential equations is old, well developed, and dense in the literature and in applications, we have been unable to find a systematic treatment of the theory's basic structure and stability properties. This book is a modest attempt to fill that void.

There are, of course, numerous treatments of the subject, but none seem to present a coherent set of results parallel to the standard treatments of stability theory given ordinary differential equations. Indeed, the student of the subject is hard put to find in the literature that the solution spaces of certain Volterra equations are identical to those for certain ordinary differential equations. Even the outstanding investigators have tended to deny such connections. For example, Miller (1971a, p. 9) states: "While it is true that all initial value problems for ordinary differential equations can be considered as Volterra integral equations, this fact is of limited importance." It is our view that this fact is of fundamental importance, and consequently, it is our goal to develop the theory of Volterra equations in such a manner that the investigator in the area of ordinary differential equations may parlay his expertise into a comprehension of Volterra equations. We hasten to add that there are indeed areas of Volterra equations that do not parallel the standard theory for ordinary differential equations. For a study of such areas, we heartily recommend the excellent treatments by Corduneanu (1991), Gripenberg *et al.* (1990), Lakshmikantham and Rao (1995), and Miller (1971a).

0.2   An Overview

It is assumed that the reader has some background in ordinary differential equations. Thus, Chapter 1 deals with numerous examples of Volterra equations reducible to ordinary differential equations. It also introduces the concept of initial functions and presents elementary boundedness results.

In Chapter 2 we point out that the structure of the solution space for the vector system

$$\mathbf{x}'(t) = A(t)\mathbf{x}(t) + \int_0^t C(t,s)\mathbf{x}(s)\,ds + \mathbf{f}(t) \tag{0.2.1}$$

is indistinguishable from that of the ordinary differential system

$$\mathbf{x}'(t) = B(t)\mathbf{x}(t) + \mathbf{g}(t)\,. \tag{0.2.2}$$

In fact, if $Z(t)$ is the $n \times n$ matrix satisfying

$$Z'(t) = A(t)Z(t) + \int_0^t C(t,s)Z(s)\,ds\,, \quad Z(0) = I\,, \tag{0.2.3}$$

and if $\mathbf{x}_p(t)$ is any solution of (0.2.1), then any solution $\mathbf{x}(t)$ of (0.2.1) on $[0, \infty)$ may be written as

$$\mathbf{x}(t) = Z(t)\big[\mathbf{x}(0) - \mathbf{x}_p(0)\big] + \mathbf{x}_p(t)\,. \tag{0.2.4}$$

Moreover, when A is a constant matrix and C is of convolution type, the solution of (0.2.1) on $[0, \infty)$ is expressed by the variation of parameters formula

$$\mathbf{x}(t) = Z(t)\mathbf{x}(0) + \int_0^t Z(t-s)\mathbf{f}(s)\,ds\,,$$

which is familiar to the college sophomore.

Chapter 2 also covers various types of stability, primarily using Liapunov's direct method. That material is presented with little background explanation, so substantial stability results are quickly obtained. Thus, by the end of Chapter 2 the reader has related Volterra equations to ordinary differential equations, has thoroughly examined the structure of the solution space, and has acquired tools for investigating boundedness and stability properties. The remainder of the book is devoted to consolidating these gains, bringing the reader to the frontiers in several areas, and suggesting certain research problems urgently in need of solution.

Chapter 3 outlines the basic existence, uniqueness, and continuation results for nonlinear ordinary differential equations. Those results and techniques are then extended to Volterra equations, making as few changes as are practical.

Chapter 4 is an in-depth account of some of the more interesting historical problems encountered in the development of Volterra equations. We trace biological growth problems from the simple Malthusian model, through the logistic equation, the predator-prey system of Lotka and Volterra, and on to Volterra's own formulation of integral equations regarding age distribution in populations. Feller's work with the renewal equation is briefly described. We then present many models of physical problems using integral equations. These problems range from electrical circuits to nuclear reactors.

Chapters 5–8 deal exclusively with Liapunov's direct method. Indeed, this book is mainly concerned with the study of stability properties of solutions of integral and integro-differential equations by means of Liapunov functionals or Liapunov-Razumikhin functions.

Chapter 5 deals with very specific Liapunov functionals yielding necessary and sufficient conditions for stability.

Chapter 6 is a basic introduction to stability theory for both ordinary differential equations and Volterra equations. Having shown the reader in Chapters 2 and 5 the power and versatility of Liapunov's direct method, we endeavor in Chapter 6 to promote a fundamental understanding of the subject. The basic theorems of ordinary differential equations are presented, proved, and discussed in terms of their history and their faults. Numerous examples of construction of Liapunov functions are given. We then show how Liapunov functionals for Volterra equations can be constructed in terms of extensions of the idea of a first integral. Theorems are proved, and examples are given concerning stability, uniform stability, asymptotic stability, uniform asymptotic stability, and perturbations.

Chapter 7 deals with the resolvent equation and its applications. These include Floquet theory, uniform asymptotic stability, and perturbations.

Chapter 8 is a brief treatment of general functional differential equations involving both bounded and unbounded delays. A main feature is the existence and stability theory synthesized and improved by Driver for functional differential equations with unbounded delay. It also contains a brief account of stability and limit sets for the equations

$$x' = F(t, x_t) \tag{0.2.5}$$

and

$$x' = f(x_t). \tag{0.2.6}$$

Much effort is devoted to certain recurring problems in earlier chapters. These may be briefly described as follows:

(i) If $V(t, \mathbf{x})$ is a scalar function whose derivative along solutions of

$$\mathbf{x}' = \mathbf{F}(t, \mathbf{x}) \tag{0.2.7}$$

is negative for $|\mathbf{x}|$ large, then it is frequently possible to conclude that solutions are bounded. Such results are of great importance in proving the existence of periodic solutions. We survey literature that tends to extend such results to Volterra and functional differential equations.

(ii) If $V(t, \mathbf{x})$ is a scalar function whose derivative along solutions of (0.2.7) is negative in certain sets, then knowledge about limit sets of solutions of (0.2.7) may be obtained, provided that $\mathbf{F}(t, \mathbf{x})$ is bounded for $\mathbf{x}$ bounded. This boundedness hypothesis is sometimes reasonable for (0.2.7), but it is ludicrous for a general functional differential equation. Yet, authors have required it for decades. We explore recent alternatives to asking $\mathbf{F}(t, \mathbf{x})$ bounded for $\mathbf{x}$ bounded in the corresponding treatment of functional differential equations.

Chapter 1

The General Problems

1.1 Introduction

We are concerned with the boundedness and stability properties of the integral equation

$$\mathbf{x}(t) = \mathbf{f}(t) + \int_0^t \mathbf{g}(t, s, \mathbf{x}(s))\, ds \qquad (1.1.1)$$

in which $\mathbf{x}$ is an n vector, $\mathbf{f} : [0, \infty) \to R^n$ is continuous, and $\mathbf{g} : \pi \times R^n \to R^n$ is continuous, where $\pi = \{(t, s) : 0 \leq s \leq t < \infty\}$.

It is unusual to ask that $\mathbf{g}$ be continuous. With considerable additional effort, one may obtain many of the results obtained here with weaker assumptions. For some such work, see Miller (1971a). The techniques we use to show boundedness will frequently require that (1.1.1) be differentiated to obtain an *integro-differential equation*

$$\mathbf{x}'(t) = \mathbf{f}'(t) + \mathbf{g}(t, t, \mathbf{x}(t)) + \int_0^t \mathbf{g}_1(t, s, \mathbf{x}(s))\, ds\,,$$

where $\mathbf{g}_1$ denotes $\partial \mathbf{g}/\partial t$ or, more generally

$$\mathbf{x}'(t) = \mathbf{h}(t, \mathbf{x}(t)) + \int_0^t \mathbf{F}(t, s, \mathbf{x}(s))\, ds\,. \qquad (1.1.2)$$

Notation. For a vector $\mathbf{x}$ and an $n \times n$ matrix A, the norm of $\mathbf{x}$ will usually be $|\mathbf{x}| = \max_i |x_i|$, whereas $|A|$ will mean $\sup_{|\mathbf{x}| \leq 1} |A\mathbf{x}|$.

5

Convention. It will greatly simplify notation if it is understood that a function written without its argument means that the function is evaluated at t. Thus (1.1.2) is

$$\mathbf{x}' = \mathbf{h}(t, \mathbf{x}) + \int_0^t \mathbf{F}(t, s, \mathbf{x}(s))\, ds\,.$$

We notice that if $\mathbf{f}$ is differentiable and $\mathbf{g}$ is independent of t, in (1.1.1), then differentiation yields an *ordinary differential equation*

$$\mathbf{x}'(t) = \mathbf{G}(t, \mathbf{x}(t))\,. \tag{1.1.3}$$

The process of going from (1.1.1) to (1.1.3) is easily reversed, as we simply write

$$\mathbf{x}(t) = \mathbf{x}(t_0) + \int_{t_0}^t \mathbf{G}(s, \mathbf{x}(s))\, ds\,.$$

To pass from (1.1.2) to (1.1.1), integrate (1.1.2) and then change the order of integration.

It is assumed that the reader has some familiarity with (1.1.3). Our procedure will generally be to state, but usually not prove, the standard result for (1.1.3) and then develop the parallel result for (1.1.1) or (1.1.2).

While investigating (1.1.1) we shall occasionally be led to examine

$$\mathbf{x}' = \mathbf{h}(t, \mathbf{x}) + \int_{t-T}^t \mathbf{F}(t, s, \mathbf{x}(s))\, ds \tag{1.1.4}$$

and

$$\mathbf{x}(t) = \mathbf{f}(t) + \int_{-\infty}^t \mathbf{g}(t, s, \mathbf{x}(s))\, ds\,. \tag{1.1.5}$$

It will turn out that results proved for (1.1.4) may be applied to the general functional differential equation with bounded delay

$$\mathbf{x}'(t) = \mathbf{H}(t, \mathbf{x}_t)\,, \tag{1.1.6}$$

where $\mathbf{x}_t$ is that segment of $\mathbf{x}(s)$ on the interval $t - h \le s \le t$ shifted back to $[-h, 0]$.

In the same way, we shall frequently see that results for (1.1.2) and (1.1.5) apply to a general functional differential equation

$$\mathbf{x}'(t) = \mathbf{K}(t, \mathbf{x}(s);\ \alpha \le s \le t)\,, \tag{1.1.7}$$

where $\alpha = -\infty$ is allowed, including

$$\mathbf{x}'(t) = \mathbf{L}\big(t, \mathbf{x}(t), \mathbf{x}(t - r(t))\big)\,, \tag{1.1.8}$$

with $r(t) \ge 0$.

One may note that Eqs. (1.1.1)–(1.1.3) are given in their order of generality.

1.2 Relations between Differential and Integral Equations

Most ordinary differential equations can be expressed as integral equations, but the reverse is not true. A given nth-order equation

$$x^{(n)}(t) = f(t, x, x', \ldots, x^{(n-1)})$$

may be expressed as a system of n first-order equations and then formally integrated. For example, if $x'' = f(t, x, x')$, then write $x = x_1$ and $x' = x_1' = x_2$, so that $x'' = x_2' = f(t, x_1, x_2)$, and the system of two first-order equations

$$\begin{bmatrix} x_1 \\ x_2 \end{bmatrix}' = \begin{bmatrix} x_2 \\ f(t, x_1, x_2) \end{bmatrix}$$

results.

And, in general, if $\mathbf{x} \in R^n$, then

$$\mathbf{x}' = \mathbf{G}(t, \mathbf{x}), \quad \mathbf{x}(t_0) = \mathbf{x}_0, \tag{1.2.1}$$

is a system of n first-order equations with initial condition (called an *initial-value problem*), written as

$$\mathbf{x}(t) = \mathbf{x}_0 + \int_{t_0}^{t} \mathbf{G}(s, \mathbf{x}(s))\, ds, \tag{1.2.2}$$

a system of n integral equations.

Thus, it is trivial to express such differential equations as integral equations. It is mainly a matter of renaming variables.

It may, however, be a surprise to find that when n is a positive integer, $f \in C^{n+1}$ on $[t_0, T)$, and g continuous, then

$$x(t) = f(t) + \int_{t_0}^{t} (t - s)^n g(s, x(s))\, ds \tag{1.2.3}$$

represents an $(n + 1)$st-order differential equation.

For example, when $n = 2$, we have

$$x'(t) = f'(t) + \int_{t_0}^t 2(t-s)g(s, x(s))\, ds\,,$$

$$x''(t) = f''(t) + \int_{t_0}^t 2g(s, x(s))\, ds\,,$$

and, finally

$$x'''(t) = f'''(t) + 2g(t, x(t))\,,$$

a third-order differential equation.

Note that $x(t_0) = f(t_0)$, $x'(t_0) = f'(t_0)$, and $x''(t_0) = f''(t_0)$, so (1.2.3) actually represents an initial-value problem and, if g is locally Lipschitz in x, we would expect a unique solution.

For a general positive integer n, we see that (1.2.3) represents an initial-value problem of order $n + 1$. Before we discuss the reverse process, let us consider a simple example in some detail. We emphasize that the form of (1.2.3) is not the only one possible for the reduction.

Example 1.2.1. Consider the scalar equation

$$x(t) = 1 + \int_0^t [-4 + e^{-(t-s)}]x(x)\, ds\,. \tag{a}$$

Differentiation yields the integro-differential equation

$$x' = -3x - \int_0^t e^{-(t-s)}x(s)\, ds\,. \tag{b}$$

Now multiply by e^t and differentiate to obtain

$$x'' + 4x' + 4x = 0 \tag{c}$$

whose general solution is

$$x(t) = c_1 e^{-2t} + c_2 t e^{-2t}\,. \tag{d}$$

Thus, (a) gives rise to (c) with two linearly independent solutions. In (b) we have $x'(0) = -3x(0)$, which, when combined with (d), yields

$$x'(t) = -2c_1 e^{-2t} + c_2 e^{-2t} - 2c_2 t e^{-2t}\,,$$
$$x'(0) = -2c_1 + c_2\,,$$
$$-3x(0) = -3c_1\,,$$

hence,

$$-2c_1 + c_2 = -3c_1$$

or

$$c_2 = -c_1 \,.$$

Thus

$$x(t) = c_1 e^{-2t} - c_1 t e^{-2t} \qquad (e)$$

is the solution of (b), and, as c_1 is arbitrary, (b) has one linearly independent solution. Finally, in (a) we have $x(0) = 1$, which, when applied to (e), yields

$$x(t) = e^{-2t} - te^{-2t} \qquad (f)$$

as the unique solution of (a).

We consider now the inverse problem for linear equations. It is worthwhile to consider $n = 2$ separately.

Let $a(t)$, $b(t)$, and $f(t)$ be continuous on an interval $[0, T)$, and consider

$$x'' + a(t)x' + b(t)x = f(t) \,, \quad x(0) = x_0 \,, \quad x'(0) = x_1 \,. \qquad (1.2.4)$$

A Liouville transformation will transform (1.2.4) to

$$u'' = -c(t)u + h(t) \,, \quad u(0) = u_0 \,, \quad u'(0) = u_1 \,, \qquad (1.2.5)$$

for $c(t)$ and $h(t)$ continuous. Integrate (1.2.5) from 0 to $t > 0$ twice obtaining successively

$$u'(t) = u_1 - \int_0^t c(s)u(s)\,ds + \int_0^t h(s)\,ds$$

and

$$u(t) = u_0 + u_1 t - \int_0^t \int_0^v c(s)u(s)\,ds\,dv + \int_0^t \int_0^v h(s)\,ds\,dv \,.$$

The integral

$$J = \int_0^t \int_0^v c(s)u(s)\,ds\,dv$$

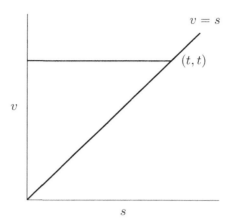

Figure 1.1: Region of integral $\int_0^t \int_0^v c(s)u(s)\,ds\,dv$.

is taken over the triangle in Fig. 1.1. We interchange the order of integration
and obtain

$$J = \int_0^t \int_s^t c(s)u(s)\,dv\,ds = \int_0^t (t-s)c(s)u(s)\,ds\,,$$

so that if we set

$$H(t) = \int_0^t \int_0^v h(s)\,ds\,dv\,,$$

then (1.2.5) becomes

$$u(t) = u_0 + u_1 t + H(t) - \int_0^t (t-s)c(s)u(s)\,ds\,. \tag{1.2.6}$$

Incidentally, the same process allows us to pass from an integro-
differential equation

$$\mathbf{x}'(t) = \mathbf{h}(t, \mathbf{x}(t)) + \int_0^t \mathbf{F}(t, s, \mathbf{x}(s))\,ds \tag{1.1.2}$$

to (1.1.1) by integrating from 0 to t and then interchanging the order of
integration and obtaining

$$\mathbf{x}(t) = \mathbf{x}(0) + \int_0^t \left[\mathbf{h}(s, \mathbf{x}(s)) + \int_s^t \mathbf{F}(u, s, \mathbf{x}(s))\,du \right] ds \tag{1.2.7}$$

under appropriate continuity conditions.

We now consider the problem of transforming a general scalar linear differential equation of order n to an integral equation.

Let f and $a_1(t), \ldots, a_n(t)$ be continuous on $[0, T)$ in

$$x^{(n)} + a_1(t)x^{(n-1)} + \cdots + a_n(t)x = f(t), \qquad (1.2.8)$$

with $x(0), x'(0), \ldots, x^{(n-1)}(0)$ given initial conditions, and set $x^{(n)}(t) = z(t)$. Then

$$x^{(n-1)}(t) = x^{(n-1)}(0) + \int_0^t z(s)\, ds\,,$$

$$x^{(n-2)}(t) = x^{(n-2)}(0) + tx^{(n-1)}(0) + \int_0^t (t-s)z(s)\, ds\,,$$

$$x^{(n-3)}(t) = x^{(n-3)}(0) + tx^{(n-2)}(0) + \frac{t^2}{2!}\,x^{(n-1)}(0)$$

$$+ \int_0^t \frac{(t-s)^2}{2}\, z(s)\, ds\,,$$

$$\vdots$$

$$\mathbf{x}(t) = x(0) + tx'(0) + \cdots + \frac{t^{n-1}}{(n-1)!}\,x^{(n-1)}(0)$$

$$+ \int_0^t \frac{(t-s)^{n-1}}{(n-1)!}\, z(s)\, ds\,.$$

If we replace these values of x and its derivatives in (1.2.8) we have a scalar integral equation for $z(t)$.

All integral equations are not reducible to differential equations; for example, let $f(t)$ be everywhere continuous and nowhere differentiable, and consider

$$x(t) = f(t) + \int_0^t [\sin(t-s)]x(s)\, ds\,. \qquad (1.2.9)$$

A different problem occurs in converting

$$x(t) = 1 + \int_0^t e^{(t-s)} \cos(t-s)x(s)\, ds \qquad (1.2.10)$$

to a differential equation.

Finally, notice that when n is a positive integer, differentiation of

$$x(t) = 1 + \int_0^t (t-s+1)^{-n}x(s)\, ds \qquad (1.2.11)$$

simply increases the exponent of $(t-s+1)^{-n}$.

Many integral equations of interest are not reducible to ordinary differential equations because the integral builds in a "memory" not present in differential equations.

The next example is familiar in the study of ordinary differential equations. Let $\mathbf{x}$ be an n vector, A an $n \times n$ constant matrix, and

$$\mathbf{x}' = A\mathbf{x} + \mathbf{f}(t, \mathbf{x}), \qquad (1.2.12)$$

where $\mathbf{f} : [0, \infty) \times R^n \to R^n$ is continuous. For a given solution $\mathbf{x}(t)$, then $\mathbf{f}(t, x(t)) \equiv \mathbf{g}(t)$ is a forcing function, and by the variation of parameters formula, we have

$$\mathbf{x}(t) = e^{At}\mathbf{x}_0 + \int_0^t e^{A(t-s)}\mathbf{f}(s, \mathbf{x}(s))\, ds, \qquad (1.2.13)$$

which is an integral equation. When the characteristic roots of A have negative real parts and $\mathbf{f}$ is small in some sense, Gronwall's inequality yields boundedness results.

Our final example concerns a control problem, often called the Problem of Lurie, concerning the $(n+1)$-dimensional system of differential equations

$$\begin{aligned}\mathbf{x}' &= A\mathbf{x} + \mathbf{B}f(y), \\ y' &= \mathbf{C}^T\mathbf{x} - rf(y)\end{aligned} \qquad (1.2.14)$$

in which A is an $n \times n$ constant matrix, $\mathbf{B}$ and $\mathbf{C}$ constant vectors, r a positive constant, and f a continuous scalar function with $yf(y) > 0$ if $y \neq 0$.

Following the work of (1.2.13) in the first equation we obtain

$$\mathbf{x}(t) = e^{At}\mathbf{x}_0 + \int_0^t e^{A(t-s)}\mathbf{B}f(y(s))\, ds,$$

which we substitute into the second equation, yielding the scalar integro-differential equation

$$y' = \mathbf{C}^T\left[e^{At}\mathbf{x}_0 + \int_0^t e^{A(t-s)}\mathbf{B}f(y(s))\, ds\right] - rf(y). \qquad (1.2.15)$$

In conclusion, although a first-order differential equation may be considered elementary by some, everyone respects a scalar integral equation.

1.3   A Glance at Initial Conditions and Existence

The standard theory of ordinary differential equations shows that if $D \subset R^n$ and $\mathbf{G}$ is continuous on $(a, b) \times D$, with $t_0 \in (a, b)$ and $\mathbf{x}_0 \in D$, then

$$\mathbf{x}' = \mathbf{G}(t, x), \quad \mathbf{x}(t_0) = \mathbf{x}_0, \tag{1.3.1}$$

has a solution. If $\mathbf{G}$ is locally Lipschitz in $\mathbf{x}$, the solution is unique. If $\mathbf{G} : [t_0, \infty) \times R^n \to R^n$ is continuous, then a solution $\mathbf{x}(t)$ on $[t_0, T)$ can be extended to $[t_0, \infty)$ unless there is a $\bar{T}$ with $\lim_{t \to \bar{T}-} |\mathbf{x}(t)| = +\infty$.

The situation for integral equations is very similar but has significant differences.

When we consider

$$\mathbf{x}(t) = \mathbf{f}(t) + \int_0^t \mathbf{g}(t, s, \mathbf{x}(s)) \, ds, \quad t \geq 0, \tag{1.3.2}$$

it is to be understood that $\mathbf{x}(0) = \mathbf{f}(0)$ and we are looking for a continuous solution $\mathbf{x}(t)$ for $t \geq 0$.

However, it may happen that $\mathbf{x}(t)$ is specified to be a certain *initial function* on an *initial interval*, say,

$$\mathbf{x}(t) = \boldsymbol{\phi}(t) \quad \text{for} \quad 0 \leq t \leq t_0 \tag{1.3.3}$$

(see Fig. 1.2). We are then looking for a solution of

$$\mathbf{x}(t) = \mathbf{f}(t) + \int_0^{t_0} \mathbf{g}(t, s, \boldsymbol{\phi}(s)) \, ds + \int_{t_0}^t \mathbf{g}(t, s, \mathbf{x}(s)) \, ds, \quad t \geq t_0. \tag{1.3.4}$$

For example, (1.3.2) may describe the population density $\mathbf{x}(t)$. A given population is observed over a time period $[0, t_0]$ and is given by $\boldsymbol{\phi}(t)$. The subsequent behavior of that density may depend greatly on $\boldsymbol{\phi}(t)$.

A change of variable will reduce the problem (1.3.4) back to one of form (1.3.2).

Let $\mathbf{x}(t + t_0) = \mathbf{y}(t)$, so that in (1.3.4) we have

$$\mathbf{x}(t + t_0) = \mathbf{f}(t + t_0) + \int_0^{t_0} \mathbf{g}(t + t_0, s, \boldsymbol{\phi}(s)) \, ds$$

$$+ \int_{t_0}^{t_0+t} \mathbf{g}(t_0 + t, s, \mathbf{x}(s)) \, ds$$

$$= \mathbf{f}(t + t_0) + \int_0^{t_0} \mathbf{g}(t + t_0, s, \boldsymbol{\phi}(s)) \, ds$$

$$+ \int_0^t \mathbf{g}(t_0 + t, u + t_0, \mathbf{x}(u + t_0)) \, du$$

or

$$\mathbf{y}(t) = \mathbf{h}(t) + \int_0^t \mathbf{g}(t_0 + t, u + t_0, \mathbf{y}(u)) \, du \qquad (1.3.5)$$

where

$$\mathbf{h}(t) = \mathbf{f}(t + t_0) + \int_0^{t_0} \mathbf{g}(t + t_0, s, \phi(s)) \, ds$$

and we want the solution for $t \geq 0$. Obviously, we need $h(0) = \phi(t_0)$.

Thus, the initial function on $[0, t_0]$ is absorbed into the forcing function, and hence, it always suffices to consider (1.3.2) with the simple condition $\mathbf{x}(0) = \mathbf{f}(0)$.

We now briefly outline an overview of the most basic existence theory. Details will be supplied in later sections.

Existence. *If $\mathbf{f}$ is continuous on $[0, \infty)$ and if $\mathbf{g}(t, s, \mathbf{x})$ is continuous for $0 \leq s \leq t < \infty$ and all $\mathbf{x}$ in R^n, then there is a number $T > 0$ and a continuous solution $\mathbf{x}(t)$ of (1.3.2) on $[0, T]$.*

Continuation. *Let the conditions in the preceding paragraph hold and let $\mathbf{x}(t)$ be a continuous, bounded solution on a right open interval $[0, S)$. Then $\mathbf{x}(t)$ can be extended to $[0, T_1]$, where $T_1 > S$.*

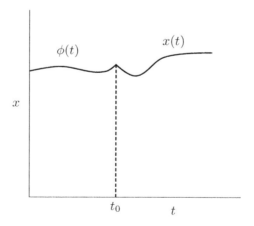

Figure 1.2: Initial function $\phi(t)$.

Existence and Uniqueness. *If a, b, and L are positive numbers, $\mathbf{f}$ is continuous on $[0, a]$, $\mathbf{g}$ is continuous on*

$$U = \left\{ (t, s, \mathbf{x}) \mid 0 \le s \le t \le a \text{ and } |\mathbf{x} - \mathbf{f}(t)| \le b \right\},$$

and $\mathbf{g}$ is Lipschitz in $\mathbf{x}$ with

$$\left| \mathbf{g}(t, s, \mathbf{x}) - \mathbf{g}(t, s, \mathbf{y}) \right| \le L|\mathbf{x} - \mathbf{y}|$$

whenever $(t, s, \mathbf{x})$ and $(t, s, \mathbf{y}) \in U$, then there is a unique solution of (1.3.2) on an interval $[0, T]$, where $T = \min[a, b/M]$ and $M = \max_U |\mathbf{g}(t, s, \mathbf{x})|$.

1.4 Building the Intuition

This exposition is directed primarily at those with a fair background in stability theory of ordinary differential equations and with an interest in stability and boundedness of Volterra equations. Accordingly, the goal of this section is to suggest connections between differential equations and Volterra equations and to give some insight into properties of Volterra equations that will have significant bearing on boundedness.

Thus, no effort is made in this section for great generality, nor is there great concern for certain fine detail. We offer instead a brief sketch of techniques, results, and relations which will be developed later. It is hoped that an intuition will develop that will assist the investigator in guessing just what is true and possible ways to prove it.

Exercise 1.4.1. Let A and B be real constants in

$$x' = Ax + \int_0^t Be^{-(t-s)} x(s) \, ds. \tag{a}$$

Integrate (a) and obtain an equation of the form

$$x(t) = x(0) + \int_0^t C(t - s) x(s) \, ds. \tag{b}$$

Differentiate (a) and obtain

$$x'' + \alpha x' + \beta x = 0. \tag{c}$$

Find the general solution of (c) and, keeping $A < 0$, deduce the range of the constant B for which all solutions of (a) will be bounded. Next, let $A \ge 0$ and determine the range of B for which all solutions are bounded.

Recall that the general solution of the linear scalar equation

$$x' = A(t)x + B(t)$$

is given by the variation of parameters formula

$$x(t) = \exp\left(\int_{t_0}^t A(s)\,ds\right)x(t_0) + \int_{t_0}^t \exp\left(\int_s^t A(u)\,du\right)B(s)\,ds\,,$$

and if A is constant and $t_0 = 0$, then

$$x(t) = Z(t)x(0) + \int_0^t Z(t-s)B(s)\,ds\,,$$

where $Z(t)$ is the solution of

$$x' = Ax\,, \quad x(0) = 1\,.$$

Exercise 1.4.2. Verify that if $Z(t)$ is a solution of the scalar equation

$$x' = hx + \int_0^t F(t-s)x(s)\,ds\,, \quad x(0) = 1\,,$$

then

$$x(t) = Z(t)x(0) + \int_0^t Z(t-s)r(s)\,ds$$

is a solution of

$$x'(t) = hx(t) + \int_0^t F(t-s)x(s)\,ds + r(t)$$

under suitable continuity assumptions. (This exercise is not trivial.)

Exercise 1.4.3. Formulate the same result for the integral equation

$$x(t) = f(t) + x(0) + \int_0^t C(t-s)x(s)\,ds\,,$$

assuming that you have a solution of

$$x(t) = x(0) + \int_0^t C(t-s)x(s)\,ds\,.$$

Work out the details. (This exercise, too, is not trivial.)

The first result is in the way of a generalization of Exercise 1.4.1. Consider the scalar equation

$$x' = -Ax + \int_0^t C(t,s)x(s)\,ds + f(t)\,, \tag{1.4.1}$$

with A constant, C continuous for $0 \le s \le t < \infty$, and f continuous for $0 \le t < \infty$.

Proposition 1.4.1. *Let $A > 0$, $|f(t)| \le M$, and $\int_0^t |C(t,s)|\,ds \le mA$ for $0 \le t < \infty$, where $0 < m < 1$ and $M > 0$. Then all solutions of (1.4.1) are bounded.*

Proof. Find $R > 0$ with $-AR + mAR + M < 0$ and $|x(0)| < R$. If $x(t)$ is an unbounded solution, then there is a first $t_1 > 0$ with $|x(t_1)| = R$. Now, if $x(t_1) = R$, then $x'(t_1) \ge 0$. But

$$x'(t_1) = -Ax(t_1) + \int_0^{t_1} C(t_1,s)x(s)\,ds + f(t_1)$$

$$\le -AR + \max_{0 \le s \le t_1} |x(s)| \int_0^{t_1} |C(t_1,s)|\,ds + M$$

$$\le -AR + RmA + M < 0\,,$$

a contradiction. Suppose, then, that $x(t_1) = -R$. Then $x'(t_1) \le 0$. But

$$x'(t_1) \ge AR - \max_{0 \le s \le t_1} |x(s)| \int_0^{t_1} |C(t_1,s)|\,ds - M$$

$$= AR - RmA - M > 0\,,$$

a contradiction. This completes the proof.

Generally, in (1.4.1) as A becomes larger and C becomes smaller, the stability is strengthened. To see this, suppose that, for fixed A and $C(t-s)$, all solutions of

$$x'(t) = -Ax + \int_0^t C(t-s)x(s)\,ds$$

are bounded. If $x(t)$ is a solution and r is a positive constant, then $x = e^{rt}y$ defines a function y tending to zero as fast as e^{-rt}. But

$$x' = re^{rt}y + e^{rt}y'$$

$$= -Ae^{rt}y + \int_0^t C(t-s)e^{rs}y(s)\,ds\,,$$

so that

$$e^{rt}y' = -(A+r)e^{rt}y + \int_0^t C(t-s)e^{rs}y(s)\,ds$$

and

$$y' = -(A+r)y + \int_0^t C(t-s)e^{-r(t-s)}y(s)\,ds\,.$$

Thus, $A \to A+r$ and $C(u) \to C(u)e^{-ru}$.

Exercise 1.4.4. Verify that

$$x(t) = 1 - \int_0^t \left[5 + \frac{3}{(t-s+1)} \right] x(s)\,ds \; + \sin t\,,$$

when differentiated, will satisfy the conditions of Proposition 1.4.1.

Proposition 1.4.2. *Let $\mathbf{f}$ be a continuous vector function on $[0,\infty)$ with $|\mathbf{f}(t)| \le M$, and suppose that $D(t,s)$ is an $n \times n$ matrix continuous for $0 \le s \le t < \infty$. If there exists $m < 1$ with $\int_0^t |D(t,s)|\,ds \le m$ for $0 \le t < \infty$, then all solutions of*

$$\mathbf{x}(t) = \mathbf{f}(t) + \int_0^t D(t,s)\mathbf{x}(s)\,ds \qquad\qquad (1.4.2)$$

are bounded.

Proof. Find $R > 0$ with $M + mR < R$. If $\mathbf{x}(t)$ is an unbounded solution, there is a $t_1 > 0$ with

$$|\mathbf{x}(t)| < R \;\text{ on }\; [0,t_1) \quad\text{and}\quad |\mathbf{x}(t_1)| = R\,.$$

Thus,

$$R = |\mathbf{x}(t_1)| = \left| \mathbf{f}(t_1) + \int_0^{t_1} D(t_1,s)\mathbf{x}(s)\,ds \right|$$

$$\le |\mathbf{f}(t_1)| + \int_0^{t_1} |D(t_1,s)|\,|\mathbf{x}(s)|\,ds$$

$$\le M + mR < R\,,$$

a contradiction.

Exercise 1.4.5. Find $k > 0$ for which the conditions of Proposition 1.4.2 hold in

$$x(t) = \sin t + \cos t + \int_0^t k\big[\sin(t-s)\big](t-s+1)^{-3}x(s)\,ds\,.$$

Both of our results are statements of boundedness. A standard technique from differential equations will yield results on solutions tending zero.

Proposition 1.4.3. *Let $f(t)$ and $C(t)$ be continuous scalar functions on $[0,\infty)$ and suppose there are positive constants A, B, and α with*

$$|f(t)| \le Ae^{-\alpha t} \quad and \quad |C(t)| \le Be^{-\alpha t}\,.$$

If $\alpha - B = \beta > 0$ and if $x(t)$ is any solution of

$$x(t) = f(t) + \int_0^t C(t-s)x(s)\,ds\,,$$

then

$$|x(t)| \le Ae^{-\beta t}.$$

Proof. We have

$$|x(t)| \le Ae^{-\alpha t} + \int_0^t Be^{-\alpha(t-s)}|x(s)|\,ds\,,$$

so that

$$e^{\alpha t}|x(t)| \le A + \int_0^t Be^{\alpha s}|x(s)|\,ds\,.$$

By Gronwall's inequality,

$$e^{\alpha t}|x(t)| \le A\exp\int_0^t B\,ds$$

or

$$|x(t)| \le A\exp\big[(B-\alpha)t\big] = A\exp(-\beta t)\,.$$

1.5   Reducible Equations

We have considered specific examples to illustrate certain properties of
Volterra equations. We also proved some specialized results on boundedness. But to conjecture provable results it is essential to have a large
collection of completely solved problems of very different types. It turns
out that such a collection is surprisingly easy to obtain.

We begin with a scalar equation of convolution type

$$x(t) = f(t) + \int_0^t C(t-s)x(s)\,ds \qquad (1.5.1)$$

in which $f^{(n)}$ and $C^{(n)}$ are continuous on the interval $[0, \infty)$.

Principle. *If $C(t)$ satisfies a linear nth-order homogeneous ordinary differential equation with constant coefficients, then (1.5.1) may be reduced
to an nth-order differential equation with constant coefficients.*

Thus, determination of stability properties of (1.5.1) rests on locating
the zeros of a polynomial. Although it is impossible to find the roots in the
general case, the Routh-Hurwitz criterion [see Gantmacher (1960, p. 194)]
allows us to decide if they all lie in the left half-plane.

Theorem 1.5.1. *All characteristic roots of a real polynomial*

$$\alpha^n + a_1\alpha^{n-1} + a_2\alpha^{n-2} + \cdots + a_n = 0 \qquad (1.5.2)$$

have negative real parts if and only if $D_k > 0$ for $k = 1, 2, \ldots, n$, where

$$D_1 = a_1\,,$$

$$D_2 = \begin{vmatrix} a_1 & a_3 \\ 1 & a_2 \end{vmatrix},$$

$$\vdots$$

$$D_k = \begin{vmatrix}
a_1 & a_3 & a_5 & \cdots & a_{2k-1} \\
1 & a_2 & a_4 & \cdots & a_{2k-2} \\
0 & a_1 & a_3 & \cdots & a_{2k-3} \\
0 & 1 & a_2 & \cdots & a_{2k-4} \\
0 & 0 & a_1 & \cdots & a_{2k-5} \\
 & 0 & 1 & \cdots & a_{2k-6} \\
 & & 0 & 0 & \\
\vdots & & \cdots & & \vdots \\
0 & & \cdots & & a_k
\end{vmatrix},$$

with $a_j = 0$ for $j > n$.

It follows readily that a_1 and a_2 must both be positive in order for

$$\alpha^2 + a_1\alpha + a_2 = 0 \tag{1.5.4}$$

to satisfy the criterion. As we shall presently require it for $n = 3$, we note that the characteristic roots of

$$\alpha^3 + a_1\alpha^2 + a_2\alpha + a_3 = 0 \tag{1.5.5}$$

will all have negative real parts just in case

$$D_1 = a_1 > 0\,, \quad D_2 = a_1 a_2 - a_3 > 0\,,$$

and

$$D_3 = \begin{vmatrix} a_1 & a_3 & 0 \\ 1 & a_2 & 0 \\ 0 & a_1 & a_3 \end{vmatrix} = a_1 a_2 a_3 - a_3^2 > 0\,.$$

As $D_3 > 0$ and $a_1 > 0$, we see that a_2 and a_3 must have the same sign. If both are negative, we may obtain a contradiction by dividing a_3 into D_3, obtaining $a_1 a_2 < a_3$.

We may now conclude that the characteristic roots of (1.5.5) all have negative real parts if and only if

$$a_i > 0 \quad \text{for } i = 1, 2, 3 \quad \text{and} \quad a_1 a_2 > a_3\,. \tag{1.5.6}$$

Exercise 1.5.1. Use the Routh-Hurwitz criterion to obtain conditions so that all characteristic roots of

$$\alpha^4 + a_1\alpha^3 + a_2\alpha^2 + a_3\alpha + a_4 = 0 \tag{1.5.7}$$

have negative real parts.

We now illustrate the principle with (1.5.1), in which we suppose that

$$C''(t) + C(t) = 0\,. \tag{1.5.8}$$

Differentiate (1.5.1) twice to obtain

$$x'(t) = f'(t) + C(0)x(t) + \int_0^t C'(t-s)x(s)\,ds$$

and

$$x''(t) = f''(t) + C(0)x'(t) + C'(0)x(t) + \int_0^t C''(t-s)x(s)\,ds\,.$$

If we add this last equation to (1.5.1) and apply (1.5.8) under the integral, we have

$$x''(t) + x(t) = f(t) + f''(t) + C(0)x'(t) + C'(0)x(t) \tag{1.5.9}$$

with the homogeneous part being

$$x'' - C(0)x' + (1 - C'(0))x = 0. \tag{1.5.10}$$

If x_1 and x_2 are linearly independent solutions of (1.5.10) and $x_p(t)$ is any solution of (1.5.9), then the solution of (1.5.1) on $[0, \infty)$ may be expressed as

$$x(t) = k_1 x_1(t) + k_2 x_2(t) + x_p(t), \tag{1.5.11}$$

with k_1 and k_2 uniquely determined, so that

$$x(0) = f(0) \quad \text{and} \quad x'(0) = f'(0) + C(0)x(0). \tag{1.5.12}$$

Exercise 1.5.2. Use the principle to find the solution of

$$x'(t) = -x(t) + \int_0^t b e^{-(t-s)} \cos(t - s)x(s)\,ds \tag{1.5.13}$$

with b constant. Determine b so that all solutions of (1.5.13) on $[0, \infty)$ will tend to zero exponentially. You will need to use (1.5.6).

Exercise 1.5.3. Use the principle and (1.5.6) to verify that all solutions of

$$x' = -2x + \int_0^t \sin(t - s)x(s)\,ds \tag{1.5.14}$$

on $[0, \infty)$ tend to zero exponentially.

Chapter 2

Linear Equations

2.1 Existence Theory

Consider the equation

$$\mathbf{x}(t) = \mathbf{f}(t) + \int_0^t B(t,s)\mathbf{x}(s)\,ds\,, \tag{2.1.1}$$

in which $\mathbf{f} : [0,\alpha) \to R^n$ is continuous and $B(t,s)$ is an $n \times n$ matrix of functions continuous for $0 \leq s \leq t < \alpha$ and $\alpha \leq \infty$.

The function $B(t,s)$ is frequently called the *kernel*. If $B(t,s)$ can be expressed as $B(t,s) = D(t-s)$, then (2.1.1) is said to be of *convolution* type.

Most writers ask less than continuity on B, but most of our work will require it as our techniques will frequently require reduction of (2.1.1) to an integro-differential equation. Thus, we will often need to also require that $\mathbf{f}$ have a derivative.

The integro-differential equations we consider have the form

$$\mathbf{x}'(t) = A(t)\mathbf{x}(t) + \int_0^t C(t,s)\mathbf{x}(s)\,ds + \mathbf{F}(t)\,, \tag{2.1.2}$$

in which $\mathbf{F} : [0,\alpha) \to R^n$ is continuous, $C(t,s)$ is an $n \times n$ matrix of functions continuous for $0 \leq s \leq t < \alpha$, and $A(t)$ is an $n \times n$ matrix of functions continuous on $[0,\alpha)$.

We now put (2.1.2) into the form of (2.1.1), so that an existence and uniqueness theorem will apply to both of them.

Equation (2.1.2) requires an initial function $\phi : [0,t_0] \to R^n$ with ϕ continuous and t_0 possibly zero. A solution of (2.1.2) is a continuous func-

23

tion $\mathbf{x}(t)$ on an interval $[t_0, T)$, such that $\mathbf{x}(t) = \phi(t)$ for $0 \le t \le t_0$. This yields

$$\mathbf{x}'(t) = A(t)\mathbf{x}(t) + \int_0^{t_0} C(t,s)\phi(s)\,ds + \mathbf{F}(t) + \int_{t_0}^t C(t,s)\mathbf{x}(s)\,ds\,.$$

As we saw before, a translation $\mathbf{y}(t) = \mathbf{x}(t + t_0)$ results in

$$\begin{aligned}
\mathbf{y}'(t) = \mathbf{x}'(t + t_0) &= A(t + t_0)\mathbf{y}(t) \\
&\quad + \int_0^{t_0} C(t_0 + t, s)\phi(s)\,ds + \mathbf{F}(t + t_0) \\
&\quad + \int_{t_0}^t C(t_0 + t, s)\mathbf{x}(s)\,ds \\
&= A(t + t_0)\mathbf{y}(t) + \int_0^t C(t_0 + t, s + t_0)\mathbf{y}(s)\,ds \\
&\quad + \int_0^{t_0} C(t_0 + t, s)\phi(s)\,ds + \mathbf{F}(t + t_0)\,,
\end{aligned}$$

which we write as

$$\mathbf{y}'(t) = \bar{A}(t)\mathbf{y}(t) + \int_0^t \bar{C}(t,s)\mathbf{y}(s)\,ds + \bar{\mathbf{F}}(t)\,,$$

again of the form (2.1.2). The initial function ϕ is absorbed into the forcing function, and the last equation then has the initial condition

$$\mathbf{y}(0) = \mathbf{x}(t_0) = \phi(t_0)\,,$$

so that an integration from 0 to t yields

$$\begin{aligned}
\mathbf{y}(t) = \phi(t_0) &+ \int_0^t \bar{A}(s)\mathbf{y}(s)\,ds + \int_0^t \bar{\mathbf{F}}(s)\,ds \\
&+ \int_0^t \int_0^u \bar{C}(u,s)\mathbf{y}(s)\,ds\,du\,. \qquad (2.1.3)
\end{aligned}$$

Interchanging the order of integration in the last term yields an equation of the form of (2.1.1). Thus, the existence and uniqueness theorem that follows applies also to (2.1.2) with a given continuous initial function.

   The uniqueness part of the proof of the result is facilitated by the following relation.

Lemma. Gronwall's Inequality $\quad$ *Let* $f, g : [0, \alpha] \to [0, \infty)$ *be continuous and let* c *be a nonnegative number. If*

$$f(t) \leq c + \int_0^t g(s) f(s) \, ds, \quad 0 \leq t < \alpha,$$

then

$$f(t) \leq c \exp \int_0^t g(s) \, ds, \quad 0 \leq t < \alpha.$$

Proof. Suppose first that $c > 0$. Divide by $c + \int_0^t g(s) f(s) \, ds$ and multiply by $g(t)$ to obtain

$$f(t) g(t) \Big/ \left[c + \int_0^t g(s) f(s) \, ds \right] \leq g(t).$$

An integration from 0 to t yields

$$\ln \left\{ \left[c + \int_0^t g(s) f(s) \, ds \right] \Big/ c \right\} \leq \int_0^t g(s) \, ds$$

or

$$f(t) \leq c + \int_0^t g(s) f(s) \, ds \leq c \exp \int_0^t g(s) \, ds.$$

If $c = 0$, take the limit as $c \to 0$ through positive values. This completes the proof.

Theorem 2.1.1. $\quad$ *Let* $0 < \alpha \leq \infty$ *and suppose that* $\mathbf{f} : [0, \alpha) \to R^n$ *is continuous and that* $B(t, s)$ *is an* $n \times n$ *matrix of functions continuous for* $0 \leq s \leq t < \alpha$. *If* $0 < T < \alpha$, *then there is one and only one solution* $\mathbf{x}(t)$ *of*

$$\mathbf{x}(t) = f(t) + \int_0^t B(t, s) \mathbf{x}(s) \, ds \tag{2.1.1}$$

on $[0, T]$.

Proof. Define a sequence of functions $\{\mathbf{x}_n(t)\}$ on $[0, T]$ by

$$\mathbf{x}_1(t) = \mathbf{f}(t)$$

$$\mathbf{x}_{n+1}(t) = \mathbf{f}(t) + \int_0^t B(t, s) \mathbf{x}_n(s) \, ds, \quad n \geq 1. \tag{2.1.4}$$

These are called *Picard's successive approximations.*

One may show by mathematical induction that each $\mathbf{x}_n(t)$ is defined on $[0, T]$ and is continuous. Let $M = \max_{0 \leq s \leq t \leq T} |B(t, s)|$ and $K = \max_{0 \leq t \leq T} |\mathbf{f}(t)|$ and consider the series

$$\mathbf{x}_1(t) + \sum_{n=1}^{\infty} \left(\mathbf{x}_{n+1}(t) - \mathbf{x}_n(t) \right) \qquad (2.1.5)$$

whose typical partial sum is $\mathbf{x}_n(t)$. We now show by induction that

$$\left| \mathbf{x}_{n+1}(t) - \mathbf{x}_n(t) \right| \leq \left[K(Mt)^n \right]/n! \,. \qquad (2.1.6)$$

It follows from (2.1.4) for $n = 1$ that

$$\left| \mathbf{x}_2(t) - \mathbf{x}_1(t) \right| = \left| \mathbf{f}(t) + \int_0^t B(t, s) \mathbf{f}(s) \, ds \ - \mathbf{f}(t) \right|$$

$$\leq \int_0^t |B(t, s) \mathbf{f}(s)| \, ds \ \leq MKt \,,$$

so that (2.1.6) is true for $n = 1$.

Assume

$$\left| \mathbf{x}_{k+1}(t) - \mathbf{x}_k(t) \right| \leq \left[K(Mt)^k \right]/k!$$

and consider

$$\left| \mathbf{x}_{k+2}(t) - \mathbf{x}_{k+1}(t) \right| = \left| \mathbf{f}(t) + \int_0^t B(t, s) \mathbf{x}_{k+1}(s) \, ds \ - \mathbf{f}(t) \right.$$

$$\left. - \int_0^t B(t, s) \mathbf{x}_k(s) \, ds \right|$$

$$\leq \int_0^t |B(t, s)| \left| \mathbf{x}_{k+1}(s) - \mathbf{x}_k(s) \right| \, ds$$

$$\leq M \int_0^t \left| \mathbf{x}_{k+1}(s) - \mathbf{x}_k(s) \right| \, ds$$

$$\leq \frac{MK}{k!} \int_0^t (Ms)^k \, ds$$

$$= \frac{M^{k+1} K t^{k+1}}{(k+1)!} \,,$$

as required.

But $K(Mt)^k/k!$ is the typical term of a Taylor series of Ke^{Mt} that converges uniformly and absolutely on $[0, T]$. Thus (2.1.5) also converges uniformly on $[0, T]$ to a continuous limit function, say $\mathbf{x}(t)$. We may, therefore, take the limit as $n \to \infty$ in (2.1.4) and pass it through the integral, obtaining

$$\mathbf{x}(t) = \mathbf{f}(t) + \int_0^t B(t, s)\mathbf{x}(s)\, ds\,,$$

so that the limit function $\mathbf{x}(t)$ is a solution of (2.1.1).

To see that $\mathbf{x}(t)$ is the only solution, suppose there are two solutions, say $\mathbf{x}(t)$ and $\mathbf{y}(t)$, on an interval $[0, T]$. Then, from (2.1.1),

$$\mathbf{x}(t) - \mathbf{y}(t) = \int_0^t B(t, s)\big[\mathbf{x}(s) - \mathbf{y}(s)\big]\, ds\,,$$

so that

$$\big|\mathbf{x}(t) - \mathbf{y}(t)\big| \leq M \int_0^t \big|\mathbf{x}(s) - \mathbf{y}(s)\big|\, ds\,.$$

This is of the form

$$|z(t)| \leq c + \int_0^t M|z(s)|\, ds$$

with $c = 0$. By Gronwall's inequality, $|z(t)| \leq ce^{Mt} = 0$. The proof is complete.

2.2   Linear Properties

Discussion of the linear properties of (2.1.1) tends to be cumbersome, whereas the linear properties of (2.1.2) are very straightforward and analogous to properties of ordinary differential equations. In fact, in the convolution case for (2.1.2) with A constant, the entire theory is almost identical to that for ordinary differential equations.

Theorem 2.2.1. *Let* $\mathbf{f}_1, \mathbf{f}_2 : [0, \alpha) \to R^n$ *be continuous and* $B(t, s)$ *be an* $n \times n$ *matrix of functions continuous for* $0 \le s \le t < \alpha$. *If* $\mathbf{x}(t)$ *and* $\mathbf{y}(t)$ *are solutions of*

$$\mathbf{x}(t) = \mathbf{f}_1(t) + \int_0^t B(t, s)\mathbf{x}(s)\, ds$$

and

$$\mathbf{y}(t) = \mathbf{f}_2(t) + \int_0^t B(t, s)\mathbf{y}(s)\, ds \,,$$

respectively, and if c_1 *and* c_2 *are real numbers, then* $c_1\mathbf{x}(t) + c_2\mathbf{y}(t)$ *is a solution of*

$$\mathbf{z}(t) = \left[c_1\mathbf{f}_1(t) + c_2\mathbf{f}_2(t)\right] + \int_0^t B(t, s)\mathbf{z}(s)\, ds \,.$$

Proof. We have

$$c_1\mathbf{x}(t) + c_2\mathbf{y}(t) = c_1\mathbf{f}_1(t) + c_2\mathbf{f}_2(t) + \int_0^t B(t, s)c_1\mathbf{x}(s)\, ds$$

$$+ \int_0^t B(t, s)c_2\mathbf{y}(s)\, ds \,,$$

and the proof is complete.

We turn now to equations

$$\mathbf{x}' = A(t)\mathbf{x} + \int_0^t C(t, s)\mathbf{x}(s)\, ds \, + \mathbf{F}(t) \qquad\qquad (2.2.1)$$

and

$$\mathbf{x}' = A(t)\mathbf{x} + \int_0^t C(t, s)\mathbf{x}(s)\, ds \,, \qquad\qquad (2.2.2)$$

with $\mathbf{F} : [0, \alpha) \to R^n$ being continuous, $A(t)$ an $n \times n$ matrix of functions continuous on $[0, \alpha)$, and $C(t, s)$ an $n \times n$ matrix of functions continuous for $0 \le s \le t < \alpha$.

Notice that if the initial condition for (2.2.1) is $\mathbf{x}(0) = \mathbf{x}_0$, then an integration from 0 to t yields

$$\mathbf{x}(t) = \mathbf{x}_0 + \int_0^t \mathbf{F}(s)\,ds + \int_0^t A(s)\mathbf{x}(s)\,ds + \int_0^t \int_0^u C(u,s)\mathbf{x}(s)\,ds\,du\,, \quad (2.2.3)$$

and upon change of order of integration, we have an equation of the form (2.1.1) with

$$\mathbf{f}(t) = \mathbf{x}_0 + \int_0^t \mathbf{F}(s)\,ds\,,$$

so that when $\mathbf{F}(t) \equiv 0$, then $\mathbf{f}(t) = \mathbf{x}_0$.

Theorem 2.2.2. *Consider* (2.2.1) *and* (2.2.2) *on* $[0, \alpha)$.

(a) *For each $\mathbf{x}_0$ there is a solution $\mathbf{x}(t)$ of* (2.2.1) *for $0 \le t < \alpha$ with $\mathbf{x}(0) = \mathbf{x}_0$.*

(b) *If $\mathbf{x}_1(t)$ and $\mathbf{x}_2(t)$ are two solutions of* (2.2.1), *then $\mathbf{x}_1(t) - \mathbf{x}_2(t)$ is a solution of* (2.2.2).

(c) *If $\mathbf{x}_1(t)$ and $\mathbf{x}_2(t)$ are two solutions of* (2.2.2) *and if c_1 and c_2 are real numbers, then $c_1\mathbf{x}_1(t) + c_2\mathbf{x}_2(t)$ is a solution of* (2.2.2).

(d) *There are n linearly independent solutions of* (2.2.2) *on $[0, \alpha)$ and any solution on $[0, \alpha)$ may be expressed as a linear combination of them.*

Proof. In view of the remarks preceding the theorem, (a) was established by Theorem 2.1.1. Parts (b) and (c) follow by direct substitution into the equations. To prove (d), consider the n constant vectors $\mathbf{e}_1, \ldots, \mathbf{e}_n$, where $\mathbf{e}_i = (0, \ldots, 0, 1_i, 0, \ldots, 0)^T$ and let $\mathbf{x}_i(t)$ be the solution with $\mathbf{x}_i(0) = \mathbf{e}_i$. For a given $\mathbf{x}_0 = (x_{01}, \ldots, x_{0n})^T$, we have $\mathbf{x}_0 = x_{01}\mathbf{e}_1 + \cdots + x_{0n}\mathbf{e}_n$, so the unique solution $\mathbf{x}(t)$ with $\mathbf{x}(0) = \mathbf{x}_0$ may be expressed as

$$\mathbf{x}(t) = x_{01}\mathbf{x}_1(t) + \cdots + x_{0n}\mathbf{x}_n(t)\,.$$

Now, the $\mathbf{x}_1(t), \ldots, \mathbf{x}_n(t)$ are linearly independent on $[0, \alpha)$; for if

$$\sum_{i=1}^n c_i\mathbf{x}_i(t) \equiv 0 \quad \text{on} \quad [0, \alpha)$$

is a nontrivial linear relation, then

$$0 = \sum_{i=1}^n c_i\mathbf{x}_i(0) = \sum_{i=1}^n c_1\mathbf{e}_i$$

is a nontrivial linear relation among the $\mathbf{e}_i$, which is impossible. This completes the proof.

Corollary.   If $\mathbf{x}_1(t), \ldots, \mathbf{x}_n(t)$ are n linearly independent solutions of (2.2.2) and if $\mathbf{x}_p(t)$ is any solution of (2.2.1), then every solution of (2.2.1) can be expressed as

$$\mathbf{x}(t) = \mathbf{x}_p(t) + c_1\mathbf{x}_1(t) + \cdots + c_n\mathbf{x}_n(t)$$

for appropriate constants $c_1, \ldots, c_n$.

Proof. If $\mathbf{x}(t)$ and $\mathbf{x}_p(t)$ are solutions of (2.2.1), then $\mathbf{x}(t) - \mathbf{x}_p(t)$ is a solution of (2.2.2) and, hence, may be expressed as

$$c_1\mathbf{x}_1(t) + \cdots + c_n\mathbf{x}_n(t).$$

This completes the proof.

2.3   Convolution and the Laplace Transform

When A is constant and C is of convolution type, then the variation of parameters formula for (2.2.1) becomes identical to that for ordinary differential equations.

Consider the systems

$$\mathbf{x}' = P\mathbf{x} + \int_0^t D(t-s)\mathbf{x}(s)\,ds + \mathbf{F}(t) \tag{2.3.1}$$

and

$$\mathbf{x}' = P\mathbf{x} + \int_0^t D(t-s)\mathbf{x}(s)\,ds, \tag{2.3.2}$$

in which P an $n \times n$ constant matrix, $D(t)$ is $n \times n$ matrix of functions continuous on $[0, \infty)$, and $\mathbf{F} : [0, \infty) \to R^n$ continuous. We suppose also that $|\mathbf{F}(t)|$ and $|D(t)|$ may be bounded by a function $Me^{\alpha t}$ for $M > 0$ and $\alpha > 0$. That is, $\mathbf{F}$ and D are said to be of exponential order.

Laplace transforms are particularly well suited to the study of convolution problems. A good (and elementary) discussion of transforms may be found in Churchill (1958). Our use here will be primarily symbolic and the necessary rudiments may be found in many elementary texts on ordinary differential equations.

The following is a list of some of the essential properties of Laplace transforms of continuous functions of exponential order [(v) requires differentiability]. The first property is a definition from which all the others may be derived very easily with the exception of (vii).

(i) If $h : [0, \infty) \to R^n$, then the Laplace transform of h is

$$L(h) = H(s) = \int_0^\infty e^{-st} h(t) \, dt \, . \tag{2.3.3}$$

(ii) If $D(t) = (d_{ij}(t))$ is a matrix (or vector), then $L(D(t)) = \big(L(d_{ij}(t))\big)$. (This is merely notation.)

(iii) If c is a constant and h_1, h_2 functions, then $L(ch_1 + h_2) = cL(h_1) + L(h_2)$.

(iv) If $D(t)$ is an $n \times n$ matrix and $\mathbf{h}(t)$ a vector function, then

$$L\left(\int_0^t D(t - s)\mathbf{h}(s) \, ds \right) = L(D)L(\mathbf{h}) \, . \tag{2.3.4}$$

(v) $L(h'(t)) = sL(h) - h(0)$.

(vi) L^{-1} is linear.

(vii) If $h_1(t)$ and $h_2(t)$ are continuous functions of exponential order and $L(h_1(t)) = L(h_2(t))$, then $h_1(t) \equiv h_2(t)$.

Theorem 2.3.1. Let $Z(t)$ be the $n \times n$ matrix whose columns are solutions of (2.3.2) with $Z(0) = I$. The solution of (2.3.1) satisfying $\mathbf{x}(0) = \mathbf{x}_0$ is

$$\mathbf{x}(t) = Z(t)\mathbf{x}_0 + \int_0^t Z(t - s)\mathbf{F}(s) \, ds \, . \tag{2.3.5}$$

Proof. Notice that $Z(t)$ satisfies (2.3.2):

$$Z'(t) = PZ(t) + \int_0^t D(t - s)Z(s) \, ds \, .$$

We first suppose that $\mathbf{F}$ and D are in $L^1[0, \infty)$. If we convert (2.3.1) into an integral equation, we have

$$\mathbf{x}(t) = \mathbf{x}(0) + \int_0^t \mathbf{F}(s) \, ds + \int_0^t \left[P + \int_s^t D(u - s) \, du \right] \mathbf{x}(s) \, ds \, ,$$

and as D and $\mathbf{F}$ are in L^1, we have

$$|\mathbf{x}(t)| \le |\mathbf{x}(0)| + K + K \int_0^t |\mathbf{x}(s)| \, ds, \quad \text{some} \quad K > 0 \quad \text{and} \quad 0 \le t < \infty.$$

By Gronwall's inequality

$$|\mathbf{x}(t)| \le \big[|\mathbf{x}(0)| + K\big]e^{Kt} .$$

Thus both $\mathbf{x}(t)$ and $Z(t)$ are of exponential order, so we can take their Laplace transforms. We have

$$Z'(t) = PZ(t) + \int_0^t D(t-s)Z(s)\,ds ,$$

and upon transforming both sides, we obtain

$$sL(Z) - Z(0) = PL(Z) + L(D)L(Z) ,$$

using (i)–(v). Thus

$$\big[sI - P - L(D)\big]L(Z) = Z(0) = I ,$$

and because the right side is nonsingular, so is $\big[sI - P - L(D)\big]$ for appropriate s. (Actually, $L(Z)$ is an analytic function of s in the half-plane Re $s \ge \alpha$, where $|Z(t)| \le Ke^{\alpha t}$.) [See Churchill (1958, p. 171).] We then have

$$L(Z) = \big[sI - P - L(D)\big]^{-1}.$$

Now, transform both sides of (2.3.1):

$$sL(\mathbf{x}) - \mathbf{x}(0) = PL(\mathbf{x}) + L(D)L(\mathbf{x}) + L(\mathbf{F})$$

or

$$\big[sI - P - L(D)\big]L(\mathbf{x}) = \mathbf{x}(0) + L(\mathbf{F}) ,$$

so that

$$\begin{aligned}
L(\mathbf{x}) &= \big[sI - P - L(D)\big]^{-1}\big[\mathbf{x}(0) + L(\mathbf{F})\big] \\
&= L(Z)\mathbf{x}(0) + L(Z)L(\mathbf{F}) \\
&= L(Z\mathbf{x}(0)) + L\left(\int_0^t Z(t-s)\mathbf{F}(s)\,ds \right) \\
&= L\left(Z(t)\mathbf{x}(0) + \int_0^t Z(t-s)\mathbf{F}(s)\,ds \right) .
\end{aligned}$$

Because $\mathbf{x}$, Z, and $\mathbf{F}$ are of exponential order and continuous, by (vii) we have the required formula. Thus, the proof is complete for D and $\mathbf{F}$ being in $L^1[0, \infty)$.

In the general case (i.e., D and $\mathbf{F}$ not in L^1), for each $T > 0$ define continuous $L^1[0, \infty)$ functions $\mathbf{F}_T$ and D_T by

$$\mathbf{F}_T(t) = \begin{cases} \mathbf{F}(t) & \text{if } 0 \le t \le T, \\ \mathbf{F}(T)\{1/[(t-T)^2+1]\} & \text{if } t > T \end{cases}$$

and

$$D_T = \begin{cases} D(t) & \text{if } 0 \le t \le T, \\ D(T)\{1/[(t-T)^2+1]\} & \text{if } t > T. \end{cases}$$

Consider (2.3.1) and

$$\mathbf{x}'(t) = P\mathbf{x}(t) + \mathbf{F}_T(t) + \int_0^t D_T(t-s)\mathbf{x}(s)\,ds\,, \qquad (2.3.1)_T$$

with $\mathbf{x}(0) = \mathbf{x}_0$ for both. Because the equations are identical on $[0, T]$, so are their solutions; this is true for each $T > 0$. Thus because (2.3.5) holds for $(2.3.1)_T$ for each $T > 0$, it holds for (2.3.1) on each interval $[0, T]$. This completes the proof.

Exercise 2.3.1.   This exercise if not trivial. Substitute (2.3.5) into (2.3.1), interchange the order of integration, and show that (2.3.5) is a solution of (2.3.1).

Although one can seldom find $Z(t)$, we shall discover certain properties that make the variation of parameters formula very useful. For example, by a change of variable

$$\int_0^t Z(t-s)\mathbf{F}(s)\,ds = \int_0^t Z(s)\mathbf{F}(t-s)\,ds\,,$$

so if we can show that

$$\int_0^\infty |Z(t)|\,dt < \infty\,,$$

it follows that, for any bounded $\mathbf{F}$, then

$$\int_0^t Z(s)\mathbf{F}(t-s)\,ds$$

is also bounded. In the study of Liapunov's direct method, one frequently finds that $\int_0^\infty |Z(t)|\,dt$ is finite. Furthermore, uniform asymptotic stability of the zero solution of (2.3.2) and $\int_0^\infty |Z(t)|\,dt < \infty$ are closely connected.

We turn now to the integral equation

$$\mathbf{x}(t) = \mathbf{f}(t) + \int_0^t B(t-s)\mathbf{x}(s)\,ds\,, \tag{2.3.6}$$

with $\mathbf{f} : [0,\infty) \to R^n$ being continuous, $B(t)$ an $n \times n$ matrix continuous on $[0,\infty)$, and both $\mathbf{f}$ and B of exponential order.

The goal is to obtain a variation of parameters formula for (2.3.6). Naturally if B and $\mathbf{f}$ are both differentiable, we could convert (2.3.6) to an equation of the form (2.3.1) and apply Theorem 2.3.1. But that seems too indirect. Such a formula should exist independently of B' and $\mathbf{f}'$. We shall see, however, that the derivative of $\mathbf{f}$ will enter, in a natural way, even when the transform of (2.3.6) is taken directly.

Theorem 2.3.2. *Let $H(t)$ be the $n \times n$ matrix satisfying*

$$H(t) = I + \int_0^t B(t-s)H(s)\,ds \tag{2.3.7}$$

and let $\mathbf{f}'(t)$ and B be continuous and of exponential order. The unique solution $\mathbf{x}(t)$ of (2.3.6) is given by

$$\mathbf{x}(t) = H(t)\mathbf{f}(0) + \int_0^t H(t-s)\mathbf{f}'(s)\,ds\,. \tag{2.3.8}$$

Proof. The Laplace transform of (2.3.7) is

$$L(H) = L(I) + L(B)L(H)$$

and, as $L(1) = s^{-1}$, $L(I) = s^{-1}I$. Thus

$$[I - L(B)]L(H) = L(I)\,,$$

and, because $L(I)$ is nonsingular, so is $[I - L(B)]$. This implies that

$$L(H) = [I - L(B)]^{-1}s^{-1}\,.$$

Now the transform of (2.3.6) is $L(\mathbf{x}) = L(\mathbf{f}) + L(B)L(\mathbf{x})$ so that

$$[I - L(B)]L(\mathbf{x}) = L(\mathbf{f})$$

or

$$L(\mathbf{x}) = [I - L(B)]^{-1}L(\mathbf{f})\,.$$

Multiply and divide by s on the right side and recall that

$$L(\mathbf{f}') = sL(\mathbf{f}) - \mathbf{f}(0).$$

This yields

$$
\begin{aligned}
L(\mathbf{x}) &= \left\{ s[I - L(B)] \right\}^{-1} sL(\mathbf{f}) \\
&= L(H)\big[L(\mathbf{f}') + \mathbf{f}(0)\big] \\
&= L(H)L(\mathbf{f}') + L(H)\mathbf{f}(0) \\
&= L\left(\int_0^t H(t-s)\mathbf{f}'(s)\,ds \right) + L\big(H(t)\mathbf{f}(0)\big) \\
&= L\left(H(t)\mathbf{f}(0) + \int_0^t H(t-s)\mathbf{f}'(s)\,ds \right),
\end{aligned}
$$

so that (2.3.8) follows from (vii). This completes the proof.

Notice that (2.3.7) represents the n integral equations

$$\mathbf{x}(t) = \mathbf{e}_i + \int_0^t B(t-s)\mathbf{x}(s)\,ds.$$

It is necessary that functions be defined on $[0, \infty)$ for the Laplace transforms to be applied. Also, $\mathbf{F}$ and D need to be of exponential order. The following exercise suggests that one may try to circumvent both problems.

Exercise 2.3.2. Suppose that $\mathbf{F}$ and D in (2.3.1) are continuous on $[0, T]$ but not defined for $t > T$. Define $\mathbf{F}$ and D on $[0, \infty)$ by asking that $\mathbf{F}(t) = \mathbf{F}(T)$ and $D(t) = D(T)$ if $t \geq 1$. Check the details to see if the variation of parameters formula will work on $[0, T]$.

Exercise 2.3.3. Continue the reasoning of Exercise 2.3.2 and suppose that it is known that $\int_0^\infty |Z(t)|\,dt < \infty$. If D is not of exponential order but $\mathbf{F}$ is bounded, can one still conclude that solutions of (2.3.1) are bounded?

We return to

$$\mathbf{x}(t) = \mathbf{f}(t) + \int_0^t B(t-s)\mathbf{x}(s)\,ds \tag{2.3.6}$$

with $\mathbf{f} : [0, \infty) \to R^n$ being continuous and B a continuous $n \times n$ matrix, both $\mathbf{f}$ and B of exponential order.

Theorem 2.3.3. *If H is defined by (2.3.7) and if H is differentiable, then the unique solution of (2.3.6) is given by*

$$\mathbf{x}(t) = \mathbf{f}(t) + \int_0^t H'(t-s)\mathbf{f}(s)\,ds. \tag{2.3.9}$$

Proof. We have

$$L(H) = \left\{ s[I - L(B)] \right\}^{-1}$$

and

$$[I - L(B)]L(x) = L(\mathbf{f}),$$

whose product yields

$$s^{-1}L(\mathbf{x}) = L(H)L(\mathbf{f})$$

or

$$L(1)L(\mathbf{x}) = L(H)L(\mathbf{f}),$$

so that

$$L\left(\int_0^t \mathbf{x}(t-s)\,ds \right) = L\left(\int_0^t H(t-s)\mathbf{f}(s)\,ds \right),$$

which implies

$$\int_0^t \mathbf{x}(s)\,ds = \int_0^t H(t-s)\mathbf{f}(s)\,ds.$$

We differentiate this to obtain (2.3.9), because $H(0) = I$. This completes the proof.

The matrices Z and H are also called resolvents, which will be discussed in Chapter 7 in some detail.

2.4   Stability

Consider the system

$$\mathbf{x}' = A(t)\mathbf{x} + \int_0^t C(t,s)\mathbf{x}(s)\,ds \qquad (2.4.1)$$

with A an $n \times n$ matrix of functions continuous for $0 \le t < \infty$ and $C(t,s)$ an $n \times n$ matrix of functions continuous for $0 \le s \le t < \infty$.

If $\phi : [0, t_0] \to R^n$ is a continuous initial function, then $\mathbf{x}(t, \phi)$ will denote the solution on $[t_0, \infty)$. If the information is needed, we may denote the solution by $\mathbf{x}(t, t_0, \phi)$. Frequently, it suffices to write $x(t)$.

Notice that $\mathbf{x}(t) \equiv \mathbf{0}$ is a solution of (2.4.1), and it is called the zero solution.

Definition 2.4.1. *The zero solution of (2.4.1) is (Liapunov) stable if, for each $\varepsilon > 0$ and each $t_0 \geq 0$, there exists $\delta > 0$ such that*

$$|\boldsymbol{\phi}(t)| < \delta \quad on \quad [0, t_0] \quad and \quad t \geq t_0$$

imply $|\mathbf{x}(t, \boldsymbol{\phi})| < \varepsilon$.

Definition 2.4.2. *The zero solution of (2.4.1) is* uniformly stable *if, for each $\varepsilon > 0$, there exists $\delta > 0$ such that*

$$t_0 \geq 0, \quad |\boldsymbol{\phi}(t)| < \delta \quad on \quad [0, t_0], \quad and \quad t \geq t_0$$

imply $|\mathbf{x}(t, \boldsymbol{\phi})| < \varepsilon$.

Definition 2.4.3. *The zero solution of (2.4.1) is* asymptotically stable *if it is stable and if for each $t_0 \geq 0$ there exists $\delta > 0$ such that $|\boldsymbol{\phi}(t)| < \delta$ on $[0, t_0]$ implies $|\mathbf{x}(t, \boldsymbol{\phi})| \to 0$ as $t \to \infty$.*

Definition 2.4.4. *The zero solution of (2.4.1) is* uniformly asymptotically stable *(U.A.S.) if it is uniformly stable and if there exists $\eta > 0$ such that, for each $\varepsilon > 0$, there is a $T > 0$ such that*

$$t_0 \geq 0, \quad |\boldsymbol{\phi}(t)| < \eta \quad on \quad [0, t_0], \quad and \quad t \geq t_0 + T$$

imply $|\mathbf{x}(t, \boldsymbol{\phi})| < \varepsilon$.

We begin with a brief reminder of Liapunov theory for ordinary differential equations. The basis idea is particularly simple. Consider a system of ordinary differential equations

$$\mathbf{x}' = \mathbf{G}(t, \mathbf{x}), \tag{2.4.2}$$

with $\mathbf{G} : [0, \infty) \times R^n \to R^n$ being continuous and $\mathbf{G}(t, \mathbf{0}) \equiv \mathbf{0}$, so that $\mathbf{x} = \mathbf{0}$ is a solution. The stability definitions apply to (2.4.2) with $\boldsymbol{\phi}(t) \equiv \mathbf{x}(t_0)$ on $[0, t_0]$.

Suppose first that there is a scalar function

$$V : [0, \infty) \times R^n \to [0, \infty)$$

having continuous first partial derivatives with respect to all variables. Suppose also that $V(t, \mathbf{x}) \to \infty$ as $|\mathbf{x}| \to \infty$ uniformly for $0 \leq t < \infty$; for example, suppose there is a continuous function $W : R^n \to [0, \infty)$ with $W(\mathbf{x}) \to \infty$ as $|\mathbf{x}| \to \infty$ and $V(t, \mathbf{x}) \geq W(\mathbf{x})$.

Notice that if $\mathbf{x}(t)$ is any solution of (2.4.2) on $[0, \infty)$, then $V(t, \mathbf{x}(t))$ is a scalar function of t, and even if $\mathbf{x}(t)$ is not explicitly known, using the chain rule and (2.4.2) it is possible to compute $V'(t, \mathbf{x}(t))$. We have

$$V'(t, \mathbf{x}(t)) = \frac{\partial V}{\partial x_1} \frac{dx_1}{dt} + \cdots + \frac{\partial V}{\partial x_n} \frac{dx_n}{dt} + \frac{\partial V}{\partial t}. \tag{a}$$

But $\mathbf{G}(t, \mathbf{x}) = \left(dx_1/dt, \ldots, dx_n/dt \right)^T$ and so (a) is actually

$$V'(t, \mathbf{x}(t)) = \mathbf{grad}\, \mathbf{V} \cdot \mathbf{G} + \partial V/\partial t. \tag{b}$$

The right-hand side of (b) consists of known functions of t and $\mathbf{x}$. If V is shrewdly chosen, many conclusions may be drawn from the properties of V'. For example, if $V'(t, \mathbf{x}(t)) \leq 0$, then $t \geq t_0$ implies $V(t, \mathbf{x}(t)) \leq V(t_0, \mathbf{x}(t_0))$, and because $V(t, \mathbf{x}) \to \infty$ as $|\mathbf{x}| \to \infty$ uniformly for $0 < t < \infty$, $\mathbf{x}(t)$ is bounded.

The object is to find a suitable V function. We now illustrate how V may be constructed in the linear constant coefficient case.

Let A be an $n \times n$ constant matrix all of whose characteristic roots have negative real parts, and consider the system

$$\mathbf{x}' = A\mathbf{x}. \tag{2.4.3}$$

All solutions tend to zero exponentially, so that the matrix

$$B = \int_0^\infty [\exp At]^T [\exp At]\, dt \tag{2.4.4}$$

is well defined, symmetric, and positive definite. Furthermore,

$$A^T B + BA = -I \tag{2.4.5}$$

because

$$
\begin{aligned}
-I &= [\exp At]^T [\exp At]\big|_0^\infty \\
&= \int_0^\infty (d/dt)[\exp At]^T [\exp At]\, dt \\
&= \int_0^\infty \left(A^T [\exp At]^T [\exp At] + [\exp At]^T [\exp At]A \right) dt \\
&= A^T B + BA.
\end{aligned}
$$

Thus, if we select V as a function of $\mathbf{x}$ alone, say

$$V(\mathbf{x}) = \mathbf{x}^T B\mathbf{x}, \tag{2.4.6}$$

then for $\mathbf{x}(t)$ a solution of (2.4.3) we have

$$
\begin{aligned}
V'(\mathbf{x}(t)) &= (\mathbf{x}^T)' B\mathbf{x} + \mathbf{x}^T B\mathbf{x}' \\
&= (\mathbf{x}')^T B\mathbf{x} + \mathbf{x}^T B\mathbf{x}' \\
&= \mathbf{x}^T A^T B\mathbf{x} + \mathbf{x}^T BA\mathbf{x} \\
&= \mathbf{x}^T (A^T B + BA)\mathbf{x} \\
&= -\mathbf{x}^T \mathbf{x}.
\end{aligned}
$$

The matrix B will be used extensively throughout the following discussions.

In some of the most elementary problems asking $V(t,\mathbf{x})$ to have continuous first partial derivatives is too severe. Instead, it suffices to ask that

$$V : [0,\infty) \times R^n \to [0,\infty) \ \text{ is continuous}$$

and   (2.4.7)

$$V \text{ satisfies a local Lipschitz condition in } \mathbf{x}.$$

Definition 2.4.5. *A function $V(t,\mathbf{x})$ satisfies a local Lipschitz condition in $\mathbf{x}$ on a subset D of $[0,\infty) \times R^n$ if, for each compact subset L of D, there is a constant $K = K(L)$ such that $(t,\mathbf{x}_1)$ and $(t,\mathbf{x}_2)$ in L imply that*

$$\left| V(t,\mathbf{x}_1) - V(t,\mathbf{x}_2) \right| \le K|\mathbf{x}_1 - \mathbf{x}_2|.$$

If V satisfies (2.4.7) then one defines the derivative of V along a solution $x(t)$ of (2.4.2) by

$$V'_{(2.4.2)}(t,\mathbf{x}) = \limsup_{h\to 0^+} \left[V\big(t+h, \mathbf{x}+h\mathbf{G}(t,\mathbf{x})\big) - V(t,\mathbf{x}) \right]/h. \quad (2.4.8)$$

Because V satisfies a local Lipschitz condition in $\mathbf{x}$, when V is independent of t [so that $V = V(\mathbf{x})$], we see that

$$|V'_{(2.4.2)}(\mathbf{x})| \le K|\mathbf{G}(t,\mathbf{x})|.$$

Next, define

$$V'(t,\mathbf{x}(t)) = \limsup_{h\to 0^+} \left[V\big(t+h, \mathbf{x}(t+h)\big) - V(t,\mathbf{x}(t)) \right]/h. \quad (2.4.9)$$

It can be shown [see T. Yoshizawa, (1966; p. 3)] that

$$V'(t,\mathbf{x}(t)) = V'_{(2.4.2)}(t,\mathbf{x}). \quad (2.4.10)$$

Moreover, from integration theory it is known that $V'(t,\mathbf{x}(t)) \le 0$ implies that $V(t,\mathbf{x}(t))$ is nonincreasing.

The next problem will be encountered frequently in the following, and it is best taken care of here. Refer to (2.4.3) and select B as in (2.4.4). Then form

$$V(\mathbf{x}) = [\mathbf{x}^T B \mathbf{x}]^{1/2}$$

and compute the derivative along a solution of (2.4.3). If $\mathbf{x} \neq 0$, then V has continuous first partial derivatives and

$$V'(\mathbf{x}) = (\mathbf{x}^T B \mathbf{x})'/2[\mathbf{x}^T B \mathbf{x}]^{1/2}$$
$$= -\mathbf{x}^T \mathbf{x}/2[\mathbf{x}^T B \mathbf{x}]^{1/2}\,.$$

Now there is a positive constant k with $|\mathbf{x}| \geq 2k[\mathbf{x}^T B \mathbf{x}]^{1/2}$, so for $\mathbf{x} \neq 0$,

$$V'(\mathbf{x}) \leq -k|\mathbf{x}|\,.$$

But we noted after (2.4.8) that

$$|V'| \leq K|G(t,\mathbf{x})| = K|A\mathbf{x}|\,,$$

so when $\mathbf{x} = 0$ we have

$$V'(\mathbf{x}) \leq 0\,.$$

Hence, for all $\mathbf{x}$ we see that

$$V'(\mathbf{x}) \leq -k|\mathbf{x}|\,.$$

The theory is almost identical for integro-differential equations, although the function $V(t,\mathbf{x})$ is generally replaced by a functional $V(t, \mathbf{x}(\cdot)) = V\big(t, \mathbf{x}(s);\ 0 \leq s \leq t\big)$. We develop this idea more fully later when we consider general functional differential equations; however, we now have sufficient material for some general results.

2.5   Liapunov Functionals and Small Kernels

We consider the system

$$\mathbf{x}' = A\mathbf{x} + \int_0^t C(t,s)\mathbf{x}(s)\,ds\,, \tag{2.5.1}$$

in which A is an $n \times n$ matrix all of whose characteristic roots have negative real parts, $C(t,s)$ an $n \times n$ matrix of functions continuous for $0 \leq s \leq t < \infty$, and

$$\int_t^\infty |C(u,s)|\,du \quad \text{is continuous for} \quad 0 \leq s \leq t < \infty\,.$$

Find a symmetric positive definite matrix B with

$$A^T B + BA = -I \,. \tag{2.5.2}$$

There are positive constants r, k, and K (not unique) with

$$|\mathbf{x}| \geq 2k[\mathbf{x}^T B\mathbf{x}]^{1/2} \,, \tag{2.5.3}$$

$$|B\mathbf{x}| \leq K[\mathbf{x}^T B\mathbf{x}]^{1/2} \,, \tag{2.5.4}$$

and

$$r|\mathbf{x}| \leq [\mathbf{x}^T B\mathbf{x}]^{1/2} \,. \tag{2.5.5}$$

A basic tool in the investigation of (2.5.1) is the functional

$$V(t, \mathbf{x}(\cdot)) = [\mathbf{x}^T B\mathbf{x}]^{1/2} + \bar{K} \int_0^t \int_t^\infty |C(u, s)| \, du \, |\mathbf{x}(s)| \, ds \,, \tag{2.5.6}$$

where $\bar{K}$ is a positive constant. This functional has continuous first partial derivatives with respect to all variables (when $\mathbf{x} \neq 0$) and it satisfies a global Lipschitz condition in $\mathbf{x}(t)$.

Let us compute the derivative of (2.5.6) along a solution $\mathbf{x}(t)$ of (2.5.1). For $\mathbf{x} \neq 0$ we have

$$V'(t, \mathbf{x}(\cdot)) = \left\{ (\mathbf{x}^T B\mathbf{x})' / 2[\mathbf{x}^T B\mathbf{x}]^{1/2} \right\}$$
$$+ \bar{K} \int_t^\infty |C(u, t)| \, du \, |\mathbf{x}| - \bar{K} \int_0^t |C(t, s)| \, |\mathbf{x}(s)| \, ds \,,$$

and because

$$(\mathbf{x}^T B\mathbf{x})' = (\mathbf{x}')^T B\mathbf{x} + \mathbf{x}^T B\mathbf{x}'$$
$$= \left[\mathbf{x}^T A^T + \int_0^t \mathbf{x}^T(s) C^T(t, s) \, ds \right] B\mathbf{x}$$
$$+ \mathbf{x}^T B \left[A\mathbf{x} + \int_0^t C(t, s)\mathbf{x}(s) \, ds \right]$$
$$= -\mathbf{x}^T \mathbf{x} + 2 \int_0^t \mathbf{x}^T(s) C^T(t, s) \, ds \, B\mathbf{x} \,,$$

by (2.5.3) and (2.5.4) we have

$$(\mathbf{x}^T B\mathbf{x})' / 2(\mathbf{x}^T B\mathbf{x})^{1/2} \leq -k|\mathbf{x}| + K \int_0^t |C(t, s)| \, |\mathbf{x}(s)| \, ds \,.$$

This yields

$$V' \leq \left[-k+\bar{K}\int_t^\infty |C(u,t)|\, du\right]|\mathbf{x}| - (\bar{K}-K)\int_0^t |C(t,s)|\,|\mathbf{x}(s)|\, ds\,. \quad (2.5.7)$$

Our basic assumption is

There exists $\bar{K} \geq K$ and $\bar{k} \geq 0$ with $\bar{k} \leq k - \bar{K}\int_t^\infty |C(u,t)|\, du\,. \quad (2.5.8)$

Theorem 2.5.1. *Let B, k, and K be defined by Eqs. (2.5.2)–(2.5.4).*

(a) *If (2.5.8) holds, the zero solution of (2.5.1) is stable.*

(b) *If (2.5.8) holds with $\bar{K} > K$ and $\bar{k} > 0$, then $\mathbf{x} = 0$ is asymptotically stable.*

(c) *If (2.5.8) holds and $\int_0^t \int_t^\infty |C(u,s)|\, du\, ds$ is bounded, then $\mathbf{x} = 0$ is uniformly stable.*

(d) *Suppose (c) holds and $\bar{K} > K$ and $\bar{k} > 0$. If for each $\rho > 0$ there exists $S > 0$ such that $P \geq S$ and $t \geq 0$ imply $\int_0^t \int_{t+P}^\infty |C(u,s)|\, du\, ds < \rho$, then $\mathbf{x} = 0$ is uniformly asymptotically stable.*

Proof of (a). Let $\varepsilon > 0$ and $t_0 \geq 0$ be given. We must find $\delta > 0$ such that

$$|\phi(t)| < \delta \quad \text{on} \quad [0,t_0] \quad \text{and} \quad t \geq t_0$$

imply $|\mathbf{x}(t,\phi)| < \varepsilon$.

Because $V'(t,\mathbf{x}(\cdot)) \leq 0$, if $|\phi(t)| < \delta$ on $[0,t_0]$, then we have

$$r|\mathbf{x}(t)| \leq V(t,\mathbf{x}(\cdot)) \leq V(t_0,\phi(\cdot))$$

$$= \left[\phi(t_0)^T B\phi(t_0)\right]^{1/2} + \bar{K}\int_0^{t_0}\int_{t_0}^\infty |C(u,s)|\, du\, |\phi(s)|\, ds$$

$$\leq (\delta/2k) + \delta\int_0^{t_0}\int_{t_0}^\infty \bar{K}|C(u,s)|\, du\, ds$$

or

$$|\mathbf{x}(t)| < \delta\left\{(1/2k) + \int_0^{t_0}\int_{t_0}^\infty \bar{K}|C(u,s)|\, du\, ds \right\}/r\,,$$

which will be smaller than ε if

$$\delta < \varepsilon r \left/ \left[(1/2k) + \int_0^{t_0}\int_{t_0}^\infty \bar{K}|C(u,s)|\, du\, ds\right]\right.\,. \quad (2.5.9)$$

This choice of $\delta = \delta(\varepsilon,t_0)$ fulfills the conditions for stability.

Proof of (c). For uniform stability, δ must be independent of t_0. If $\int_0^t \int_t^\infty |C(u,s)| \, du \, ds \leq M$ for $0 \leq t < \infty$ and some $M > 0$, then (2.5.9) may be replaced by

$$\delta < \varepsilon r / \left[(1/2k) + \bar{K}M) \right], \qquad (2.5.10)$$

yielding uniform stability.

Proof of (b). We have

$$V'(t, \mathbf{x}(\cdot)) \leq -\bar{k}|\mathbf{x}| - (\bar{K} - K) \int_0^t |C(t,s)| \, |\mathbf{x}(s)| \, ds \,,$$

and hence, there is a $\mu > 0$ with

$$V'(t, \mathbf{x}(\cdot)) \leq -\mu \left[|\mathbf{x}| + |\mathbf{x}'| \right], \qquad (2.5.11)$$

which is a fundamental relation. If $\mathbf{x}(t)$ is a solution of (2.5.1), then

$$r|\mathbf{x}(t)| \leq V(t, \mathbf{x}(\cdot)) \leq V(t_0, \boldsymbol{\phi}(\cdot)) - \mu \int_{t_0}^t |\mathbf{x}(s)| \, ds \; - \mu \int_{t_0}^t |\mathbf{x}'(s)| \, ds \,.$$

If Euclidean length is used for $|\mathbf{x}|$, then the second integral is arc length. Let

$$\mathbf{x}[a,b] \quad \text{denote arc length of } \mathbf{x}(t) \text{ on } [a,b] \,. \qquad (2.5.12)$$

Then we have

$$r|\mathbf{x}(t)| \leq V(t_0, \boldsymbol{\phi}(\cdot)) - \mu \int_{t_0}^t |\mathbf{x}(s)| \, ds \; - \mu \mathbf{x}[t_0, t] \,. \qquad (2.5.13)$$

Because $|\mathbf{x}(t)| \geq 0$, we have $\int_{t_0}^\infty |\mathbf{x}(s)| \, ds < \infty$, which implies that there is a sequence $\{t_n\} \to \infty$ with $|\mathbf{x}(t_n)| \to 0$. Also, $\mathbf{x}[t_0, t]$ is bounded. Thus, $|\mathbf{x}(t)| \to 0$. Because (a) is satisfied, the proof of (b) is complete.

Proof of (d). By (b), $\mathbf{x} = 0$ is uniformly stable. Find $\delta > 0$ such that $|\boldsymbol{\phi}| < \delta$ implies $|\mathbf{x}(t, \boldsymbol{\phi})| < 1$. Take $\eta = \delta$ and let $\varepsilon > 0$ be given. We then must find T such that

$$t_0 \geq 0, \ |\boldsymbol{\phi}(t)| < \delta \quad \text{on} \quad [0, t_0] \,, \quad \text{and} \quad t \geq t_0 + T$$

imply $|\mathbf{x}(t, \boldsymbol{\phi})| < \varepsilon$.

The proof has three distinct parts.

(i)   Find $L > 0$ and $\rho > 0$ with $(\varepsilon/2kL) + \rho\bar{K} + (\bar{K}\varepsilon M/L) < r\varepsilon$. For that ρ find S in (d). We show that if $|\mathbf{x}(t)| < \varepsilon/L$ on an interval of length S, then $|\mathbf{x}(t)| < \varepsilon$ always.

Suppose $|\mathbf{x}(t)| < \varepsilon/L$ on an interval $[t_1, t_1 + P]$ with $P \geq S$. Then at $t = t_1 + P$ we have (as $|\mathbf{x}(t)| < 1$)

$$r|\mathbf{x}(t)| \leq V(t, \mathbf{x}(\cdot)) = [\mathbf{x}^T B\mathbf{x}]^{1/2} + \int_0^{t_1} \int_{t_1+P}^\infty \bar{K}|C(u,s)|\, du\, |\mathbf{x}(s)|\, ds$$

$$+ \int_{t_1}^{t_1+P} \int_{t_1+P}^\infty \bar{K}|C(u,s)|\, du\, |\mathbf{x}(s)|\, ds$$

$$\leq (\varepsilon/2kL) + \rho\bar{K} + \bar{K}M\varepsilon/L < r\varepsilon.$$

As $V' \leq 0$, we have

$$r|x(t)| \leq V(t, \mathbf{x}(\cdot)) \leq V(t_1 + P, \mathbf{x}(\cdot)) < r\varepsilon$$

for $t \geq t_1 + P$, so that $|\mathbf{x}(t)| < \varepsilon$ if $t \geq t_1 + P$.

(ii)   Note that there is a $P_1 > 0$ such that the inequality $|\mathbf{x}(t)| \geq \varepsilon/2L$ on an interval of length P_1 must fail because

$$0 \leq r|\mathbf{x}(t)| \leq V(t, \mathbf{x}(\cdot)) \leq (1/2k) + M - \mu\int_{t_0}^t |\mathbf{x}(s)|\, ds.$$

(iii)   There is an N such that $|\mathbf{x}(t)|$ moves from $\varepsilon/2L$ to ε/L at most N times because

$$0 \leq V(t, \mathbf{x}(\cdot)) \leq (1/2k) + M - \mu\mathbf{x}[t_0, t].$$

Thus, on each interval of length $S + P_1$, either $|\mathbf{x}(t)|$ remains smaller than ε/L for S time units or moves from ε/L to $\varepsilon/2L$. The motion from ε/L to $\varepsilon/2L$ happens at most N times. Thus if $t > t_0 + N(S + P_1)$, then we will have $|\mathbf{x}(t)| < \varepsilon$ always. Then taking $T = N(S + P_1)$ completes the proof.

Remark. The corollary to Theorem 2.6.1 will show that the conclusion in (b) demonstrates uniform asymptotic stability in the convolution case.

Exercise 2.5.1. Consider the scalar equation

$$x' = -x + \int_0^t \alpha(t, s)(t - s + 1)^{-n} x(s)\, ds$$

for $n > 1$ and $\alpha(t, s)$ a continuous scalar function satisfying $|\alpha(t, s)| \leq d$ for some $d > 0$. Determine conditions on d and n to ensure that each part of

Theorem 2.5.1 is satisfied. That is, give different conditions for each part of the theorem. Pay careful attention to (d) and notice how Part (i) of the proof would be accomplished.

Exercise 2.5.2. Consider

$$x' = -x + \int_0^t d(t - s + 1)^{-n} x(s)\, ds + \sin t,$$

with d and n positive constants. Determine d and n such that the variation of parameters formula yields all solutions bounded.

There is also a variation of parameters formula for

$$\mathbf{x}' = A\mathbf{x} + \int_0^t C(t, s)\mathbf{x}(s)\, ds + \mathbf{F}(t), \qquad (2.5.14)$$

namely,

$$\mathbf{x}(t) = R(t, 0)\mathbf{x}(0) + \int_0^t R(t, s)\mathbf{F}(s)\, ds, \qquad (2.5.15)$$

where $R(t, s)$ is called the *resolvent* and is an $n \times n$ matrix that satisfies

$$\partial R(t, s)/\partial s = -R(t, s)A - \int_s^t R(t, u)C(u, s)\, du, \qquad (2.5.16)$$

for $0 \le s \le t$ and $R(t, t) = I$.

When $C(t, s)$ is of convolution type so is $R(t, s)$, and in fact, $R(t, s) = Z(t - s)$, where $Z(t)$ is the $n \times n$ matrix satisfying

$$Z'(t) = AZ(t) + \int_0^t C(t - s)Z(s)\, ds \qquad (2.5.17)$$

and $Z(0) = I$.

We found conditions for which $\int_0^\infty |\mathbf{x}(t)|\, dt < \infty$ for each solution of (2.5.1), so that $\int_0^\infty |Z(t)|\, dt < \infty$. Thus, in the convolution case, a bounded $\mathbf{F}$ in (2.5.14) produced bounded solutions.

But in the general case of (2.5.14), we have too little evidence of the integrability of $R(t, s)$. Thus we are motivated to consider Liapunov's direct method for the forced equation (2.5.14).

Extensive treatment of the resolvent may be found in Miller (1971a), in a series of papers by Miller (see also the references mentioned in Chapter 7), and in papers by Grossman and Miller appearing in the *Journal of Differential Equations* from 1969 to mid-1970s. Additional results and references are found in Grimmer and Seifert (1975). The following is one of their results, but the proof presented here is different.

Theorem 2.5.2. *Let A by an $n \times n$ constant matrix all of whose characteristic roots have negative real parts, let $C(t, s)$ be continuous for $0 \le s \le t < \infty$, and let $\mathbf{F} : [0, \infty) \to R^n$ be bounded and continuous. Suppose B satisfies $A^T B + BA = -I$ and α^2 and β^2 are the smallest and largest eigenvalues of B, respectively. If $\int_0^t |BC(t, s)|\, ds \le M$ for $0 \le t < \infty$ and $2\beta M/\alpha < 1$, then all solutions of $(2.5.14)$ are bounded.*

Proof. If the theorem is false, there is a solution $\mathbf{x}(t)$ with $\limsup_{t \to \infty} \mathbf{x}^T(t) B\mathbf{x}(t) = +\infty$. Thus, there are values of t with $|\mathbf{x}(t)|$ as large as we please and $\left[\mathbf{x}^T(t) B\mathbf{x}(t)\right]' \ge 0$, say, at $t = S$, and $\mathbf{x}^T(t) B\mathbf{x}(t) \le \mathbf{x}^T(S) B\mathbf{x}(S)$ if $t \le S$. Hence, at $t = S$ we have

$$
\begin{aligned}
\left[\mathbf{x}^T(t) B\mathbf{x}(t)\right]' &= -\mathbf{x}^T(t)\mathbf{x}(t) + \int_0^t 2\mathbf{x}^T(s) C^T(t, s) B\mathbf{x}(t)\, ds \\
&\quad + 2\mathbf{F}^T(t) B\mathbf{x}(t) \\
&\ge 0
\end{aligned}
$$

or

$$
\begin{aligned}
\mathbf{x}^T(S)\mathbf{x}(S) &\le \int_0^S 2\mathbf{x}^T(s) C^T(S, s) B\mathbf{x}(S)\, ds + 2\mathbf{F}^T(S) B\mathbf{x}(S) \\
&\le 2|\mathbf{x}(S)| \int_0^S |BC(S, s)|\, |\mathbf{x}(s)|\, ds + 2\mathbf{x}^T(S) B\mathbf{F}(S) \\
&\le 2|\mathbf{x}(S)| \int_0^S |BC(S, s)| \left[\left(\mathbf{x}^T(s) B\mathbf{x}(s)\right)^{1/2}/\alpha\right] ds \\
&\quad + 2\mathbf{x}^T(S) B\mathbf{F}(S) \\
&\le (2/\alpha)\, |\mathbf{x}(S)| \left(\mathbf{x}^T(S) B\mathbf{x}(S)\right)^{1/2} \int_0^S |BC(S, s)|\, ds \\
&\quad + 2\mathbf{x}^T(S) B\mathbf{F}(S) \\
&\le (2/\alpha)\, |\mathbf{x}(S)|\, \beta\, |\mathbf{x}(S)| M + 2\mathbf{x}^T(S) B\mathbf{F}(S) \\
&= (2\beta M/\alpha)\, |\mathbf{x}(S)|^2 + 2\mathbf{x}^T(S) B\mathbf{F}(S)\,.
\end{aligned}
$$

As $2\beta M/\alpha < 1$ we have a contradiction for $|\mathbf{x}(S)|$ sufficiently large.

The proof of the last theorem is a variant of what is known as the Liapunov-Razumikhin technique, which uses a Liapunov function (rather than a functional) to show boundedness and stability results for a functional differential equation. An introduction to the method for general functional differential equations is found in Driver (1962).

Detailed adaptations of the Razumikhin method to Volterra equations may be found in Grimmer and Seifert (1975) and in Grimmer (1979). Most of those results are discussed in Chapter 8. Halanay and Yorke (1971) argue very strongly for the merits of this method over the method of Liapunov functionals.

Notice that the main conditions in the last two theorems are very different. In Theorem 2.5.1 we mainly ask that $\int_t^\infty |C(u,t)|\, du$ be small, where the first coordinate is integrated. But in Theorem 2.5.2 we ask that $\int_0^t |BC(t,s)|\, ds$ be small, where the second coordinate is integrated.

Under certain conditions on $C(t,s)$ it is possible to obtain a differential inequality when considering (2.5.6), (2.5.7), and (2.5.1). That is, we differentiate $V(t, \mathbf{x}(\cdot))$ along a solution of (2.5.1) and attempt to find a scalar function $\eta(t) > 0$ with

$$V'(t, \mathbf{x}(\cdot)) \le \eta(t) V(t, \mathbf{x}(\cdot)) \,. \tag{2.5.18}$$

When that situation occurs, owing to the global Lipschitz condition in $\mathbf{x}(t)$ that V satisfies, it turns out that the derivative of V along a solution of the forced equation (2.5.14) results in the inequality

$$V'(t, \mathbf{x}(\cdot)) \le -\eta(t) V(t, \mathbf{x}(\cdot)) + K|\mathbf{F}(t)| \,. \tag{2.5.19}$$

It then follows that for a solution $\mathbf{x}(t, \boldsymbol{\phi})$ on $[t_0, \infty)$

$$V(t, \mathbf{x}(\cdot)) \le V(t_0, \phi) \exp\left[-\int_{t_0}^t \eta(s)\, ds \right]$$
$$+ K \int_{t_0}^t |\mathbf{F}(s)| \left\{ \exp\left[-\int_s^t \eta(u)\, du \right] \right\} ds \,, \tag{2.5.20}$$

which can ensure boundedness, depending on the properties of η and F.

Equation (2.5.20) becomes a substitute variation of parameters formula for (2.5.14), acting in place of (2.5.15). In fact, it may be superior to (2.5.15) in many ways even if much is known about $R(t,s)$. To see this, recall that, for a system of ordinary differential equations

$$\mathbf{x}' = P(t)\mathbf{x} + \mathbf{Q}(t)$$

with $P(t)$ not constant, if $Z(t)$ is the $n \times n$ matrix satisfying

$$Z'(t) = P(t)Z(t) \,, \quad Z(0) = I \,,$$

then the variation of parameters formula is

$$\mathbf{x}(t) = Z(t)\mathbf{x}(0) + \int_0^t Z(t)Z^{-1}(s)\mathbf{Q}(s)\, ds \,.$$

Even if $Z(t)$ is bounded, $Z^{-1}(s)$ may be very badly behaved. One usually needs to ask that

$$\int_0^t \operatorname{tr} P(s)\, ds \geq -M > -\infty$$

to utilize that variation of parameters formula; and this condition may imply that $Z(t) \nrightarrow 0$ as $t \to \infty$. In that case, the hope of concluding that bounded $\mathbf{Q}$ produces bounded $\mathbf{x}(t)$ vanishes.

To achieve (2.5.19) we examine

$$V' \leq \left[-k+\bar{K}\int_t^\infty |C(u,t)|\, du\right]|x|-(\bar{K}-K)\int_0^t |C(t,s)|\,|\mathbf{x}(s)|\, ds \qquad (2.5.7)$$

once more and observe that we require a function

$$\lambda : [0,\infty) \to [0,1]$$

with

$$|C(t,s)| \geq \lambda(t)\int_t^\infty |C(u,s)|\, du\,, \qquad (2.5.21)$$

for $0 \leq s \leq t < \infty$. For, in that case, if $\bar{K} > K$ and

$$\bar{k} \leq k - \bar{K}\int_t^\infty |C(u,t)|\, du$$

with $\bar{k}$ positive, then from (2.5.7) we have

$$V' \leq -\bar{k}|\mathbf{x}| - (\bar{K}-K)\lambda(t)\int_0^t\int_t^\infty |C(u,s)|\, du\, |\mathbf{x}(s)|\, ds$$
$$\leq -2k\bar{k}\lambda(t)[\mathbf{x}^T B\mathbf{X}]^{1/2}$$
$$\quad - [(\bar{K}-K)/\bar{K}\,]\lambda(t)\bar{K}\int_0^t\int_t^\infty |C(u,s)|\, du\, |\mathbf{x}(s)|\, ds$$
$$\leq -\eta(t)V(t,\mathbf{x}(\cdot))\,,$$

where

$$\eta(t) = \lambda(t)\min\left[2k\bar{k};\ (\bar{K}-K)/\bar{K}\,\right]. \qquad (2.5.22)$$

These calculations prove the following result.

Theorem 2.5.3. *Suppose that the conditions of Theorem 2.5.1 (b) hold and that (2.5.21) and (2.5.22) are satisfied. If $\mathbf{x}(t, \phi)$ is a solution of (2.5.14) on $[t_0, \infty)$ and if V is defined by (2.5.6), then*

$$V'(t, \mathbf{x}(\cdot)) \leq -\eta(t)V(t, \mathbf{x}(\cdot)) + K|\mathbf{F}(t)|,$$

and therefore,

$$V(t, \mathbf{x}(\cdot)) \leq V(t_0, \phi) \exp\left[-\int_{t_0}^{t} \eta(s)\, ds\right]$$
$$+ K \int_{t_0}^{t} |\mathbf{F}(s)| \left\{ \exp\left[-\int_{s}^{t} \eta(u)\, du\right] \right\} ds.$$

Exercise 2.5.3. Verify that (2.5.19) holds.

Exercise 2.5.4. Consider the scalar equation

$$x'(t) = -x(t) + \int_{0}^{t} C(t, s)x(s)\, ds + \alpha \cos t,$$

where $|C(t, s)| \leq c_1\{\exp[-h(t-s)]\}$ with c_1 and h being positive constants. Find conditions on h and c_1 to ensure that the conditions of Theorem 2.5.3 are satisfied and $\eta(t)$ is constant. Your conditions may take the form

$$\eta(t) = h(\beta - 1)/\beta \leq 1 \quad \text{for some} \quad \beta > 1$$

and

$$\beta c_1/h \leq \alpha \quad \text{for some} \quad \alpha < 1.$$

In the convolution case there is a natural way to search for $\eta(t)$.

Exercise 2.5.5. Consider the vector system

$$\mathbf{x}' = A\mathbf{x} + \int_{0}^{t} D(t-s)\mathbf{x}(s)\, ds \tag{2.5.23}$$

in which the characteristic roots of A have negative real parts and $|D(t)| > 0$ on $[0, \infty)$. Let B, k, and K be defined as before and suppose that there is a $d > K$ and $k_1 > 0$ with

$$k > k_1 \geq d \int_{t}^{\infty} |D(u-t)|\, du, \quad 0 \leq t < \infty.$$

Prove the following result.

Theorem 2.5.4. *If there is a continuous and nonincreasing scalar function* $\lambda : [0, \infty) \to (0, \infty)$ *with*

$$|D(v)| \geq \lambda(v) \int_v^\infty |D(u)| \, du \,,$$

then there is a constant $q > 0$ *such that for* $\mathbf{x}(t)$ *a solution of (2.5.23) and*

$$V(t, \mathbf{x}(\cdot)) = [\mathbf{x}^T B \mathbf{x}]^{1/2} + d \int_0^t \int_t^\infty |D(u - s)| \, du \, |\mathbf{x}(s)| \, ds$$

we have

$$V'(t, \mathbf{x}(\cdot)) \leq -q\lambda(t) V(t, \mathbf{x}(\cdot)) \,.$$

In our discussion of the variation of parameters formula for an ordinary differential equation

$$\mathbf{x}' = P(t)\mathbf{x} + \mathbf{Q}(t)$$

with P not constant, but P and Q continuous on $[0, \infty)$, we looked at $Z(t)Z^{-1}(s)$ where

$$Z'(t) = P(t)Z(t), \quad Z(0) = I \,.$$

Jacobi's identity [or the Wronskian theorem [see Hale (1969; pp. 90–91)] states that $\det Z(t) = \exp \int_0^t \operatorname{tr} P(s) \, ds$, so that $\det Z(t)$ never vanishes.

However, if $Z(t)$ is the principal matrix solution of

$$\mathbf{x}' = A\mathbf{x} + \int_0^t B(t - s)\mathbf{x}(s) \, ds \qquad (2.5.24)$$

with A constant and B continuous, then $\det Z(t)$ may vanish for many values of t.

Theorem 2.5.5. *Suppose that (2.5.24) is a scalar equation with* $A \leq 0$ *and* $B(t) \leq 0$ *on* $[0, \infty)$. *If there exists* $t_1 > 0$ *such that*

$$\int_{t_1}^t \int_0^{t_1} B(u - s) \, ds \, du \to -\infty \quad \text{as} \quad t \to \infty \,,$$

then there exists $t_2 > 0$ *such that if* $x(0) = 1$, *then* $x(t_2) = 0$.

Proof. If the theorem is false, then $x(t)$ has a positive minimum, say x_1, on $[0, t_1]$. Then for $t \geq t_1$ we have

$$x'(t) \leq \int_0^{t_1} B(t-s)x(s)\,ds + \int_{t_1}^t B(t-s)x(s)\,ds$$
$$\leq \int_0^{t_1} B(t-s)x_1\,ds$$

implying, upon integration, that

$$x(t) \leq x_1 + \int_{t_1}^t \int_0^{t_1} B(u-s)x_1\,ds\,du \to -\infty$$

as $t \to \infty$, a contradiction. This completes the proof.

2.6 Uniform Asymptotic Stability

We noticed in Theorem 2.5.1 that every solution $\mathbf{x}(t)$ of (2.5.1) may satisfy

$$\int_0^\infty |\mathbf{x}(t)|\,dt < \infty$$

(that is, $\mathbf{x}$ is $L^1[0, \infty)$) under considerably milder conditions than those required for uniform asymptotic stability. However, in the convolution case

$$\mathbf{x}' = A\mathbf{x} + \int_0^t D(t-s)\mathbf{x}(s)\,ds \tag{2.6.1}$$

with $D(t)$ continuous on $[0, \infty)$ and A being an $n \times n$ constant matrix, then

$$\int_0^\infty |D(t)|\,dt < \infty \quad \text{and} \quad \int_0^\infty |\mathbf{x}(t)|\,dt < \infty \tag{2.6.2}$$

is equivalent to uniform asymptotic stability of (2.6.1). This is a result of Miller (1971b), and we present part of it here.

Theorem 2.6.1. *If each solution $\mathbf{x}(t)$ of (2.6.1) on $[0, \infty)$ is $L^1[0, \infty)$, if $D(t)$ is $L^1[0, \infty)$, and if A is a constant $n \times n$ matrix, then the zero solution of (2.6.1) is uniformly asymptotically stable.*

Proof. If $Z(t)$ is the $n \times n$ matrix with $Z(0) = I$ and

$$Z'(t) = AZ(t) + \int_0^t D(t-s)Z(s)\,ds\,,$$

then $Z(t)$ is $L^1[0, \infty)$.

Let $\mathbf{x}(t, t_0, \boldsymbol{\phi})$ be a solution of (2.6.1) on $[t_0, \infty)$. Then

$$\mathbf{x}'(t, t_0, \boldsymbol{\phi}) = A\mathbf{x}(t, t_0, \boldsymbol{\phi}) + \int_0^{t_0} D(t-s)\boldsymbol{\phi}(s)\,ds + \int_{t_0}^t D(t-s)\mathbf{x}(s, t_0, \boldsymbol{\phi})\,ds\,,$$

so that

$$\mathbf{x}'(t + t_0, t_0, \boldsymbol{\phi}) = A\mathbf{x}(t + t_0, t_0, \boldsymbol{\phi}) + \int_0^{t_0} D(t + t_0 - s)\boldsymbol{\phi}(s)\,ds$$

$$+ \int_0^t D(t - s)\mathbf{x}(t_0 + s, t_0, \boldsymbol{\phi})\,ds$$

or $\mathbf{x}(t + t_0, t_0, \boldsymbol{\phi})$ is a solution of the nonhomogeneous equation

$$\mathbf{x}' = A\mathbf{x} + \int_0^t D(t - s)\mathbf{x}(s)\,ds + \int_0^{t_0} D(t + t_0 - s)\boldsymbol{\phi}(s)\,ds\,,$$

which we write as

$$\mathbf{y}' = A\mathbf{y} + \int_0^t D(t - s)\mathbf{y}(s)\,ds + \mathbf{F}(t) \tag{2.6.3}$$

with $\mathbf{y}(0) = \mathbf{x}(t_0, t_0, \boldsymbol{\phi}) = \boldsymbol{\phi}(t_0)$ and

$$\mathbf{F}(t) = \int_0^{t_0} D(t + t_0 - s)\boldsymbol{\phi}(s)\,ds\,. \tag{2.6.4}$$

By the variation of parameters formula [see Eq. (2.3.5) in Theorem 2.3.1] we have

$$\mathbf{y}(t) = Z(t)\boldsymbol{\phi}(t_0) + \int_0^t Z(t - s)\mathbf{F}(s)\,ds$$

or

$$\mathbf{x}(t + t_0, t_0, \boldsymbol{\phi}) = Z(t)\boldsymbol{\phi}(t_0) + \int_0^t Z(t - s)\left\{ \int_0^{t_0} D(s + t_0 - u)\boldsymbol{\phi}(u)\,du \right\}ds\,,$$

so that

$$\mathbf{x}(t + t_0, t_0, \boldsymbol{\phi}) = Z(t)\boldsymbol{\phi}(t_0)$$

$$+ \int_0^t Z(t - s)\left\{ \int_0^{t_0} D(s + u)\boldsymbol{\phi}(t_0 - u)\,du \right\}ds\,. \tag{2.6.5}$$

Next, notice that, because A is constant and $Z(t)$ is $L^1[0, \infty)$, then $AZ(t)$ is $L^1[0, \infty)$. Also, the convolution of two functions in $L^1[0, \infty)$ is

$L^1[0, \infty)$, as may be seen by Fubini's theorem [see Rudin (1966, p. 156).]
Thus $\int_0^t D(t-s)Z(s)\,ds$ is $L^1(0, \infty)$, and hence, $Z'(t)$ is $L^1[0, \infty)$.

Now, because $Z'(t)$ is $L^1[0, \infty)$, it follows that $Z(t)$ has a limit as $t \to \infty$. But, because $Z(t)$ is $L^1[0, \infty)$, the limit is zero.

Moreover, the convolution of an $L^1[0, \infty)$ function with a function tending to zero as $t \to \infty$ yields a function tending to zero as $t \to \infty$. (*Hint:* Use the dominated convergence theorem.) Thus

$$Z'(t) = AZ(t) + \int_0^t D(t-s)Z(s)\,ds \to 0$$

as $t \to \infty$.

Examine (2.6.5) again and review the definition of uniform asymptotic stability (Definition 2.4.4). We must show that $|\phi(t)| < \eta$ on $[0, t_0]$ implies that $\mathbf{x}(t + t_0, t_0, \phi) \to 0$ independently of t_0.

Now in (2.6.5) we see that $Z(t)\phi(t_0) \to 0$ independently of t_0. The second term is bounded by

$$\eta \int_0^t |Z(t-s)| \int_0^{t_0} |D(s+u)|\,du\,ds \leq \eta \int_0^t |Z(t-s)| \int_s^\infty |D(v)|\,dv\,ds\,,$$

and that is the convolution of an L^1 function with a function tending to zero as $t \to \infty$ and, hence, is a (bounded) function tending to zero as $t \to \infty$. Thus,

$$\mathbf{x}(t + t_0, t_0, \phi) \to 0 \quad \text{as} \quad t \to \infty$$

independently of t_0. The proof is complete

Corollary 1. *If the conditions of Theorem 2.5.1(b) hold and if $C(t, s)$ is of convolution type, then the zero solution of (2.5.1) is uniformly asymptotically stable.*

Proof. Under the stated conditions, we saw that solutions of (2.5.1) were $L^1[0, \infty)$.

Corollary 2. *Consider*

$$\mathbf{x}' = A\mathbf{x} + \int_0^t D(t-s)\mathbf{x}(s)\,ds \tag{2.6.1}$$

with A being an $n \times n$ constant matrix and D continuous on $[0, \infty)$. Suppose that each solution of (2.6.1) with initial condition $\mathbf{x}(0) = \mathbf{x}_0$ tends to zero as $t \to \infty$. If there is a function $\lambda(s)\varepsilon L^1[0, \infty)$ with $\int_0^{t_0} |D(s+u)|\,du \leq \lambda(s)$ for $0 \leq t_0 < \infty$ and $0 \leq s < \infty$, then the zero solution of (2.6.1) is uniformly asymptotically stable.

Proof. We see that $Z(t) \to 0$ as $t \to \infty$, and in (2.6.5), then, we have

$$\left| \mathbf{x}(t + t_0, t_0, \boldsymbol{\phi}) \right| \leq |Z(t)\boldsymbol{\phi}(t_0)| + \max_{0 \leq s \leq t_0} |\boldsymbol{\phi}(s)| \int_0^t |Z(t - s)|\lambda(s) \, ds \,.$$

The integral is the convolution of an L^1 function with a function tending to zero as $t \to \infty$ and, hence, tends to zero. Thus $\mathbf{x}(t + t_0, t_0, \boldsymbol{\phi}) \to 0$ as $t \to \infty$ uniformly in t_0. This completes the proof.

Example 2.6.1. Let $D(t) = (t+1)^{-n}$ for $n > 2$. Then

$$\int_0^{t_0} D(s + u) \, du = \int_0^{t_0} (s + u + 1)^{-n} \, du$$

$$= \frac{(s + u + 1)^{-n+1}}{-n+1} \bigg|_0^{t_0} \leq \frac{(s+1)^{-n+1}}{n-1} \,,$$

which is L^1.

Recall that for a linear system

$$\mathbf{x}' = A(t)\mathbf{x} \tag{2.6.6}$$

with $A(t)$ an $n \times n$ matrix and continuous on $[0, \infty)$, the following are equivalent:

(i) All solutions of (2.6.6) are bounded.
(ii) the zero solution is stable.

The following are also equivalent under the same conditions:

(i) All solutions of (2.6.6) tend to zero.
(ii) The zero solution is asymptotically stable.

However, when $A(t)$ is T-periodic, then the following are equivalent:

(i) All solutions of (2.6.6) are bounded.
(ii) The zero solution is uniformly stable.

Also, $A(t)$ periodic implies the equivalence of:

(i) All solutions of (2.6.6) tend to zero.
(ii) The zero solution is uniformly asymptotically stable.
(iii) All solutions of

$$\mathbf{x}' = A(t)\mathbf{x} + \mathbf{F}(t) \tag{2.6.7}$$

are bounded for each bounded and continuous $\mathbf{F} : [0, \infty) \to R^n$.

Property (iii) is closely related to Theorem 2.6.1. Also, the result is true with $|A(t)|$ bounded instead of periodic. But with A periodic, the result is simple, because, from Floquet theory, there is a nonsingular T-periodic matrix P and a constant matrix R with $Z(t) = P(t)e^{Rt}$ being an $n \times n$ matrix satisfying (2.6.6). By the variation of parameters formula each solution $\mathbf{x}(t)$ of (2.6.7) on $[0, \infty)$ may be expressed as

$$\mathbf{x}(t) = Z(t)\mathbf{x}(0) + \int_0^t Z(t)Z^{-1}(s)\mathbf{F}(s)\,ds.$$

In particular, when $\mathbf{x}(0) = 0$, then

$$\mathbf{x}(t) = \int_0^t P(t)[e^{R(t-s)}]P^{-1}(s)\mathbf{F}(s)\,ds.$$

Now $P(t)$ and $P^{-1}(s)$ are continuous and bounded. One argues that if $\mathbf{x}(t)$ is bounded for each bounded $\mathbf{F}$, then the characteristic roots of R have negative real parts; but, it is more to the point that

$$\int_0^\infty |P(t)e^{Rt}|\,dt < \infty.$$

Thus, one argues from (iii) that solutions of (2.6.6) are $L^1[0, \infty)$ and then that the zero solution of (2.6.6) is uniformly asymptotically stable.

We shall shortly (proof of Theorem 2.6.6) see a parallel argument for (2.6.1).

The preceding discussion is a special case of a result by Perron for $|A(t)|$ bounded. A proof may be found in Hale (1969; p. 152).

Problem 2.6.1. Examine (2.6.5) and decide if:

(a) boundedness of all solutions of (2.6.1) implies that $x = 0$ is stable.

(b) whenever all solutions of (2.6.1) tend to zero then the zero solution of (2.6.1) is asymptotically stable.

We next present a set of equivalent statements for a scalar Volterra equation of convolution type in which A is constant and $D(t)$ positive. An n-dimensional counterpart is given in Theorem 2.6.6.

Theorem 2.6.2. *Let A be a positive real number, $D : [0, \infty) \to (0, \infty)$ continuous, $\int_0^\infty D(t)\,dt < \infty$, $-A + \int_0^\infty D(t)\,dt \neq 0$, and*

$$x' = -Ax + \int_0^t D(t - s)x(s)\,ds. \tag{2.6.8}$$

The following statements are equivalent.

(a) *All solutions tend to zero.*

(b) $-A + \int_0^\infty D(t)\,dt < 0.$

(c) *Each solution is $L^1[0, \infty)$.*

(d) *The zero solution is uniformly asymptotically stable.*

(e) *The zero solution is asymptotically stable.*

Proof. We show that each statement implies the succeeding one and, of course, (e) implies (a).

Suppose (a) holds, but $-A + \int_0^{t_0} D(t)\,dt > 0$. Choose t_0 so large that $\int_0^\infty D(t)\,dt > A$ and let $\phi(t) \equiv 2$ on $[0, t_0]$. Then we claim that $x(t, \phi) > 1$ on $[t_0, \infty)$. If not, then there is a first t_1 with $x(t_1) = 1$, and therefore, $x'(t_1) \leq 0$. But

$$x'(t_1) = -Ax(t_1) + \int_0^{t_1} D(t_1 - s)x(s)\,ds$$

$$= -A + \int_0^{t_1} D(s)x(t_1 - s)\,ds$$

$$\geq -A + \int_0^{t_1} D(s)\,ds$$

$$> -A + \int_0^{t_0} D(s)\,ds > 0,$$

a contradiction. Thus (a) implies (b).

Let (b) hold and define

$$V(t, x(\cdot)) = |x| + \int_0^t \int_t^\infty D(u - s)\,du\,|x(s)|\,ds,$$

so that if $x(t)$ is a solution of (2.6.8), then

$$V'(t, x(\cdot)) \leq -A|x| + \int_0^t D(t - s)\,|x(s)|\,ds$$

$$+ \int_t^\infty D(u - t)\,du\,|x| - \int_0^t D(t - s)\,|x(s)|\,ds$$

$$= \left[-A + \int_t^\infty D(u - t)\,du \right] |x|$$

$$= \left[-A + \int_0^\infty D(u)\,du \right] |x|$$

$$= -\alpha|x| \quad \text{for some} \quad \alpha > 0.$$

An integration yields

$$0 \leq V(t, x(\cdot)) \leq V(t_0, \phi) - \alpha \int_{t_0}^{t} |x(s)| \, ds \,,$$

as required. Thus, (b) implies (c).

Now Theorem 2.6.1 shows that (c) implies (d).

Clearly (d) implies (e), and the proof is complete.

To this point we have depended on the strength of A to overcome the effects of $D(t)$ in

$$\mathbf{x}' = A\mathbf{x} + \int_0^t D(t-s)\mathbf{x}(s) \, ds \qquad (2.6.1)$$

to produce boundedness and stability. We now turn from that view and consider a system

$$\mathbf{x}' = A(t)\mathbf{x} + \int_0^t C(t, s)\mathbf{x}(s) \, ds + \mathbf{F}(t) \,, \qquad (2.6.9)$$

with A being an $n \times n$ matrix and continuous on $[0, \infty)$, $C(t, s)$ continuous for $0 \leq s \leq t < \infty$ and $n \times n$, and $\mathbf{F} : [0, \infty) \to R^n$ bounded and continuous. Suppose that

$$G(t, s) = -\int_t^{\infty} C(u, s) \, du \qquad (2.6.10)$$

is defined and continuous for $0 \leq s \leq t < \infty$. Define a matrix Q on $[0, \infty)$ by

$$Q(t) = A(t) - G(t, t) \qquad (2.6.11)$$

and require that

$$Q \text{ commutes with its integral} \qquad (2.6.12)$$

(as would be the case if A were constant and C of convolution type) and that

$$\left| \exp \int_u^t Q(s) \, ds \right| \leq M \exp \left[-\alpha(t - u) \right] \qquad (2.6.13)$$

for $0 \leq u \leq t$ and some positive constants M and α.

Here, when L is a square matrix, then e^L is defined as the usual power series (of matrices). Also, when $Q(t)$ commutes with its integral, then $\exp \int_{t_0}^{t} Q(s)\, ds$ is a solution matrix of

$$\mathbf{x}' = Q(t)\mathbf{x}\,.$$

Moreover,

$$Q(t) \exp \left[\int_{t_0}^{t} Q(s)\, ds \right] = \left\{ \exp \left[\int_{t_0}^{t} Q(s)\, ds \right] \right\} Q(t)\,.$$

Notice that (2.6.9) may be written as

$$\mathbf{x}' = \big[A(t) - G(t,t)\big]\mathbf{x} + \mathbf{F}(t) + (d/dt) \int_{0}^{t} G(t,s)\mathbf{x}(s)\, ds\,. \qquad (2.6.14)$$

If we subtract $Q\mathbf{x}$ from both sides, left multiply by $\exp\left[-\int_0^t Q(s)\, ds\,\right]$, and group terms, then we obtain

$$\left\{ \exp \left[-\int_0^t Q(s)\, ds\, \mathbf{x}(t) \right] \right\}'$$
$$= \left\{ \exp \left[-\int_0^t Q(s)\, ds \right] \right\} \left[(d/dt) \int_0^t G(t,s)\mathbf{x}(s)\, ds + \mathbf{F}(t) \right]\,.$$

Let ϕ be a given continuous initial function on $[0, t_0]$. Integrate the last equation from t_0 to t and obtain

$$\left\{ \exp \left[-\int_0^t Q(s)\, ds \right] \right\} \mathbf{x}(t)$$
$$= \left\{ \exp \left[-\int_0^{t_0} Q(s)\, ds \right] \right\} \mathbf{x}(t_0)$$
$$+ \int_{t_0}^t \left\{ \exp \left[-\int_0^u Q(s)\, ds \right] \right\} \mathbf{F}(u)\, du$$
$$+ \int_{t_0}^t \left\{ \exp \left[-\int_0^u Q(s)\, ds \right] \right\} \times (d/du) \int_0^u G(u,s)\mathbf{x}(s)\, du$$

If we integrate the last term by parts, we obtain

$$\left\{ \exp\left[-\int_0^t Q(s)\,ds \right] \right\} \mathbf{x}(t)$$

$$= \left\{ \exp\left[-\int_0^{t_0} Q(s)\,ds \right] \right\} \mathbf{x}(t_0)$$

$$+ \int_{t_0}^t \left\{ \exp\left[-\int_0^u Q(s)\,ds \right] \right\} \mathbf{F}(u)\,du$$

$$+ \left\{ \exp\left[-\int_0^t Q(s)\,ds \right] \right\} \int_0^t G(t,s)\mathbf{x}(s)\,ds$$

$$- \left\{ \exp\left[-\int_0^{t_0} Q(s)\,ds \right] \right\} \int_0^{t_0} G(t_0,s)\mathbf{x}(s)\,ds$$

$$+ \int_{t_0}^t Q(u)\left\{ \exp\left[-\int_0^u Q(s)\,ds \right] \right\} \int_0^u G(u,s)\mathbf{x}(s)\,ds\,du\,.$$

Left multiply by $\exp\left[\int_0^t Q(s)\,ds \right]$, take norms, and use (2.6.13) to obtain

$$|\mathbf{x}(t)| \leq \left[M|\mathbf{x}(t_0)| + \int_0^{t_0} |G(t_0,s)||\boldsymbol{\phi}(s)|\,ds \right] \exp\left[-\alpha(t-t_0) \right]$$

$$+ \int_{t_0}^t Me^{-\alpha(t-u)}|\mathbf{F}(u)|\,du + \int_0^t |G(t,s)|\,|\mathbf{x}(s)|\,ds \qquad (2.6.15)$$

$$+ \int_{t_0}^t |Q(u)|Me^{-\alpha(t-u)} \int_0^u |G(u,s)|\,|\mathbf{x}(s)|\,ds\,du\,.$$

Theorem 2.6.3. If $\mathbf{x}(t)$ is a solution of (2.6.9), if $|Q(t)| \leq D$ on $[0,\infty)$ for some $D > 0$, and if $\sup_{0 \leq t < \infty} \int_0^t |G(t,s)|\,ds \leq \beta$, then for β sufficiently small, $\mathbf{x}(t)$ is bounded.

Proof. For the given t_0 and $\boldsymbol{\phi}$, because $\mathbf{F}$ is bounded there is a $K_1 > 0$ with

$$M|\mathbf{x}(t_0)| + \int_0^{t_0} \big|G(t_0,s)\boldsymbol{\phi}(s)\big|\,ds$$

$$+ \sup_{t_0 \leq t < \infty} \int_{t_0}^t M\big\{ \exp\left[-\alpha(t-u) \right] \big\}\,|\mathbf{F}(u)|\,du < K_1\,.$$

From this and (2.6.15) we obtain

$$|\mathbf{x}(t)| \le K_1 + \int_0^t |G(t,s)|\,|\mathbf{x}(s)|\,ds$$

$$+ \int_{t_0}^t DM \exp\big[-\alpha(t-u)\big] \int_0^u |G(u,s)|\,|\mathbf{x}(s)|\,ds\,du$$

$$\le K_1 + \beta \sup_{0 \le s \le t} |\mathbf{x}(s)| + (DM\beta/\alpha) \sup_{0 \le s \le t} |\mathbf{x}(s)|$$

$$= K_1 + \beta\big[1 + (DM/\alpha)\big] \sup_{0 \le s \le t} |\mathbf{x}(s)|\,.$$

Let β be chosen so that $\beta\big[1 + (DM/\alpha)\big] = m < 1$, yielding

$$|\mathbf{x}(t)| \le K_1 + m \sup_{0 \le s \le t} |\mathbf{x}(s)|\,.$$

Let $K_2 > \max_{0 \le t \le t_0} |\phi(t)|$ and $K_1 + mK_2 < K_2$. If $|\mathbf{x}(t)|$ is not bounded, then there is a first $t_1 > t_0$ with $|\mathbf{x}(t_1)| = K_2$. Then

$$K_2 = |\mathbf{x}(t_1)| \le K_1 + mK_2 < K_2\,,$$

a contradiction. This completes the proof.

Exercise 2.6.1. Let $a > 0$ and

$$x' = -\int_0^t a(t - s + 1)^{-3} x(s)\,ds + F(t)\,.$$

Work through the entire sequence of steps from (2.6.10) to (2.6.15). Then state Theorem 2.6.3 for this equation, let $F(t) = \sin t$ and $\phi(t) = 1$ on $[0, t_0]$, and follow the proof of Theorem 2.6.3 to find M, K_1, D, α, K_2, and β.

Exercise 2.6.2. Interchange the order of integration in the last term of (2.6.15), assume $|G(t,s)| \le Le^{-\gamma(t-s)}$ for L and γ positive, and use Gronwall's inequality to bound $|\mathbf{x}(t)|$ under appropriate restrictions on the constants.

Theorem 2.6.4. In (2.6.9) let $\mathbf{F}(t) \equiv 0$, $|Q(t)| \le D$ on $[0, \infty)$ and $\int_0^t |G(t,s)|\,ds \le \beta$. If β is sufficiently small, then the zero solution is uniformly stable.

Proof. Let $\varepsilon > 0$ be given. We wish to find $\delta > 0$ such that

$$t_0 \ge 0\,, \quad |\phi(t)| < \delta \text{ on } [0, t_0]\,, \quad \text{and} \quad t \ge t_0$$

imply $|\mathbf{x}(t, \phi)| < \varepsilon$.

Let $\delta < \varepsilon$ with δ yet to be determined. If $|\phi(t)| < \delta$ on $[0, t_0]$, then

$$M|\mathbf{x}(t_0)| + \int_0^{t_0} \left| G(t_0, s)\phi(s) \right| ds \le (M + \beta)\delta.$$

From (2.6.15) (with $\mathbf{F} \equiv 0$),

$$|\mathbf{x}(t)| \le (M + \beta)\delta + \beta \sup_{0 \le s \le t} |\mathbf{x}(s)| + (\beta DM/\alpha) \sup_{0 \le s \le t} |\mathbf{x}(s)|$$

$$= (M + \beta)\delta + \beta\big[1 + (DM/\alpha)\big] \sup_{0 \le s \le t} |\mathbf{x}(s)|.$$

First, pick β so that $\beta\big[1 + (DM/\alpha)\big] \le \frac{3}{4}$. Then pick δ so that $(M + \beta)\delta + \frac{3}{4}\varepsilon < \varepsilon$. If $|\phi(t)| < \delta$ on $[0, t_0]$ and if there is a first $t_1 > t_0$ with $|\mathbf{x}(t_1)| = \varepsilon$, we have

$$\varepsilon = |\mathbf{x}(t_1)| < (M + \beta)\delta + \frac{3}{4}|\mathbf{x}(t_1)| \le \varepsilon,$$

a contradiction. Thus, the zero solution is uniformly stable. The proof is complete.

Naturally, one believes that the conditions of Theorem 2.6.3 imply that the unforced equation (2.6.9) is uniformly asymptotically stable. We would expect to give a proof parallel to that of Perron showing that the resolvent satisfies $\sup_{0 \le t < \infty} \int_0^t |R(t, s)|\, ds < \infty$, where $R(t, s)$ is defined in (2.5.16). We would then hope to use the variation of parameters formula (2.5.15) and prove a theorem similar to Theorem 2.6.1 showing that $R(t, s)$ in $L^1[0, \infty)$ implies uniform asymptotic stability. Now we proceed with the convolution case.

Theorem 2.6.5.   *Let the conditions of Theorem 2.6.3 hold and let $\mathbf{F} \equiv 0$. Suppose also that A is constant and $C(t, s) = D(t - s)$. If $\sup_{0 \le t < \infty} \int_0^t |G(t - s)|\, ds \le \beta$, then for β sufficiently small the zero solution of (2.6.9) is uniformly asymptotically stable.*

Proof. By Theorem 2.6.3 all solutions of (2.6.9) are bounded for bounded $\mathbf{F}$. If $Z'(t) = AZ(t) + \int_0^t D(t - s)Z(s)\, ds$ with $Z(0) = I$, then by the variation of parameters formula a solution $\mathbf{x}(t)$ of (2.6.9) on $[0, \infty)$ is written as

$$\mathbf{x}(t) = Z(t)\mathbf{x}(0) + \int_0^t Z(t - s)\mathbf{F}(s)\, ds.$$

For $\mathbf{x}(0) = 0$ this is

$$\mathbf{x}(t) = \int_0^t Z(t - s)\mathbf{F}(s)\, ds,$$

which is bounded for every bounded $\mathbf{F}$. One may repeat the proof of Perron's theorem [Hale (1969, p. 152)] for ordinary differential equations to conclude that

$$\int_0^\infty |Z(t)|\, dt < \infty\,.$$

By Theorem 2.6.1 the zero solution is uniformly asymptotically stable. The proof is complete.

Remark. Theorems 2.6.3–2.6.5 are found in Burton (1983a).

We return now to the n-dimensional system

$$\mathbf{x}' = A\mathbf{x} + \int_0^t D(t-s)\mathbf{x}(s)\, ds\,, \tag{2.6.1}$$

with A constant and D continuous. Our final result of this section is a set of equivalences for systems similar to Theorems 2.6.2 for scalar equations. These two results may be found in Burton and Mahfoud (1983), together with examples showing a certain amount of sharpness.

Let $Z(t)$ be the $n \times n$ matrix satisfying

$$Z'(t) = AZ(t) + \int_0^t D(t-s)Z(s)\, ds\,, \quad Z(0) = I\,. \tag{2.6.16}$$

Theorem 2.6.6. *Suppose there is a constant $M > 0$ such that $0 \le t_0 < \infty$ and $0 \le t < \infty$ we have*

$$\int_0^t \int_0^{t_0} |D(u+v)|\, du\, dv \le M\,. \tag{2.6.17}$$

Then the following statements are equivalent.

(a) *$Z(t) \to 0$ as $t \to \infty$.*
(b) *All solutions $\mathbf{x}(t, t_0, \boldsymbol{\phi})$ of (2.6.1) tend to zero as $t \to \infty$.*
(c) *The zero solution of (2.6.1) is uniformly asymptotically stable.*
(d) *$Z(t)$ is in $L^1[0, \infty)$ and $Z(t)$ is bounded.*
(e) *Every solution $\mathbf{x}(t, 0, \mathbf{x}_0)$ of*

$$\mathbf{x}' = A\mathbf{x} + \int_0^t D(t-s)\mathbf{x}(s)\, ds + \mathbf{F}(t) \tag{2.6.18}$$

on $[0, \infty)$ is bounded for every bounded and continuous $\mathbf{F} : [0, \infty) \to R^n$.

(f) *The zero solution of (2.6.1) is asymptotically stable.*

Furthermore, the following is a second set of equivalents:

(g) $Z(t)$ *is bounded.*

(h) *All solutions* $\mathbf{x}(t, t_0, \phi)$ *of (2.6.1) are bounded.*

(i) *The zero solution of (2.6.1) is uniformly stable.*

(j) *The zero solution of (2.6.1) is stable.*

Proof. Let (a) hold. Then a solution $\mathbf{x}(t, t_0, \phi)$ of (2.6.1) may be considered as a solution of

$$\mathbf{x}' = A\mathbf{x} + \int_0^{t_0} D(t-s)\phi(s)\,ds + \int_{t_0}^t D(t-s)\mathbf{x}(s)\,ds$$

for $t \geq t_0$ with the second term on the right treated as a forcing term. If we translate the equation by $\mathbf{y}(t) = \mathbf{x}(t + t_0)$, we obtain

$$\mathbf{y}'(t) = A\mathbf{y}(t) + \int_0^t D(t-s)\mathbf{y}(s)\,ds + \int_0^{t_0} D(t+t_0-s)\phi(s)\,ds\,.$$

We may now apply the variation of parameters formula and write

$$\mathbf{y}(t) = Z(t)\phi(t_0) + \int_0^t Z(t-u) \int_0^{t_0} D(u+t_0-s)\phi(s)\,ds\,du\,.$$

The substitution $s = t_0 - v$ yields

$$\mathbf{y}(t) = Z(t)\phi(t_0) + \int_0^t Z(t-u) \int_0^{t_0} D(u+v)\phi(t_0-v)\,dv\,du\,.$$

Because $|\phi(t)| \leq K$, $K > 0$, on $[0, t_0]$, we have

$$|\mathbf{y}(t)| \leq K|Z(t)| + K \int_0^t |Z(t-u)| \int_0^{t_0} |D(u+v)|\,dv\,du\,.$$

The last term is the convolution of an $L^1[0, \infty)$ function ($\int_0^{t_0} |D(u+v)|\,dv$) with a function tending to zero ($Z(t)$), and so it tends to zero. Thus, (a) implies (b).

Suppose that (b) holds. Then, in particular, all solutions of the form $\mathbf{x}(t, 0, \mathbf{x}_0)$ tend to zero, which implies that $Z(t) \to 0$ as $t \to \infty$. Now

$$\int_0^t \int_0^{t_0} |D(u + v)|\, du\, dv \leq M$$

uniformly in t_0. Thus the integral

$$\int_0^t |Z(t - u)| \int_0^{t_0} |D(u + v)|\, dv\, du$$

tends to zero uniformly in t_0 and hence $\mathbf{x}(t, t_0, \boldsymbol{\phi})$ tends to zero uniformly in t_0 for bounded $\boldsymbol{\phi}$. Thus, (b) implies (c).

Let (c) hold. Then Miller's result implies that $Z(t)$ is $L^1[0, \infty)$. Also, the uniform asymptotic stability implies that $Z(t)$ is bounded. Hence, (d) holds.

Suppose (d) is satisfied. Then solutions $\mathbf{x}(t, 0, \mathbf{x}_0)$ of (2.6.18) on $[0, \infty)$ are expressed as

$$\mathbf{x}(t) = Z(t)\mathbf{x}(0) + \int_0^t Z(t - s)\mathbf{F}(s)\, ds\,.$$

Because $Z(t)$ is $L^1[0, \infty)$ and bounded and because $\mathbf{F}$ is bounded, then $\mathbf{x}(t)$ is bounded. Hence, (e) holds.

Suppose (e) is satisfied. Then the argument in the proof of Perron's theorem [see Hale (1969, p. 152)] yields $Z(t)$ being $L^1[0, \infty)$. This, in turn, implies uniform asymptotic stability. Of course, uniform asymptotic stability implies asymptotic stability, so (e) implies (f).

Certainly, (f) implies (a). This completes the proof of the first set of equivalences.

Let (g) hold. The variation of parameters formula implies that $\mathbf{x}(t, \boldsymbol{\phi})$ is bounded. Thus, (h) holds.

Suppose (h) is satisfied. Then $|Z(t)| \leq P$ and $\int_0^t \int_0^{t_0} |D(u + v)|\, du\, dv \leq M$ imply that whenever $|\boldsymbol{\phi}(t)| < \delta$ on $[0, t_0]$ we have

$$|\mathbf{x}(t + t_0, \boldsymbol{\phi})| \leq P|\boldsymbol{\phi}(t_0)| + \delta \int_0^t |Z(t - u)| \int_0^{t_0} |C(u + v)|\, dv\, du$$

$$\leq P\delta + \delta PM < \varepsilon\,,$$

provided that $\delta < \varepsilon/(P + PM)$. Hence, $\mathbf{x} = 0$ is uniformly stable. Thus, (h) implies (i).

Certainly, (i) implies (j).

Finally, if $\mathbf{x} = 0$ is stable, then $Z(t)$ is bounded, so (j) implies (g). This completes the proof of the theorem.

2.7   Reducible Equations Revisited

In Section 1.5 we dealt with Volterra equations reducible to ordinary differential equations. That discussion led us to a large class of solvable equations. But that is only a small part of a general theory of equations that can be reduced to Volterra equations with $L^1[0, \infty)$ kernels.

Though all this can be done for vector equations, it is convenient to consider the scalar equations

$$x(t) = f(t) + \int_0^t C(t-s)x(s)\,ds \tag{2.7.1}$$

or

$$x' = Ax + \int_0^t C(t-s)x(s)\,ds + f(t) \tag{2.7.2}$$

in which f and C have n continuous derivatives on $[0, \infty)$ and A is constant.

Definition 2.7.1. *Equation (2.7.1) or (2.7.2) is said to be*

(a) reducible *if $C(t)$ is a solution of a linear nth-order ordinary differential equation with constant coefficients*

$$L(y) = a_0 y^{(n)} + a_1 y^{(n-1)} + \cdots + a_n y = F(t) \tag{2.7.3}$$

with F continuous on $[0, \infty)$ and

$$\int_0^\infty |F(t)|\,dt < \infty\,; \tag{2.7.4}$$

(b) *V-reducible if it is reducible and if*

$$\int_0^t \int_t^\infty |F(u-s)|\,du\,ds \quad \text{exists on} \quad [0, \infty)\,; \tag{2.7.5}$$

(c) *t-reducible if it is reducible and if*

$$\int_0^\infty |tF(t)|\,dt < \infty\,; \tag{2.7.6}$$

(d) uniformly reducible *if it is reducible and if there exists $M > 0$ with*

$$\int_0^t \int_0^{t_0} |F(u+v)|\,du\,dv \le M \tag{2.7.7}$$

for $0 \le t < \infty$ and $0 \le t_0 < \infty$;

(e) completely reducible *if reducible and*

$$F(t) \equiv 0\,. \tag{2.7.8}$$

We have discussed (e) in Section 1.5. The vast majority of stability results for (2.7.2) concern one of the reducible forms. In most cases the results are stated for $n = 0$, so that one is directly assuming at least one of (2.7.4)–(2.7.7) with F replaced by C, so that (2.7.3) is just $L(y) = y = C(t)$.

Of course, when (2.7.1) or (2.7.2) is reducible, then we operate on it using (2.7.3) to obtain a higher-order integro-differential equation with F as the new kernel. Thus, (2.7.4) is the basic assumption for Miller's results on uniform asymptotic stability (U.A.S.); (2.7.5) was used in Theorem 2.5.1 and elsewhere; (2.7.6) is a basic (but unstated) assumption of Brauer (1978) in deriving certain results on U.A.S.; and (2.7.7) was just used in Theorem 2.6.6.

We now give examples that will be of interest in later chapters.

Example 2.7.1. Let A and α be constants and suppose $\int_0^\infty |C'(v)|\, dv < \infty$. If we differentiate the scalar equation

$$x' = Ax + \int_0^t \big[\alpha + C(t-s)\big]x(s)\, ds \,, \tag{2.7.9}$$

we obtain

$$x'' = Ax' + \big[\alpha + C(0)\big]x + \int_0^t C'(t-s)x(s)\, ds \,, \tag{2.7.10}$$

which we may write as

$$x' = y$$

$$y' = \big[\alpha + C(0)\big]x + Ay + \int_0^t C'(t-s)x(s)\, ds \,,$$

or in matrix form as

$$\begin{pmatrix} x \\ y \end{pmatrix}' = \begin{pmatrix} 0 & 1 \\ \alpha + C(0) & A \end{pmatrix}\begin{pmatrix} x \\ y \end{pmatrix} + \int_0^t \begin{pmatrix} 0 & 0 \\ C'(t-s) & 0 \end{pmatrix}\begin{pmatrix} x(s) \\ y(s) \end{pmatrix} ds \,,$$

which we finally express as

$$\mathbf{X}' = B\mathbf{X} + \int_0^t D(t-s)\mathbf{X}(s)\, ds \,, \tag{2.7.11}$$

in which D is in $L^1[0, \infty)$ and (2.7.5) is satisfied so that (2.7.9) is V-reducible. Now, it is possible to investigate (2.7.11) by means of Theorem 2.5.1 (and others) because (2.7.9) is not covered by Theorem 2.5.1.

Let us look more closely at (2.7.10). The integral is viewed as a small perturbation of

$$x'' - Ax' - [\alpha + C(0)]x = 0 \tag{2.7.12}$$

because $C'(t)$ is in $L^1[0, \infty)$. Also $-A$ is the coefficient of the "damping," whereas $-[\alpha + C(0)]$ is the coefficient of the "restoring force." From well-established theory of ordinary differential equations we expect (2.7.10) to be stable if $-A > 0$, $-[\alpha + C(0)] > 0$, and $D(t)$ is small.

Example 2.7.2. Consider the scalar equation

$$x' = Ax + \int_0^t B \ln(t - s + a)x(s)\, ds \tag{2.7.13}$$

with $a > 1$, $A < 0$, and $b < 0$. Differentiate and obtain

$$x'' = Ax' + b(\ln a)x + \int_0^t b(t - s + a)^{-1}x(s)\, ds$$

and

$$x''' = Ax'' + b(\ln a)x' + (b/a)x - \int_0^t b(t - s + a)^{-2}x(s)\, ds.$$

Now the kernel is $L^1[0, \infty)$ so we express it as a system

$$x' = y$$
$$y' = z$$
$$z' = (b/a)x + b(\ln a)y + Az - \int_0^t b(t - s + a)^{-2}x(s)\, ds,$$

which may be written as

$$\mathbf{X}' = B\mathbf{X} + \int_0^t D(t - s)\mathbf{X}(s)\, ds$$

with D in $L^1[0, \infty)$. By the Routh-Hurwitz criterion, the characteristic roots of B will have negative real parts if $|aA \ln a| > 1$. We expect stability if b is small enough.

Exercise 2.7.1. Consider the scalar equation

$$x' = Ax + \int_0^t b[\cos(t - s)](t - s + a)^{-1}x(s)\, ds. \tag{2.7.14}$$

Can (2.7.14) be reduced to an integro-differential equation with L^1 kernel?

These problems and theorems give us a good start in our understanding of stability. We will see various parts of them again in Chapters 5, 6, 7, and 8. The problems with arc length will become central. The constructions for Liapunov functionals will generalize in a natural way to systems. Chapter 3 will provide details on existence, uniqueness, and continuation which were simply assumed in this introductory chapter.

Chapter 3

Existence Properties

3.1 Definitions, Background, and Review

From our point of view there is a close parallel between the existence theory of ordinary differential equations and that of integral equations. Indeed, ordinary differential equations are frequently converted to integral equations to prove existence results. We hasten to add, however, that some writers consider integral equations under such general conditions that the similarities are lost.

Consistent with our aim to make a very gentle transition from differential equations to integral equations, we first state the standard results for ordinary differential equations and briefly sketch the concept of proof as a motivation, a comparison, and, sometimes, a contrast with integral equations.

Definition 3.1.1. *Let $\{f_n(t)\}$ be a sequence of functions from an interval $[a, b]$ to real numbers.*

(a) $\{f_n(t)\}$ is uniformly bounded *on $[a, b]$ if there exists M such that*

$$n \text{ a positive integer and } t \in [a, b]$$

imply $|f_n(t)| \leq M$.

(b) $\{f_n(t)\}$ is equicontinuous *if for any $\varepsilon > 0$ there exists $\delta > 0$ such that*

$$\left[n \text{ a positive integer, } t_1 \in [a, b], \ t_2 \in [a, b], \text{ and } |t_1 - t_2| < \delta \right]$$

imply $|f_n(t_1) - f_n(t_2)| < \varepsilon$.

Part (b) is sometimes called uniformly equicontinuous. Also, some writers consider a family of functions (possibly uncountable) instead of a sequence. Presumably, one uses the axiom of choice to obtain a sequence from the family.

Lemma 3.1.1.   Ascoli-Arzela   *If $\{f_n(t)\}$ is a uniformly bounded and equicontinuous sequence of real functions on an interval $[a, b]$, then there is a subsequence that converges uniformly on $[a, b]$ to a continuous function.*

Proof. Because the rational numbers are countable, we may let $t_1, t_2, \ldots$ be a sequence of all rational numbers on $[a, b]$ taken in any fixed order. Consider the sequence $\{f_n(t_1)\}$. This sequence is bounded, so it contains a convergent subsequence, say, $\{f_n^1(t_1)\}$, with limit $\phi(t_1)$. The sequence $\{f_n^1(t_2)\}$ also has a convergent subsequence, say, $\{f_n^2(t_2)\}$, with limit $\phi(t_2)$. If we continue in this way, we obtain a sequence of sequences (there will be one sequence for each value of m):

$$f_n^m(t), \quad m = 1, 2, \ldots; \quad n = 1, 2, \ldots,$$

each of which is a subsequence of all the preceding ones, such that for each m we have

$$f_n^m(t_m) \to \phi(t_m) \quad \text{as} \quad n \to \infty.$$

We select the diagonal. That is, consider the sequence of functions

$$F_k(t) = f_k^k(t).$$

It is a subsequence of the given sequence and is, in fact, a subsequence of each of the sequences $\{f_n^m(t)\}$, for n large. As $f_n^m(t_m) \to \phi(t_m)$, it follows that $F_k(t_m) \to \phi(t_m)$ as $k \to \infty$ for each m.

We now show that $\{F_k(t)\}$ converges uniformly on $[a, b]$. Let $\varepsilon_1 > 0$ be given, and let $\varepsilon = \varepsilon_1/3$. Choose δ with the property described in the definition of equicontinuity for the number ε. Now, divide the interval $[a, b]$ into p equal parts, where p is any integer larger than $(b - a)/\delta$. Let ξ_j be a rational number in the jth part $(j = 1, \ldots, p)$; then $\{F_k(t)\}$ converges at each of these points. Hence, for each j there exists an integer M_j such that $|F_r(\xi_j) - F_s(\xi_j)| < \varepsilon$ if $r > M_j$ and $s > M_j$. Let M be the largest of the numbers M_j.

If t is in the interval $[a, b]$, it is in one of the p parts, say the jth, so $|t - \xi_j| < \delta$ and $|F_k(t) - F_k(\xi_j)| < \varepsilon$ for every k. Also, if $r > M \geq M_j$ and $s > M$, then $|F_r(\xi_j) - F_s(\xi_j)| < \varepsilon$. Hence, if $r > M$ and $s > M$, then

$$
\begin{aligned}
&|F_r(t) - F_s(t)| \\
&\quad = \big|\big(F_r(t) - F_r(\xi_j)\big) + \big(F_r(\xi_j) - F_s(\xi_j)\big) - \big(F_s(t) - F_s(\xi_j)\big)\big| \\
&\quad \leq |F_r(t) - F_r(\xi_j)| + |F_r(\xi_j) - F_s(\xi_j)| + |F_s(t) - F_s(\xi_j)| \\
&\quad < 3\varepsilon = \varepsilon_1 .
\end{aligned}
$$

By the Cauchy criterion for uniform convergence, the sequence $\{F_k(t)\}$ converges uniformly to some function $\phi(t)$. As each $F_k(t)$ is continuous, so is $\phi(t)$. This completes the proof.

The lemma is, of course, also true for vector functions. Suppose that $\{F_n(t)\}$ is a sequence of functions from $[a, b]$ to R^p, for instance, $\mathbf{F}_n(t) = \big(f_n(t)_1, \ldots, f_n(t)_p\big)$. [The sequence $\{\mathbf{F}_n(t)\}$ is uniformly bounded and equicontinuous if all the $\{f_n(t)_j\}$ are.] Pick a uniformly convergent subsequence $\{f_{kj}(t)_1\}$ using the lemma. Consider $\{f_{kj}(t)_2\}$ and use the lemma to obtain a uniformly convergent subsequence $\{f_{kjr}(t)_2\}$. Continue and conclude that $\{\mathbf{F}_{kjr\cdots s}(t)\}$ is uniformly convergent.

We are now in a position to state the fundamental existence theorem for the initial-value problem for ordinary differential equations.

Theorem 3.1.1.   *Let $(t_0, \mathbf{x}_0) \in R^{n+1}$ and suppose there are positive constants a, b, and M such that $D = \big\{(t, \mathbf{x}) : |t - t_0| \leq a,\ |\mathbf{x} - \mathbf{x}_0| \leq b\big\}$, $\mathbf{G} : D \to R^n$ is continuous, and $|\mathbf{G}(t, \mathbf{x})| \leq M$ if $(t, \mathbf{x}) \in D$. Then there is at least one solution $x(t)$ of*

$$
\mathbf{x}' = \mathbf{G}(t, \mathbf{x}), \quad \mathbf{x}(t_0) = \mathbf{x}_0 , \tag{3.1.1}
$$

and $\mathbf{x}(t)$ is defined for $|t - t_0| \leq T$ with $T = \min[a,\ b/M]$.

Sketch of Proof. Let j be a fixed positive integer and divide $[t_0, t_0 + T]$ into j equal subintervals, say, $t_0 < t_1 < \cdots < t_n = t_0 + T$, with $t_{k+1} - t_k = \delta_j = T/j$. (We construct the solution only on $[t_0, t_0 + T]$; the rest being similar, its construction is left to the reader.) Define a continuous function $\mathbf{x}_j(t)$ on $[t_0, t_0 + T]$ as follows. Let

$$
\mathbf{x}_j(t) = \mathbf{x}_0 + (t - t_0)\mathbf{G}(t_0, \mathbf{x}_0) \quad \text{for} \quad t_0 \leq t \leq t_1 ,
$$

and define

$$
\mathbf{x}_j(t_1) = \mathbf{y}_1 .
$$

Then

$$\mathbf{x}_j(t) = \mathbf{y}_1 + (t - t_1)\mathbf{G}(t_1, \mathbf{y}_1) \quad \text{on} \quad [t_1, t_2]$$

and

$$\mathbf{x}_j(t_2) = \mathbf{y}_2 .$$

We may continue in this way and define a piecewise linear function on $[t_0, t_0 + T]$. Because of the properties of M and T, the graph will not leave D.

One then shows that $\{\mathbf{x}_j(t)\}$ is uniformly bounded and equicontinuous. This follows from the fact that each $\mathbf{x}_j(t)$ stays in D and the slopes are bounded by M. Thus, there is a uniformly convergent subsequence. The differential equation is then converted to an integral equation containing the approximating sequence $\{\mathbf{x}_j(t)\}$:

$$\mathbf{x}_j(t) = \mathbf{x}(t_0) + \int_{t_0}^{t} \left\{ \mathbf{G}(s, \mathbf{x}_j(s)) + \Delta_j(s) \right\} ds ,$$

where

$$\Delta_j(t) = \begin{cases} \mathbf{x}_j(t) - \mathbf{G}(t, \mathbf{x}_j(t)) & \text{if } t \neq t_i , \\ 0 & \text{if } t = t_i . \end{cases}$$

By the uniform convergence, the limit may be passed through the integral to show the existence of a solution.

This is called the Cauchy-Euler-Peano method, and we shall see that it requires drastic changes for integral equations.

Exercise 3.1.1. Given the scalar initial-value problem

$$x' = x^{1/3}, \quad x(0) = 0 ,$$

notice that $x(t) \equiv 0$ satisfies the problem. Separate variables and obtain a second solution. Note that there is a whole "funnel" of solutions to the problem.

Nonuniqueness of solutions of the initial-value problem is most easily "cured" by requiring that $\mathbf{G}$ satisfy a local Lipschitz condition in the second argument.

Definition 3.1.2.   *Let $U \subset R^{n+1}$ and $\mathbf{G} : U \to R^n$. We say that $\mathbf{G}$ satisfies a local Lipschitz condition with respect to $\mathbf{x}$ if for each compact subset M of U there is a constant K such that $(t, \mathbf{x}_i)$ in M implies*

$$\left| \mathbf{G}(t, \mathbf{x}_1) - \mathbf{G}(t, \mathbf{x}_2) \right| \leq K |\mathbf{x}_1 - \mathbf{x}_2| .$$

Frequently we need to allow t on only one side of t_0.

Exercise 3.1.2.   Show that $G(t, x) = x^{1/3}$ does not satisfy a local Lipschitz condition when

$$U = \left\{ (t, x) : -\infty < t < \infty , \; |x| < 1 \right\} .$$

Exercise 3.1.3.   Let A be an $n \times n$ real constant matrix and $\mathbf{G}(t, x) = A\mathbf{x}$, $\mathbf{x} \in R^n$. Show that there is a $K > 0$ with

$$\left| \mathbf{G}(t, \mathbf{x}_1) - \mathbf{G}(t, \mathbf{x}_2) \right| \leq K |\mathbf{x}_1 - \mathbf{x}_2|$$

for all $\mathbf{x}_i \in R^n$.

In Chapter 2 (Section 2.1) we stated and proved Gronwall's inequality. It was used to obtain uniqueness of solutions of the linear initial-value problem. We use it here in the same way, and state it (without proof) for convenient reference.

Theorem 3.1.2.   *Let f and g be continuous nonnegative scalar functions on $[a, b]$ and let $A \geq 0$. If*

$$f(t) \leq A + \int_a^t g(s) f(s) \, ds$$

for $a \leq t \leq b$, then

$$f(t) \leq A \exp \int_a^t g(s) \, ds , \quad a \leq t \leq b . \tag{3.1.2}$$

Theorem 3.1.3.   *Let the conditions of Theorem 3.1.1 hold and suppose there is a constant L such that $(t, \mathbf{x}_i) \in D$ implies*

$$\left| \mathbf{G}(t, \mathbf{x}_1) - \mathbf{G}(t, \mathbf{x}_2) \right| \leq L |\mathbf{x}_1 - \mathbf{x}_2| .$$

Then (3.1.1) has one and only one solution.

Proof. Theorem 3.1.1 yields one solution. If $\mathbf{x}(t)$ and $\mathbf{y}(t)$ are two solutions with $(t, \mathbf{x}(t)) \in D$ and $(t, \mathbf{y}(t)) \in D$ for $t_0 \leq t \leq T_1 \leq t_0 + T$, then

$$\mathbf{x}(t) = \mathbf{x}_0 + \int_{t_0}^{t} \mathbf{G}(s, \mathbf{x}(s)) \, ds$$

and

$$\mathbf{y}(t) = \mathbf{x}_0 + \int_{t_0}^{t} \mathbf{G}(s, \mathbf{y}(s)) \, ds \,,$$

so that

$$|\mathbf{x}(t) - \mathbf{y}(t)| \leq \int_{t_0}^{t} L|\mathbf{x}(s) - \mathbf{y}(s)| \, ds \,,$$

yielding

$$|\mathbf{x}(t) - \mathbf{y}(t)| \leq 0 e^{L(t - t_0)}$$

by Gronwall's inequality.

Continual dependence of solutions on initial conditions is easily obtained in exactly the same way. For if $\mathbf{x}(t)$ and $\mathbf{y}(t)$ are solutions of (3.1.1) on $[t_0, T_1]$, then

$$|\mathbf{x}(t) - \mathbf{y}(t)| \leq |\mathbf{x}(t_0) - \mathbf{y}(t_0)| + L \int_{t_0}^{t} |\mathbf{x}(s) - \mathbf{y}(s)| \, ds \,,$$

so that

$$|\mathbf{x}(t) - \mathbf{y}(t)| \leq |\mathbf{x}(t_0) - \mathbf{y}(t_0)| e^{L(t - t_0)} .$$

Under the conditions of Theorem 3.1.3, the existence of solutions may be proved as in Chapter 2 (Theorem 2.1.1) by using Picard's successive approximations.

Sketch of Alternative Proof of Theorem 3.1.3 . Write (3.1.1) as

$$\mathbf{x}(t) = \mathbf{x}_0 + \int_{t_0}^{t} \mathbf{G}(s, \mathbf{x}(s)) \, ds \,,$$

and define a sequence inductively by

$$\mathbf{x}_1(t) = \mathbf{x}_0$$
$$\mathbf{x}_{n+1}(t) = \mathbf{x}_0 + \int_{t_0}^{t} \mathbf{G}(s, \mathbf{x}_n(s)) \, ds \,, \quad n \geq 1 \,.$$

Notice that $\mathbf{x}_{n+1}(t)$ is the typical partial sum of the series

$$\mathbf{x}_1(t) + \sum_{n=1}^{\infty} \left(\mathbf{x}_{n+1}(t) - \mathbf{x}_n(t) \right)$$

and that series converges uniformly and absolutely to a function $\mathbf{x}(t)$ by comparison with a series for $Ae^{k(t-t_0)}$. Thus, in the definition of $\mathbf{x}_{n+1}(t)$, we take the limit through the integral and obtain

$$\mathbf{x}(t) = \mathbf{x}_0 + \int_{t_0}^{t} \mathbf{G}(s, \mathbf{x}(s)) \, ds \, .$$

This is almost identical to the proof of existence of solutions of linear Volterra equations given in Chapter 2. We shall see that few changes are necessary for nonlinear Volterra equations.

The successive approximation technique has been traced to Liouville (1838, p. 19) and, in its general form, to Picard (1890, p. 217).

Banach (1932) recognized that Picard's result was actually a fixed point theorem, because the operator

$$T(\mathbf{x})(t) = \mathbf{x}_0 + \int_{t_0}^{t} \mathbf{G}(s, \mathbf{x}(s)) \, ds$$

has a fixed point $\mathbf{x}(t)$. A good setting for this is a complete metric space.

Definition 3.1.3.   *A pair $(\mathcal{S}, \rho)$ is a metric space if $\mathcal{S}$ is a set and $\rho : \mathcal{S} \times \mathcal{S} \to [0, \infty)$ such that when y, z, and u are in $\mathcal{S}$ then*

(a) *$\rho(y, z) \geq 0$, $\rho(y, y) = 0$, and $\rho(y, z) = 0$ implies $y = z$,*
(b) *$\rho(y, z) = \rho(z, y)$, and*
(c) *$\rho(y, z) \leq \rho(y, u) + \rho(u, z)$.*

The metric space is complete if every Cauchy sequence in $(\mathcal{S}, \rho)$ has a limit in that space.

Definition 3.1.4.   *Let $(\mathcal{S}, \rho)$ be a metric space and $A : \mathcal{S} \to \mathcal{S}$. the operator A is a contraction operator if there is an $\alpha \in (0, 1)$ such that $x \in \mathcal{S}$ and $y \in \mathcal{S}$ imply*

$$\rho[A(x), A(y)] \leq \alpha \rho(x, y) \, .$$

Theorem 3.1.4.   Contractive Mapping Principle   *Let $(\mathcal{S}, \rho)$ be a complete metric space and $A : \mathcal{S} \to \mathcal{S}$ a contraction operator. Then there is a unique $\phi \in \mathcal{S}$ with $A(\phi) = \phi$. Furthermore, if $\psi \in \mathcal{S}$ and if $\{\psi_n\}$ is defined inductively by $\psi_1 = A(\psi)$ and $\psi_{n+1} = A(\psi_n)$, then $\psi_n \to \phi$, the unique fixed point. That is, the equation $A(\phi) = \phi$ has one and only one solution.*

Proof. Let $x_0 \in \mathcal{S}$ and define a sequence $\{x_n\}$ in $\mathcal{S}$ by $x_1 = A(x_0)$, $x_2 = A(x_1) \stackrel{\text{def}}{=} A^2 x_0, \ldots, x_n = A x_{n-1} = A^n x_0$. To see that $\{x_n\}$ is a Cauchy sequence, note that if $m > n$, then

$$
\begin{aligned}
\rho(x_n, x_m) &= \rho(A^n x_0, A^m x_0) \\
&\leq \alpha \rho(A^{n-1} x_0, A^{m-1} x_0) \\
&\quad\vdots \\
&\leq \alpha^n \rho(x_0, x_{m-n}) \\
&\leq \alpha^n \{\rho(x_0, x_1) + \rho(x_1, x_2) + \cdots + \rho(x_{m-n-1}, x_{m-n})\} \\
&\leq \alpha^n \{\rho(x_0, x_1) + \alpha \rho(x_0, x_1) + \cdots + \alpha^{m-n-1} \rho(x_0, x_1)\} \\
&= \alpha^n \rho(x_0, x_1)\{1 + \alpha + \cdots + \alpha^{m-n-1}\} \\
&\leq \alpha^n \rho(x_0, x_1)\{1/(1 - \alpha)\}\,.
\end{aligned}
$$

Because $\alpha < 1$, the right side tends to zero as $n \to \infty$. Thus, $\{x_n\}$ is a Cauchy sequence, and because $(\mathcal{S}, \rho)$ is complete, it has a limit $x \in \mathcal{S}$. Now A is certainly continuous, so

$$
A(x) = A(\lim_{n\to\infty} x_n) = \lim_{n\to\infty} A(x_n) = \lim_{n\to\infty} x_{n+1} = x\,,
$$

and x is a fixed point. To see that x is unique, consider $A(x) = x$ and $A(y) = y$. Then

$$
\rho(x, y) = \rho\big(A(x), A(y)\big) \leq \alpha \rho(x, y)\,,
$$

and because $\alpha < 1$, we conclude that $\rho(x, y) = 0$, so that $x = y$. This completes the proof.

With the contractive mapping principle, the proof of Theorem 3.1.3 becomes very simple, as we shall see in the next section.

3.2   Existence and Uniqueness

Let $\mathbf{x}$, $\mathbf{f}$, and $\mathbf{g}$ be n vectors and let

$$
\mathbf{x}(t) = \mathbf{f}(t) + \int_0^t \mathbf{g}(t, s, x(s))\, ds\,. \tag{3.2.1}
$$

Recall from Chapter 1 [see Eq. (1.1.2)] that an integro-differential equation with initial conditions can be put into the form of (3.2.1).

Theorem 3.2.1. *Let a, b, and L be positive numbers, and for some fixed $\alpha \in (0,1)$ define $c = \alpha/L$. Suppose*

(a) $\mathbf{f}$ *is continuous on* $[0, a]$,

(b) $\mathbf{g}$ *is continuous on*

$$U = \left\{ (t, s, \mathbf{x}) : 0 \le s \le t \le a \text{ and } |\mathbf{x} - \mathbf{f}(t)| \le b \right\},$$

(c) $\mathbf{g}$ *satisfies a Lipschitz condition with respect to* $\mathbf{x}$ *on* U

$$\left| \mathbf{g}(t, s, \mathbf{x}) - \mathbf{g}(t, s, \mathbf{y}) \right| \le L|\mathbf{x} - \mathbf{y}|$$

if $(t, s, \mathbf{x})$, $(t, s, \mathbf{y}) \in U$.

If $M = \max_U |\mathbf{g}(t, s, \mathbf{x})|$, *then there is a unique solution of* (3.2.1) *on* $[0, T]$, *where* $T = \min[a, b/M, c]$.

Proof. Let $\mathcal{S}$ be the space of continuous functions from $[0, T] \to R^n$ with $\psi \in \mathcal{S}$ if

$$\|\psi - \mathbf{f}\| \stackrel{\text{def}}{=} \max_{0 \le t \le T} |\psi(t) - \mathbf{f}(t)| \le b.$$

(The norm defines the metric ρ.) Define an operator $A : \mathcal{S} \to \mathcal{S}$ by

$$\mathbf{A}(\psi)(t) = \mathbf{f}(t) + \int_0^t \mathbf{g}(t, s, \psi(s)) \, ds.$$

To see that $\mathbf{A} : \mathcal{S} \to \mathcal{S}$ notice that ψ continuous implies $\mathbf{A}(\psi)$ continuous and that

$$\|\mathbf{A}(\psi) - \mathbf{f}\| = \max_{0 \le t \le T} |\mathbf{A}(\psi)(t) - \mathbf{f}(t)|$$

$$= \max_{0 \le t \le T} \left| \int_0^t \mathbf{g}(t, s, \psi(s)) \, ds \right| \le MT \le b.$$

To see that A is a contraction mapping, notice that if ϕ and $\psi \in \mathcal{S}$, then

$$\|\mathbf{A}(\phi) - \mathbf{A}(\psi)\| = \max_{0 \le t \le T} \left| \int_0^t \mathbf{g}(t, s, \phi(s)) \, ds - \int_0^t \mathbf{g}(t, s, \psi(s)) \, ds \right|$$

$$\le \max_{0 \le t \le T} \int_0^t \left| \mathbf{g}(t, s, \phi(s)) - \mathbf{g}(t, s, \psi(s)) \right| \, ds$$

$$\le \max_{0 \le t \le T} L \int_0^t |\phi(s) - \psi(s)| \, ds$$

$$\le T \max_{0 \le t \le T} L|\phi(s) - \psi(s)|$$

$$= TL\|\phi - \psi\| \le cL\|\phi - \psi\| = \alpha\|\phi - \psi\|.$$

Thus, by Theorem 3.1.4, there is a unique function $\mathbf{x} \in \mathcal{S}$ with

$$\mathbf{A}(\mathbf{x})(t) = \mathbf{x}(t) = \mathbf{f}(t) + \int_0^t \mathbf{g}(t, s, \mathbf{x}(s)) \, ds \,.$$

This completes the proof.

The interval $[0, T]$ can be improved by using a different norm. The conclusion can be strengthened to $T = \min[a, b/M]$. On an interval $[0, T]$ one may ask that

$$\|\phi\| = \max_{0 \le s \le T} A e^{-Bs} |\phi(s)|$$

for A and B positive constants [see Hale (1969, p. 20)].

Using a continuation result, we shall find that for most theoretical purposes a short interval of existence of a solution is as good as a longer one. There are, however, times when one needs a long interval.

Example 3.2.1. Consider a system of ordinary differential equations

$$\mathbf{x}' = \mathbf{F}(t, \mathbf{x})$$

with $\mathbf{F} : (-\infty, \infty) \times R^n \to R^n$ continuous. Suppose that solutions are unique. Also suppose that there exists $T > 0$ with $\mathbf{F}(t + T, \mathbf{x}) = \mathbf{F}(t, \mathbf{x})$ for all $(t, \mathbf{x})$ and that $\mathbf{F}(-t, \mathbf{x}) = -\mathbf{F}(t, \mathbf{x})$ for all $(t, \mathbf{x})$. If for each $\mathbf{x}_0$ the successive approximations $\{\mathbf{X}_n\}$ defined inductively by

$$\mathbf{X}_0 = \mathbf{x}_0$$

and

$$\mathbf{X}_{n+1} = \mathbf{x}_0 + \int_0^t \mathbf{F}(s, \mathbf{X}_n(s)) \, ds$$

converge uniformly on $[0, T]$, then every solution of the initial-value problem

$$\mathbf{x}' = \mathbf{F}(t, \mathbf{x}) \,, \quad \mathbf{x}(0) = \mathbf{x}_0$$

is T-periodic.

Proof. Proceed by verifying that $\mathbf{X}_1$ is even and T-periodic. Thus, $\mathbf{F}(t, \mathbf{X}_1(t))$ is odd and T-periodic. If follows by induction that $\mathbf{X}_n$ is even and T-periodic for each n. By the uniform convergence we may pass the limit through the integral and conclude that the solution is T-periodic. This completes the proof.

Technical problems arise if the successive approximations converge on an interval smaller than $[0, T]$.

If we look at the proof of the contraction theorem (Theorem 3.1.4) and our existence result (Theorem 3.2.1), we see that the latter simply uses Picard's successive approximations, which we outline in the sketch of the alternative proof of Theorem 3.1.3 for ordinary differential equations. However, there are interesting differences in the sets over which we are working. For ordinary differential equations we look at the set in Fig. 3.1, whereas for integral equations we look at the set in Fig. 3.2.

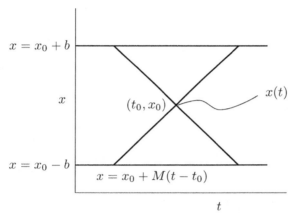

Figure 3.1: Set for ordinary differential equations.

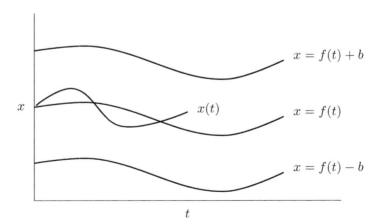

Figure 3.2: Set for integral equations.

For existence without uniqueness, we ask that $\mathbf{f}$ and $\mathbf{g}$ be continuous and use an integral version of the Cauchy-Euler-Peano method.

We emphatically point out that this is not the usual treatment, and it avoids many of the interesting theoretical problems in integral equations. Ordinarily, much less than the continuity of $\mathbf{g}$ is required. For such discussions, consult the excellent book by Miller (1971a). There are two reasons for our choice of treatment. First, this is a book on stability and our main methods will require that the integral equation be converted to an integro-differential equation; and for that it is most convenient that $\mathbf{g}$ be continuous. The second reason is our wish to present the theory as nearly as possible as if it were ordinary differential equations, so the investigator from that area may parlay existing expertise into a knowledge of Volterra equations in a very direct manner.

Theorem 3.2.2.   *Let* $\mathbf{f} : [0, a] \to R^n$ *and* $\mathbf{g} : U \to R^n$ *both be continuous, where*

$$U = \left\{ (t, s, \mathbf{x}) : 0 \le s \le t \le a \text{ and } |\mathbf{x} - \mathbf{f}(t)| \le b \right\}.$$

Then there is a continuous solution of

$$\mathbf{x}(t) = \mathbf{f}(t) + \int_0^t \mathbf{g}(t, s, \mathbf{x}(s)) \, ds \tag{3.2.1}$$

on $[0, T]$, *where* $T = \min[a, b/M]$ *and* $M = \max_U |\mathbf{g}(t, s, \mathbf{x})|$.

Proof. We construct a sequence of continuous functions on $[0, T]$ such that

$$\mathbf{x}_1(t) = \mathbf{f}(t),$$

and if j is a fixed integer, $j > 1$, then $\mathbf{x}_j(t) = \mathbf{f}(t)$ when $t \in [0, T/j]$ and

$$\mathbf{x}_j(t) = \mathbf{f}(t) + \int_0^{t-(T/j)} \mathbf{g}\big(t - (T/j), s, \mathbf{x}_j(s)\big) \, ds$$

for $T/j \le t \le T$.

Let us examine this definition. If j is a fixed integer, $j > 1$, then

$$\mathbf{x}_j(t) = \mathbf{f}(t) \quad \text{on} \quad [0, T/j]$$

and

$$\mathbf{x}_j(t) = \mathbf{f}(t) + \int_0^{t-(T/j)} \mathbf{g}\big(t - (T/j), s, \mathbf{f}(s)\big) \, ds$$

for $t \in [T/j, 2T/j]$. At $t = 2T/j$, the upper limit is T/j, so the integrand was defined on $[T/j, 2T/j]$, thereby defining $\mathbf{x}_j(t)$ on $[T/j, 2T/j]$. Now, for $t \in [2T/j, 3T/j]$ we still have

$$\mathbf{x}_j(t) = \mathbf{f}(t) + \int_0^{t-(T/j)} \mathbf{g}\big(t - (T/j), s, \mathbf{x}_j(s)\big)\, ds\,,$$

and the upper limit goes to $2T/j$ on that interval, so $\mathbf{x}_j(s)$ is defined, and hence the integral is well defined. This process is continued to $[(j-1)T/j, T]$, obtaining $\mathbf{x}_j(t)$ on $[0, T]$ for each j. A Lipschitz condition will be avoided because $\mathbf{x}_j(t)$, $j > 1$, is independent of all other $\mathbf{x}_k(t)$, $j \neq k$.

Notice that

$$|\mathbf{x}_j(t) - \mathbf{f}(t)| \leq \int_0^{t-(T/j)} \big|\mathbf{g}\big(t - (T/j), s, \mathbf{x}_j(s)\big)\big|\, ds$$
$$\leq M(t - (T/j)) \leq M(b/M) = b\,.$$

This sequence $\{\mathbf{x}_j(t)\}$ is uniformly bounded because

$$|\mathbf{x}_j(t)| \leq |\mathbf{f}(t)| + b \leq \max_{0 \leq t \leq T} |\mathbf{f}(t)| + b\,.$$

To see that $\{\mathbf{x}_j(t)\}$ is an equicontinuous sequence, let $\varepsilon > 0$ be given, let n be an arbitrary integer, let t and v be in $[0, T]$ with $t > v$, and consider

$$|\mathbf{x}_n(t) - \mathbf{x}_n(v)| \leq |\mathbf{f}(t) - \mathbf{f}(v)|$$
$$+ \left| \int_0^{v-(T/n)} \big[\mathbf{g}\big(t - (T/n), s, \mathbf{x}_n(s)\big) \right.$$
$$\left. - \mathbf{g}\big(v - (T/n), s, \mathbf{x}_n(s)\big) \big]\, ds \right|$$
$$+ \left| \int_{v-(T/n)}^{t-(T/n)} \mathbf{g}\big(t - (T/n), s, \mathbf{x}_n(s)\big)\, ds \right|$$
$$\leq |\mathbf{f}(t) - \mathbf{f}(v)| + M|t - v|$$
$$+ \int_0^{v-(T/n)} \big|\mathbf{g}\big(t - (T/n), s, \mathbf{x}_n(s)\big)$$
$$- \mathbf{g}\big(v - (T/n), s, \mathbf{x}_n(s)\big)\big|\, ds\,.$$

By the uniform continuity of $\mathbf{f}$, there is a $\delta_1 > 0$ such that $|t - v| < \delta_1$ implies $|\mathbf{f}(t) - \mathbf{f}(v)| < \varepsilon/3$.

By the uniform continuity of $\mathbf{g}$ on U, there is a $\delta_2 > 0$ such that $|t - v| < \delta_2$ implies

$$T\big|\mathbf{g}\big(t - (T/n), s, \mathbf{x}_n(s)\big) - \mathbf{g}\big(v - (T/n), s, \mathbf{x}_n(s)\big)\big| < \varepsilon/3\,.$$

Let $\delta = \min[\delta_1, \delta_2, \varepsilon/3M]$. This yields equicontinuity. Apply Lemma 3.1.1.

3.3 Continuation of Solutions

Theorems 3.2.1 and 3.2.2 are local existence results. They guarantee a solution on an interval $[0, T]$ with T obtained as follows:

(i) $\mathbf{f}$ is continuous on $[0, a]$,

(ii) $\mathbf{g}$ is continuous on

$$U = \big\{(t, s, \mathbf{x}) : 0 \le s \le t \le a\,, \ |\mathbf{f}(t) - \mathbf{x}| \le b\big\}\,,$$

(iii) $|\mathbf{g}(t, s, \mathbf{x})| \le M$ on U, and

(iv) $T = \min[a, b/M]$.

Suppose, however, that $\mathbf{f}$ is continuous on $[0, \infty)$ and $\mathbf{g}$ is continuous for $0 \le s \le t < \infty$ and all $\mathbf{x} \in R^n$. Then the number a drops out of the definition of T and M need not exist. In some way we must manufacture both a and b to apply the existence theorem. We wonder just how large the interval of existence can be made. Presumably, if we take a or b too large, then M becomes so large that b/M is small.

To be definite and to start the search for the interval of definition, let $a = b = \pi$ and find M. This yields T, and we have at least one well-defined solution, say $\boldsymbol{\phi}(t)$, on $[0, T]$. Thus, we would like to start a solution at $(T, \boldsymbol{\phi}(T))$ and apply the existence theorem again. But that theorem starts solutions at $t = 0$. Thus, we must translate our equation

$$\mathbf{x}(t) = \mathbf{f}(t) + \int_0^t \mathbf{g}(t, s, \mathbf{x}(s))\,ds$$

by letting $\mathbf{y}(t) = \mathbf{x}(t + T)$ (as in Chapter 1 with initial functions), so that

$$\mathbf{y}(t) = \mathbf{x}(t + T)$$

$$= \mathbf{f}(t + T) + \int_0^{t+T} \mathbf{g}(t + T, s, \mathbf{x}(s))\, ds$$

$$= \mathbf{f}(t + T) + \int_0^T \mathbf{g}(t + T, s, \boldsymbol{\phi}(s))\, ds + \int_T^{t+T} \mathbf{g}(t + T, s, \mathbf{x}(s))\, ds$$

$$\overset{\text{def}}{=} \mathbf{h}(t) + \int_0^t \mathbf{g}(t + T, s + T, \mathbf{y}(s))\, ds$$

$$\overset{\text{def}}{=} \mathbf{h}(t) + \int_0^t \mathbf{q}(t, s, \mathbf{y}(s))\, ds\,.$$

The function $\mathbf{h} : [0, \infty) \to R^n$ is continuous and $\mathbf{q}$ is continuous for $0 \leq s \leq t < \infty$ and $\mathbf{y} \in R^n$. Thus, the existence theorem may be applied to

$$\mathbf{y}(t) = \mathbf{h}(t) + \int_0^t \mathbf{q}(t, s, \mathbf{y}(s))\, ds\,.$$

One may take $a = b = \pi$ again, but $\mathbf{q}$ is a translation of $\mathbf{g}$, so a new M, say, M_1, will be obtained and an interval $[0, T_1]$, with $T_1 = \min[\pi, \pi/M_1]$, will result. Any solution $\mathbf{y}(t)$ on $[0, T_1]$, translated to the right by $\mathbf{y}(t-T)$, so as to be defined on $[T, T + T_1]$, will be called a *continuation* of $\boldsymbol{\phi}$. Naturally, there may be many continuations of $\boldsymbol{\phi}$ if solutions are not unique. A nightmare of that sort for ordinary differential equations may be seen in Hartman (1964, p. 19).

At least in theory, we may continue the process to obtain continuations of $\boldsymbol{\phi}$ on intervals $T, T_1, T_2, \ldots$.

With slight modifications the process can be used if $\mathbf{f}$ is defined on $[0, \infty)$ and $\mathbf{g}$ is defined and continuous for $0 \leq s \leq t < \infty$ and $\mathbf{x} \in D$, where D is an open and connected subset of R^n. The number b will change each time, becoming the distance from the new $\mathbf{f}(t)$ to the boundary of D.

Now, there are two possibilities.

The intervals $T, T_1, T_2, \ldots$, when translated, may exhaust $[0, \infty)$. This would happen if, for example, $|\mathbf{g}(t, s, \mathbf{x})|$ is bounded and $D = R^n$, because M and M_1 would be the same. [Is it also true if $|\mathbf{g}(t, s, \mathbf{x})| \leq \alpha + \beta |\mathbf{x}|$?]

If $\mathbf{g}$ is continuous for $0 \leq s \leq t < \infty$ and $\mathbf{x} \in D$, then $\mathbf{x}(t)$ may approach the boundary of D and T_n may approach zero too quickly. If $D = R^n$, then $\mathbf{x}(t)$ may become unbounded as t approaches some number L from the left.

One thing is clear, however, from our translation argument: if $D = R^n$, then bounded solutions can be continued to $t = \infty$, because the bounds M_i on $|\mathbf{g}|$ are then bounded on any interval $[0, K]$. The next result formalizes this.

Theorem 3.3.1. *Let* $\mathbf{f} : [0, \infty) \to R^n$ *and* $\mathbf{g} : U \to R^n$ *be continuous where* $U = \{(t, s, \mathbf{x}) : 0 \leq s \leq t < \infty, \ \mathbf{x} \in R^n\}$. *If* $\mathbf{x}(t)$ *is a solution of*

$$\mathbf{x}(t) = \mathbf{f}(t) + \int_0^t \mathbf{g}(t, s, \mathbf{x}(s)) \, ds \qquad (3.3.1)$$

on an interval $[0, T)$ *and if there is a constant* P *with* $|\mathbf{x}(t) - \mathbf{f}(t)| \leq P$ *on* $[0, T)$, *then there is a* $\bar{T} > T$ *such that* $\mathbf{x}(t)$ *can be continued to* $[0, \bar{T}]$.

Proof. If we can show that $\lim_{t \to T^-} \mathbf{x}(t)$ exists, then an application of the existence theorem starting at $t = T$ will complete the proof.

Let $\{t_n\}$ be a monotonic increasing sequence with limit T. We shall show that $\{\mathbf{x}(t_n)\}$ is a Cauchy sequence. Now

$$|\mathbf{x}(t_m) - \mathbf{x}(t_n)| \leq |\mathbf{f}(t_m) - \mathbf{f}(t_n)|$$
$$+ \left| \int_0^{t_m} \mathbf{g}(t_m, s, \mathbf{x}(s)) \, ds - \int_0^{t_n} \mathbf{g}(t_n, s, \mathbf{x}(s)) \, ds \right|,$$

and, because $\mathbf{f}$ is continuous and $t_n \to T$, the first term on the right tends to zero as $n, m \to \infty$. Also, if $t_m > t_n$, then

$$\left| \int_0^{t_m} \mathbf{g}(t_m, s, \mathbf{x}(s)) \, ds - \int_0^{t_n} \mathbf{g}(t_n, s, \mathbf{x}(s)) \, ds \right|$$
$$\leq \left| \int_0^{t_n} \left[\mathbf{g}(t_m, s, \mathbf{x}(s)) \, ds - \mathbf{g}(t_n, s, \mathbf{x}(s)) \right] ds \right|$$
$$+ \left| \int_{t_n}^{t_m} \mathbf{g}(t_m, s, \mathbf{x}(s)) \, ds \right|.$$

The function $\mathbf{g}$ is uniformly continuous on the set

$$\tilde{U} = \left\{ (t, s, \mathbf{x}) : 0 \leq s \leq t \leq T, \ |\mathbf{x} - \mathbf{f}(t)| \leq P \right\},$$

so for a given $\varepsilon > 0$ there exists $\delta > 0$ such that $|(t_m, s, \mathbf{x}) - (t_n, s, \mathbf{x})| < \delta$ implies $|\mathbf{g}(t_m, s, \mathbf{x}) - \mathbf{g}(t_n, s, \mathbf{x})| < \varepsilon$. Thus, in the last integral inequality, the first term on the right tends to zero as $n, m \to \infty$; because $\mathbf{g}$ is bounded on $\tilde{U}$ and $t_m, t_n \to T$, the second integral also tends to zero.

Hence $\{\mathbf{x}(t_n)\}$ is a Cauchy sequence with limit, say $\mathbf{x}(T)$, making $\mathbf{x}(t)$ continuous on $[0, T]$. Now, translate by $\mathbf{y}(t) = \mathbf{x}(t + T)$ and apply the existence theorem again. That will complete the proof.

From this it readily follows that f continuous on $[0, \infty)$ and $\mathbf{g}$ continuous for $0 \leq s \leq t < \infty$ and all $\mathbf{x} \in R^n$ implies that solutions that remain bounded are continuable to all of $[0, \infty)$.

There is one important difference in the continuation problem of integral equations and that of differential equations.

A solution $\mathbf{x}(t)$ of (3.3.1) can be continued to $[0, \infty)$ unless there is a T with $\mathbf{x}(t)$ defined on $[0, T)$ and $\limsup_{t \to T^-} |\mathbf{x}(t)| = +\infty$, assuming that $\mathbf{f}$ is continuous on $[0, \infty)$ and $\mathbf{g}$ is continuous for $0 \leq s \leq t < \infty$ and $\mathbf{x} \in R^n$.

But for

$$\mathbf{x}' = \mathbf{G}(t, \mathbf{x}), \quad \mathbf{x}(0) = \mathbf{x}_0 \tag{3.3.2}$$

with $\mathbf{G}$ continuous on $[0, \infty) \times R^n$, a solution $\mathbf{x}(t)$ can be continued to all of $[0, \infty)$ unless there is a T with $\mathbf{x}(t)$ defined on $[0, T)$ and $\lim_{t \to T^-} |\mathbf{x}(t)| = +\infty$.

The reason for this is well worth explaining. Suppose that $\mathbf{x}(t)$ is a solution of (3.3.2) on $[0, T)$ with $\limsup_{t \to T^-} |\mathbf{x}(t)| = +\infty$, but $|\mathbf{x}(t)| \not\to \infty$ as $t \to T^-$. Then there is an $R > 0$ and a sequence t_n tending monotonically upward to T with $|\mathbf{x}(t_n)| \leq R$. Because $\limsup_{t \to T^-} |\mathbf{x}(t)| = +\infty$ and $\mathbf{x}(t)$ is continuous, there is a sequence $\{\delta_n\}$ of positive numbers with $|\mathbf{x}(t_n + \delta_n)| = 2R$ and $|\mathbf{x}(t)| \leq 2R$ on $[t_n, t_n + \delta_n]$. Now $\mathbf{G}(t, \mathbf{x})$ is continuous on $[0, T] \times \{x \in R^n : |\mathbf{x}| \leq 2R\}$, and hence, there is an M with $|\mathbf{G}(t, \mathbf{x})| \leq M$ on that set. Moreover,

$$R \leq \left| \mathbf{x}(t_n + \delta_n) - \mathbf{x}(t_n) \right|$$

$$= \left| \int_{t_n}^{t_n + \delta_n} \mathbf{G}(s, \mathbf{x}(s)) \, ds \right| \leq M \delta_n,$$

so that $\delta_n \geq R/M$. This implies that $T = \infty$. That is, the speed of $|\mathbf{x}(t)|$ in moving from R to $2R$ is bounded by M as the argument of $\mathbf{G}$ is confined to a compact set, and hence, at least R/M time units elapse during the journey.

But the situation is different for (3.3.1). If we use the same notation with a solution $\mathbf{x}(t)$ of (3.3.1), we have

$$\mathbf{x}(t_n + \delta_n) = \mathbf{f}(t_n + \delta_n) + \int_0^{t_n + \delta_n} \mathbf{g}(t_n + \delta_n, s, \mathbf{x}(s)) \, ds$$

and

$$\mathbf{x}(t_n) = \mathbf{f}(t_n) + \int_0^{t_n} \mathbf{g}(t_n, s, \mathbf{x}(s)) \, ds,$$

so that

$$
\begin{aligned}
R &\le |\mathbf{x}(t_n + \delta_n) - \mathbf{x}(t_n)| \\
&= \left| \mathbf{f}(t_n + \delta_n) - \mathbf{f}(t_n) + \int_0^{t_n + \delta_n} \mathbf{g}(t_n + \delta_n, s, \mathbf{x}(s))\, ds \right. \\
&\quad \left. - \int_0^{t_n} \mathbf{g}(t_n, s, \mathbf{x}(s))\, ds \right|.
\end{aligned}
$$

There is no way to exclude the values of $\mathbf{x}(s)$ when $|\mathbf{x}(s)| > 2R$ from the consideration. Thus, these integrands may become arbitrarily large even though $0 \le t \le T$ and $|\mathbf{x}(s)| \le 2R$ for $t_n \le s \le t_n + \delta_n$.

There are (very complicated) examples showing that $\limsup_{t \to T^-} |\mathbf{x}(t)| = +\infty$, but $|\mathbf{x}(t)| \not\to \infty$ as $t \to T^-$ for various delay equations. See, for example, Herdman (1980) or Myshkis (1951).

The ability to look at general classes of integral equations and decide if solutions are continuable to all $[0, \infty)$ is indispensable. In the theory of ordinary differential equations one of the primary tools of this investigation is Liapunov functions. It turns out that much less is required of the Liapunov function to prove continuation of solutions than was required in Chapter 2 for boundedness.

The following is the main result of this type for ordinary differential equations. If $\mathbf{G}$ is fairly nice, then its converse is also true. It can be traced back to Conti, Kato and Strauss, and Yoshizawa.

Theorem 3.3.2. *Let* $\mathbf{G} : [0, \infty) \times R^n \to R^n$ *be continuous, and suppose there is a function* $V : [0, \infty) \times R^n \to [0, \infty)$ *having continuous first partial derivatives. If on* $[0, \infty) \times R^n$ *we have*

$$
V'(t, \mathbf{x}) = \mathbf{grad}\, \mathbf{V} \cdot \mathbf{G} + \partial V/\partial t \le 0, \tag{3.3.3}
$$

and if for each $T > 0$ *the relation*

$$
V(t, \mathbf{x}) \to \infty \quad as \quad |\mathbf{x}| \to \infty \quad uniformly\ for \quad 0 \le t \le T \tag{3.3.4}
$$

holds, then every solution of (3.3.2) *can be continued to* $+\infty$.

Proof. If the theorem is false, then there is a solution $\mathbf{x}(t)$ of (3.3.2) defined on some interval $[t_0, T)$ with

$$
\lim_{t \to T^-} |\mathbf{x}(t)| = +\infty. \tag{3.3.5}
$$

According to (3.3.3), $V'(t, \mathbf{x}(t)) \le 0$ so $V(t, \mathbf{x}(t)) \le V(t_0, \mathbf{x}(t_0))$ on $[t_0, T)$. But by (3.3.4) and (3.3.5) we have $V(t, \mathbf{x}(t)) \to \infty$ as $t \to T^-$, a contradiction. This completes the proof.

The drawback of this result is, of course, the necessity of finding the function V. We illustrate such a function in the proof of the following classical result known as the Conti-Wintner theorem.

Theorem 3.3.3. *Let* $\mathbf{G} : [0, \infty) \times R^n \to R^n$ *be continuous and suppose there are continuous functions* $\lambda : [0, \infty) \to [0, \infty)$ *and* $\omega : [0, \infty) \to [1, \infty)$ *with* $\int_1^\infty [ds/\omega(s)] = +\infty$. *If* $|\mathbf{G}(t, \mathbf{x})| \le \lambda(t)\omega(|\mathbf{x}|)$, *then any solution of*

$$\mathbf{x}' = \mathbf{G}(t, \mathbf{x}), \quad \mathbf{x}(t_0) = \mathbf{x}_0 \tag{3.3.2}'$$

can be continued to $[t_0, \infty)$.

Proof. Let $\mathbf{x}(t)$ be a solution of (3.3.2) and define

$$V(t, \mathbf{x}) = \left\{ \int_0^{|\mathbf{x}|} [ds/\omega(s)] + 1 \right\} \exp\left[-\int_0^t \lambda(s)\,ds \right].$$

(A differentiable norm may be chosen for $x \ne 0$. Because we are concerned with $\lim_{t \to T^-} |\mathbf{x}(t)| = +\infty$, we will not be bothered with the nondifferentiability of $|\mathbf{x}|$ at 0.) We find that

$$V'_{(3.3.2)}(t, \mathbf{x}) \le -\lambda(t)V(t, \mathbf{x}) + \left[|\mathbf{G}(t, \mathbf{x})/\omega(|\mathbf{x}|)| \right] \exp\left[-\int_0^t \lambda(s)\,ds \right]$$

$$\le -\lambda(t)V(t, \mathbf{x}) + \lambda(t)\exp\left[-\int_0^t \lambda(s)\,ds \right] \le 0.$$

The result now follows from Theorem 3.3.2.

When ω is nondecreasing the result may be extended to (3.3.1).

Definition 3.3.1. *Let* $h : [0, \infty) \to (-\infty, \infty)$ *and for*

$$U = \left\{ (t, s, x) | 0 \le s \le t < \infty, \ x \in R \right\},$$

let $p : U \to (-\infty, \infty)$. *Let* $x(t)$ *be a continuous solution of the scalar equation*

$$x(t) = h(t) + \int_0^t p(t, s, x(s))\,ds \tag{3.3.6}$$

on $[0, A]$ *with the property that if* $y(t)$ *is any other solution, then as long as* $y(t)$ *exists and* $t \le A$ *we have* $y(t) \le x(t)$. *Then* $x(t)$ *is called the* maximal solution *of (3.3.6). Minimal solutions are defined by asking* $y(t) \ge x(t)$.

Of course, if solutions are unique, then the unique solution is the maximal and minimal solution.

Much can be proved concerning maximal solutions, and we shall repeat little of it here. The interested reader may consult Hartman (1964, pp. 25–31) for some interesting properties of ordinary differential equations and integral equations. Extensive results of this type are also found in Lakshmikantham and Leela (1969, e.g., pp. 11–31).

Theorem 3.3.4. *Let the maximal solution $x(t)$ of the scalar equation*

$$x(t) = B + \int_0^t p(s, x(s)) \, ds$$

exist on $[0, A]$, where B is constant, and let $p : [0, A] \times R \to R$ be continuous and nondecreasing in x when $0 \leq t \leq A$. If $y(t)$ is a continuous scalar function on $[0, A]$ satisfying

$$y(t) \leq y_0 + \int_0^t p(s, y(s)) \, ds, \quad y_0 \leq B,$$

then $y(t) \leq x(t)$ on $[0, A]$.

Proof. Let

$$Y(t) = y_0 + \int_0^t p(s, y(s)) \, ds,$$

so that $y(t) \leq Y(t)$ and

$$Y'(t) = p(t, y(t)) \leq p(t, Y(t))$$

by monotonicity. Hence,

$$Y'(t) \leq p(t, Y(t)), \quad Y(0) = y_0$$

and $Y(t) \leq x(t)$ according to (1964, p. 26), or see Theorem 8.1.1 for $p \geq 0$. This completes the proof.

Unfortunately, the monotonicity in x cannot be dropped for integral inequalities as it can be for differential inequalities. For this reason the Conti-Wintner theorem for integral equations is not as satisfactory as for differential equations.

Theorem 3.3.5. *Let the conditions of Theorem 3.3.1 hold ($\mathbf{f}$ and $\mathbf{g}$ are continuous) and suppose that for each $T > 0$ there is a constant $K(T) > 0$ and a continuous function $\omega : [0, \infty) \to [1, \infty)$ with ω nondecreasing, $|\mathbf{g}(t, s, \mathbf{x})| \leq K(T)\omega(|\mathbf{x}|)$ if $0 \leq s \leq t \leq T$, and $\int_1^\infty [ds/\omega(s)] = \infty$. If $\mathbf{x}(t)$ is a solution of (3.3.1) on any interval $[0, \alpha)$, then it is bounded and, hence, is continuable to $[0, \infty)$.*

Proof. Because $\mathbf{x}(t)$ is defined on $[0, \alpha)$, take $T = \alpha$ and have

$$|\mathbf{x}(t)| \leq |\mathbf{f}(t)| + \int_0^t |\mathbf{g}(t, s, \mathbf{x}(s))| \, ds$$
$$\leq M + K(\alpha) \int_0^t \omega(|\mathbf{x}(s)|) \, ds \,,$$

where $|\mathbf{f}(t)| \leq M$ on $[0, \alpha]$. Because ω is monotone, $|\mathbf{x}(t)|$ is bounded by the maximal solution of

$$y = M + \int_0^t K(\alpha)\omega(y(s)) \, ds \,,$$

or of the initial-value problem

$$y' = K(\alpha)\omega(y), \quad y(0) = M \,.$$

Separating variables yields

$$\int_M^{y(t)} [ds/\omega(s)] = K(\alpha)t \,.$$

Because $\int_M^\infty [ds/\omega(s)] = \infty$, $y(\alpha)$ exists; hence $x(\alpha)$ exists. This completes the proof.

The next result is not as strong as Theorem 3.3.5, but the proof is very simple and ω is specified.

Theorem 3.3.6. *Let $\mathbf{g} : [0, \infty) \times [0, \infty) \times R^n \to R^n$ be continuous, and suppose that for each $T > 0$ there is a continuous scalar function $M(s, T)$ with $|\mathbf{g}(t, s, \mathbf{x})| \leq M(s, T)(1 + |\mathbf{x}|)$ if $0 \leq s \leq t \leq T$. Let $\mathbf{f} : [0, \infty) \to R^n$ be continuous. If $\mathbf{x}(t)$ is a solution of*

$$\mathbf{x}(t) = \mathbf{f}(t) + \int_0^t \mathbf{g}(t, s, \mathbf{x}(s)) \, ds \tag{3.3.7}$$

on some interval $[0, \alpha)$, then it is bounded, and, hence, is continuable to all of $[0, \infty)$.

Proof. Let $\mathbf{x}(t)$ be a solution on $[0, \alpha)$ and let

$$|\mathbf{f}(t)| + \int_0^\alpha M(s, \alpha)\, ds \leq Q\,.$$

Then for $0 \leq t < \alpha$, we have

$$|\mathbf{x}(t)| \leq |\mathbf{f}(t)| + \int_0^t M(s, \alpha)(1 + |\mathbf{x}(s)|)\, ds$$

$$\leq Q + \int_0^t M(s, \alpha)|\mathbf{x}(x)|\, ds\,, \qquad (3.3.8)$$

so that

$$|\mathbf{x}(t)| \leq Q \exp\left[\int_0^t M(s, \alpha)\, ds\right]$$

by Gronwall's inequality. This completes the proof.

Theorem 3.3.6 showed that if $\mathbf{g}$ is small enough, then the sign of $\mathbf{g}$ has nothing to do with continuation. The next result shows that if the "signs are right," then the growth of $\mathbf{g}$ has nothing to do with continuation. We then show that if $\mathbf{g}$ grows too fast and if the "signs are wrong," then solutions cannot always be continued. Examples of the counterparts for differential equations are

$$\begin{aligned} x' &= x && \text{(continuable)}\,, \\ x' &= -x^3 && \text{(continuable)}\,, \end{aligned}$$

and

$$x' = x^3 \qquad\qquad \text{(not continuable)}\,.$$

Theorem 3.3.7. *Let $f : [0, \infty) \to (-\infty, \infty)$, let $C(t, s)$ be a scalar function defined for $0 \leq s \leq t < \infty$, and let $f'(t)$, $C_t(t, s)$ and $C(t, s)$ be continuous on their domains. Suppose $g : (-\infty, \infty) \to (-\infty, \infty)$ with $xg(x) > 0$ if $x \neq 0$, and for each $T > 0$ we have $C(t, t) + \int_t^T |C_u(u, t)|\, du \leq 0$. Then each solution of*

$$x(t) = f(t) + \int_0^t C(t, s)g(x(s))\, ds \qquad (3.3.9)$$

can be continued for all future times.

Proof. We show that if a solution $x(t)$ is defined on $[0, \alpha)$, it is bounded. Let $|f'(t)| \leq M$ on $[0, \alpha]$ and define

$$V(t, x(\cdot)) = e^{-Mt} \left[1 + |x(t)| + \int_0^t \int_t^\alpha |C_u(u, s)| \, du \, |g(x(s))| \, ds \right],$$

so that

$$V'(t, x(\cdot)) \leq e^{-Mt} \left[-M - M|x| + |f'(t)| + C(t, t)|g(x)| \right.$$

$$+ \int_0^t |C_t(t, s)| \, |g(x(s))| \, ds + \int_t^\alpha |C_u(u, t)| \, du \, |g(x)|$$

$$\left. - \int_0^t |C_t(t, s)| \, |g(x(s))| \, ds \right]$$

$$\leq 0 \,.$$

This shows that V, and hence $|x(t)|$, is bounded on $[0, \alpha)$. This completes the proof.

Example 3.3.1. Let $C(t, s) = -(t - s + 1)^{-2}$, $g(x) = x^3$, and $f(t) = e^t$. Then (3.3.9) becomes

$$x(t) = e^t - \int_0^t (t - s + 1)^{-2} x^3(s) \, ds \,.$$

We have

$$C(t, t) + \int_t^T |C_u(u, t)| \, du = -1 + \int_t^T 2(u - t + 1)^{-3} \, du$$

$$= -1 - (u - t + 1)^{-2} \Big|_t^T$$

$$= -1 - (T - t + 1)^{-2} + 1 \leq 0 \,.$$

Thus, solutions are continuable by Theorem 3.3.7.

It is clear that Theorem 3.3.7 actually concerns an integro-differential equation when we differentiate (3.3.9) and obtain

$$x'(t) = f'(t) + C(t, t)g(x) + \int_0^t C_t(t, s)g(x(s)) \, ds \,. \tag{3.3.10}$$

Exercise 3.3.1. Generalize (3.3.10) to

$$x'(t) = h(t) + A(t)g(x) + \int_0^t B(t, s)r(x(s)) \, ds \,. \tag{3.3.11}$$

Examine the statement and proof of Theorem 3.3.7, and place conditions on (3.3.11) so that Theorem 3.3.7 can be stated and proved for (3.3.11).

Now consider the second-order scalar differential equation

$$x'' + a(t)f(x) = 0,\tag{3.3.12}$$

where $a : [0, \infty) \to (-\infty, \infty)$, $f : (-\infty, \infty) \to (-\infty, \infty)$, both are continuous, and $xf(x) > 0$ for $x \neq 0$. Write

$$F(x) = \int_0^x f(s)\, ds.$$

The following is a fundamental continuation result for (3.3.12) that we wish to generalize, in some sense, to encompass integral equations.

Theorem 3.3.8. *Suppose $a(t_1) < 0$ for some $t_1 > 0$. If either*

(a) $\int_0^{+\infty} [1 + F(x)]^{-1/2}\, dx < \infty$ *or*

(b) $\int_0^{-\infty} [1 + F(x)]^{-1/2}\, dx > -\infty,$

then (3.3.12) has solutions not continuable to $+\infty$. Moreover, if $a(t) < 0$ on an interval $[t_1, t_2)$, then (3.3.12) has a solution $x(t)$ defined at t_1 satisfying $\lim_{t \to T^-} |x(t)| = +\infty$ for some T satisfying $t_1 < T \leq t_2$ if and only if either (a) or (b) holds.

Proof. Because $a(t_1) < 0$ and a is continuous, there are positive numbers δ, m, and M such that $-M \leq a(t) \leq -m < 0$ if $t_1 \leq t \leq t_1 + \delta$. In (3.3.12) let $x' = y$ to obtain the system

$$x' = y, \quad y' = -a(t)f(x).\tag{3.3.12'}$$

Assume that condition (a) holds. Denote by $(x(t), y(t))$ a solution satisfying $x(t_1) = 0$ with $y(t_1)$ large and to be determined. As long as $(x(t), y(t))$ is defined on $[t_1, t_1 + \delta)$ we have both $x(t)$ and $y(t)$ monotonically increasing. From (3.3.12)' we have $yy' = -a(t)f(x)x'$, which, upon integration, yields

$$y^2(t) - y^2(t_1) = -2 \int_{t_1}^t a(s)f(x(s))x'(s)\, ds.$$

Because $x(t)$ is increasing, there is a $\bar{t}$ satisfying $t_1 < \bar{t} < t$ such that

$$y^2(t) - y^2(t_1) = -2a(\bar{t}) \int_{x(t_1)}^{x(t)} f(u)\, du.$$

And, because $x(t_1) = 0$, we obtain

$$y^2(t) - y^2(t_1) = -2a(\bar{t})F(x(t)),$$

as long as $(x(t), y(t))$ is defined and $t_1 < t < t_1 + \delta$. Thus

$$y(t) = \left[y^2(t_1) - 2a(\bar{t})F(x(t))\right]^{1/2},$$

so

$$\left[y^2(t_1) + 2mF(x(t))\right]^{1/2} \leq y(t) \leq \left[y^2(t_1) + 2MF(x(t))\right]^{1/2}.$$

Because $x' = y$, we have

$$x'(t) \geq \left[y^2(t_1) + 2mF(x(t))\right]^{1/2}$$

or

$$\left[y^2(t_1) + 2mF(x)\right]^{-1/2} dx \geq dt.$$

Integrating both sides from t_1 to t and recalling that $x(t_1) = 0$, we have

$$\int_0^{x(t)} \left[y^2(t_1) + 2mF(s)\right]^{-1/2} ds \geq t - t_1.$$

Because (a) holds, we may choose $y^2(t_1)$ so large that the integral is smaller than δ. It then follows that $x(t) \to \infty$ before t reaches $t_1 + \delta$. This completes the proof of the first part of the theorem when (a) holds. If (b) holds, then a similar proof is carried out in Quadrant III of the xy plane. The details showing that the integral can be made smaller than δ and the proof of the second part of the theorem can be found in Burton and Grimmer (1971). That paper also contains results on the uniqueness of the zero solution that may be extended to integral equations.

We return now to our integral equation and show that if a grows too fast and if $C(t, s)$ becomes positive at one point, then there are solutions with finite escape time.

It is convenient to introduce an initial function ϕ on an initial interval $[0, a]$ and show that the solution generated by this initial function has finite escape time. As discussed in Chapter 1, it is possible to translate the equation by $y(t) = x(t + a)$, so that the initial function becomes a forcing function.

Theorem 3.3.9. *Consider the scalar equation*

$$x(t) = x_0 + \int_0^t C(t,s)g(x(s))\,ds\,, \tag{3.3.13}$$

where g is continuous and positive for $x > 0$ and $C(t,s)$ and $C_t(t,s)$ are continuous on $0 \le s \le t \le t_1$ for $t_1 > 0$. Suppose also

(a) *there exist $\varepsilon > 0$ and $c_0 > 0$ with $C(t,s) \ge c_0$ if $t_1 - \varepsilon \le s \le t \le t_1$,*

(b) *$g(x)/x \to \infty$ as $x \to \infty$, and*

(c) *$\int_1^\infty [dx/g(x)] < \infty$.*

Then there is a $t_2 \in (0, t_1)$ and an initial function $\phi : [0, t_2] \to [0, \infty)$ such that a solution $x(t, \phi)$ has finite escape time.

Proof. Because $C(t,s) \ge c_0$ if $t_1 - \varepsilon \le s \le t \le t_1$, there is a $K > 0$ with $|C_t(t,s)| \le KC(t,s)$ for $t_1 - \varepsilon \le s \le t \le t_1$. Also, there is an $R > 0$ with

$$c_0 g(x) - Kx \overset{\text{def}}{=} h(x) > 0 \quad \text{for} \quad x \ge R\,.$$

Note that $g(x)/x \to \infty$ as $x \to \infty$ implies that $h(x) \ge Mg(x)$ for some $M > 0$ and x large. Thus $\int_R^\infty [dx/h(x)] < \infty$, and we may choose $R_1 > R$ with $\int_{R_1}^\infty [dx/h(x)] < \varepsilon/2$.

Now, pick $x_0 = R_1$. Define an initial function ϕ with $\phi(t) \equiv 0$ on $[0, t_1 - \varepsilon]$ and let ϕ be linear from $(t_1 - \varepsilon, 0)$ to $(t_1 - (\varepsilon/2), x_0)$. Then as long as the solution $x(t) = x(t, \phi)$ exists on $t_1 - (\varepsilon/2) \le t < t_1$, it satisfies

$$x'(t) = C(t,t)g(x) + \int_0^t C_t(t,s)g(x(s))\,ds$$

$$\ge C(t,t)g(x) - K\int_0^t C(t,s)g(x(s))\,ds\,,$$

because $|C_t(t,s)| \le KC(t,s)$ for $t_1 - \varepsilon \le s \le t \le t_1$. But from (3.3.13), we have

$$x(t) - x_0 = \int_0^t C(t,s)g(x(s))\,ds\,,$$

so that on $t_1 - (\varepsilon/2) \le t \le t_1$ we have

$$x'(t) \ge C(t,t)g(x) - K[x - x_0]$$
$$\ge c_0 g(x) - Kx = h(x) > 0\,.$$

Thus $dx/h(x) \geq dt$, so

$$\int_{x_0}^{x} [ds/h(s)] \geq t - (t_1 - \varepsilon/2) \,,$$

or

$$\varepsilon/2 > \int_{R_1}^{\infty} [dx/h(x)] = \int_{x_0}^{\infty} [ds/h(s)] \geq \int_{x_0}^{x} [ds/h(s)]$$
$$\geq t - (t_1 - (\varepsilon/2)) \,.$$

Thus, if $x(t)$ exists to $t = t_1$, then $\varepsilon/2 > \varepsilon/2$, a contradiction.

Roughly speaking, this theorem tells us that if $C(t,t) > 0$ at some $t = t_1$, if $g(x) > 0$ for $x > 0$, and if $\int_1^{\infty}[ds/g(s)] < \infty$, then

$$x(t) = f(t) + \int_0^t C(t,s)g(x(s))\,ds$$

has solutions with finite escape time.

Exercise 3.3.2. Study the statement and proof of Theorem 3.3.9.

(a) State and prove a similar result for

$$x(t) = f(t) + \int_0^t g(t,s,x(s))\,ds \,.$$

(b) Do the same for

$$x' = h(t,x) + \int_0^t g(t,s,x(s))\,ds \,.$$

3.4 Continuity of Solutions

In Chapter 1 we saw that the innocent-appearing $\mathbf{f}(t)$ in

$$\mathbf{x}(t) = \mathbf{f}(t) + \int_0^t \mathbf{g}(t,s,\mathbf{x}(s))\,ds \qquad (3.4.1)$$

may, in fact, be filled with complications. It may contain constants $\mathbf{x}'(0)$, $\mathbf{x}''(0), \ldots, \mathbf{x}^{(n)}(0)$, all of which are arbitrary, or (even worse) a continuous initial function $\phi : [0, t_0] \to R^n$, where both ϕ and t_0 are arbitrary. Recall that for a given initial function ϕ we write

$$\mathbf{x}(t) = \mathbf{f}(t) + \int_0^{t_0} \mathbf{g}(t, s, \phi(s)) \, ds + \int_{t_0}^t \mathbf{g}(t, s, \mathbf{x}(s)) \, ds \,, \qquad (3.4.2)$$

and ask for a solution of the latter equation for $t \geq t_0$. To change it into the form of (3.4.1) we let

$$\mathbf{y}(t) = \mathbf{x}(t + t_0) = \mathbf{f}(t + t_0) + \int_0^{t_0} \mathbf{g}(t + t_0, s, \phi(s)) \, ds$$

$$+ \int_{t_0}^{t_0 + t} \mathbf{g}(t_0 + t, s, \mathbf{x}(s)) \, ds \,,$$

and define

$$\mathbf{h}(t) = \mathbf{f}(t + t_0) + \int_0^{t_0} \mathbf{g}(t + t_0, s, \phi(s)) \, ds \,. \qquad (3.4.3)$$

Then

$$\int_{t_0}^{t_0 + t} \mathbf{g}(t_0 + t, s, \mathbf{x}(s)) \, ds = \int_0^t \mathbf{g}(t_0 + t, u + t_0, \mathbf{x}(u + t_0)) \, du$$

$$= \int_0^t \mathbf{g}(t_0 + t, s + t_0, \mathbf{y}(s)) \, ds \,,$$

and if

$$\mathbf{g}(t_0 + t, s + t_0, \mathbf{y}(s)) \stackrel{\text{def}}{=} \mathbf{G}(t, s, \mathbf{y}(s)) \,, \qquad (3.4.4)$$

then (3.4.2) becomes

$$\mathbf{y}(t) = \mathbf{h}(t) + \int_0^t \mathbf{G}(t, s, \mathbf{y}(s)) \, ds \,, \qquad (3.4.5)$$

an equation of the form of (3.4.1). In particular, ϕ is consolidated with the forcing function $\mathbf{f}$.

Consider (3.4.3) and note how $\mathbf{h}(t)$ will change as t_0 and ϕ change. We shall want to see precisely how a solution $\mathbf{x}(t)$ of (3.4.2) will change as t_0 and ϕ change.

For future reference we note the role of t in the integrand in (3.4.3). Physical problems frequently demand that a solution $x(t)$ of (3.4.1) take into account its past history; thus, the term ϕ appears in the integral. However, events of long ago tend to fade from memory, and hence, in (3.4.3) as $t \to \infty$, the term $\int_0^{t_0} \mathbf{g}(t + t_0, s, \phi(s)) \, ds$ may reasonably be expected to tend to zero and, frequently, even be $L^1[0, \infty)$.

Example 3.4.1. Consider the scalar equation

$$x(t) = 1 + \int_0^t q(t,s)e^{-(t-s)}x(s)\,ds\,,$$

where $|q(t,s)| \le 1$. Let $\phi : [0,1] \to [-L,L]$ for some $L > 0$. Then

$$h(t) - 1 = \int_0^1 q(t+1,s)e^{-(t+1-s)}\phi(s)\,ds$$

$$= e^{-(t+1)} \int_0^1 q(t+1,s)e^s\phi(s)\,ds$$

and

$$|h(t) - 1| \le Le^{(-t+1)} \int_0^1 e^s\,ds$$

$$= Le^{-(t+1)}(e-1)\,.$$

In the literature, theorems on the continual dependence of solutions on initial conditions (and parameters) take many different forms. Our treatment here will be quite brief. Much can be found on the subject in Miller (1971a).

Basically, we want to say that if $\mathbf{x}(t)$ and $\mathbf{y}(t)$ are solutions of

$$\mathbf{x}(t) = \mathbf{f}(t) + \int_0^t \mathbf{g}(t,s,\mathbf{x}(s))\,ds \qquad (3.4.1)$$

and

$$\mathbf{y}(t) = \mathbf{f}_1(t) + \int_0^t \mathbf{g}(t,s,\mathbf{y}(s))\,ds \qquad (3.4.6)$$

on an interval $[0,T]$, with $|\mathbf{f}_1(t) - \mathbf{f}(t)|$ small on $[0,T]$, then for $\mathbf{g}$ Lipschitz we also have $|\mathbf{x}(t) - \mathbf{y}(t)|$ small on $[0,T]$.

It is worthwhile to state the result locally with a local Lipschitz condition, but the details tend to obscure the basic simplicity.

Theorem 3.4.1. *Let* $\mathbf{f}, \mathbf{f}_1 : [0,a] \to R^n$ *and* $\mathbf{g} : U \to R^n$ *be continuous with*

$$U = \big\{(t,s,\mathbf{x}) : 0 \le s \le t \le a,\ \mathbf{x} \in R^n\big\}\,.$$

Suppose there exists $L > 0$ *with* $|\mathbf{g}(t,s,\mathbf{x}) - \mathbf{g}(t,s,\mathbf{y})| \le L|\mathbf{x} - \mathbf{y}|$ *on* U. *Let* $\mathbf{x}(t)$ *and* $\mathbf{y}(t)$ *be solutions of (3.4.1) and (3.4.6), respectively, on an interval* $[0,T]$ *and let* $\delta = \max_{0 \le t \le T} |\mathbf{f}(t) - \mathbf{f}_1(t)|$. *Then*

$$|\mathbf{x}(t) - \mathbf{y}(t)| \le \delta e^{Lt} \quad \text{for} \quad 0 \le t \le T\,.$$

Proof. From (3.4.1) and (3.4.6) we have

$$|\mathbf{x}(t) - \mathbf{y}(t)| \leq |\mathbf{f}(t) - \mathbf{f}_1(t)| + \int_0^t \big|\mathbf{g}(t,s,\mathbf{x}(s)) - \mathbf{g}(t,s,\mathbf{y}(s))\big|\, ds$$

$$\leq \delta + L \int_0^t |\mathbf{x}(s) - \mathbf{y}(s)|\, ds \,,$$

so the result now follows from Gronwall's inequality.

Exercise 3.4.1. State and prove Theorem 3.4.1 using a local Lipschitz condition. Begin with $\mathbf{f}$ continuous on $[0,a]$ and $\mathbf{g}$ continuous on

$$U = \big\{(t,s,\mathbf{x}) : (0 \leq s \leq t \leq a \,,\, |\mathbf{x} - \mathbf{f}(t)| \leq b\big\}\,.$$

By the existence theorem, a solution $\mathbf{x}(t)$ of (3.4.1) will exist on $[0,\alpha]$ with $\alpha = \min[a, b/M]$. Next, let $\mathbf{f}_1$ be continuous on $[0,a]$ and satisfy $|\mathbf{f}(t) - \mathbf{f}_1(t)| \leq b/3$. By the existence theorem, a solution of (3.4.6) will exist on $[0,\beta]$ for a definite $\beta > 0$, say $\beta \leq \alpha$. Now let

$$\bar{U} = \big\{(t,s,\mathbf{x}) : 0 \leq s \leq t \leq \beta \,,\, |\mathbf{x} - \mathbf{f}(t)| \leq b\big\}$$

and let $\mathbf{g}$ satisfy a Lipschitz condition on $\bar{U}$ with constant L. Formulate and prove the local form of Theorem 3.4.1.

From Theorem 3.4.1 and the exercise we see that the continuity of $\mathbf{f}$ and $\mathbf{g}$, together with a local Lipschitz condition in $\mathbf{x}$ on $\mathbf{g}$, yield existence, uniqueness, and continual dependence of solutions on initial conditions. (It is, perhaps, surprising to learn that the uniqueness of solutions implies a property of this general kind.)

Before proceeding with integral equations, it is worth noting the standard results for ordinary differential equations. For simplicity we consider only an autonomous equation.

Throughout the remainder of this section we assume a strong growth condition on the equation simply to avoid protracted arguments concerning continuation of solutions. The reader will recognize that (3.4.8) can be replaced by a Conti-Wintner equation.

We consider the system

$$\mathbf{x}' = \mathbf{g}(\mathbf{x})\,, \quad \mathbf{x}(0) = \xi_0\,, \tag{3.4.7}$$

in which $\mathbf{g} : R^n \to R^n$ is continuous and for some $K > 0$ we have

$$|\mathbf{g}(x)| \leq K(1 + |\mathbf{x}|)\,, \quad \text{all } \mathbf{x} \in R^n\,. \tag{3.4.8}$$

Notation. $\mathbf{g}_n \rightrightarrows \mathbf{F}$ means that $\mathbf{g}_n$ converges uniformly to $\mathbf{F}$.

Theorem 3.4.2. *Consider a sequence of functions* $\mathbf{g}_k : R^n \to R^n$ *being continuous and satisfying* $|\mathbf{g}_k(\mathbf{x})| \leq K(1 + |\mathbf{x}|)$ *on* R^n. *Suppose*

(a) *for each compact set* $B \subset R^n$, *then* $\mathbf{g}_k(x) \rightrightarrows \mathbf{g}(x)$ *on* B;

(b) *for each* k, $\boldsymbol{\psi}_k(t)$ *is a solution of*

$$\boldsymbol{\psi}' = \mathbf{g}_k(\boldsymbol{\psi}), \quad \boldsymbol{\psi}_k(0) = \boldsymbol{\xi}_k$$

on $[0, \infty)$ *with* $\boldsymbol{\xi}_k \to \boldsymbol{\xi}_0$.

Then for any $T > 0$ *there is a subsequence* $k_j \to \infty$ *with* j *such that*

$$\boldsymbol{\psi}_{k_j}(t) \rightrightarrows \boldsymbol{\psi}(t) \quad \text{on} \quad [0, T] \text{as} \quad j \to \infty,$$

and $\boldsymbol{\psi}(t)$ *is a solution of*

$$\boldsymbol{\xi}' = \mathbf{g}(\boldsymbol{\xi}), \quad \boldsymbol{\psi}(0) = \boldsymbol{\xi}_0.$$

Proof. Write (b) as

$$\boldsymbol{\psi}_k(t) = \boldsymbol{\xi}_k + \int_0^t \mathbf{g}_k(\boldsymbol{\psi}_k(s)) \, ds,$$

so that on $[0, T]$ we have

$$|\boldsymbol{\psi}_k(t)| \leq |\boldsymbol{\xi}_k| + KT + \int_0^t K|\boldsymbol{\psi}_k(s)| \, ds.$$

By Gronwall's inequality we obtain $|\boldsymbol{\psi}_k(t)| \leq (|\boldsymbol{\xi}_k| + KT)e^{KT}$ on $[0, T]$. As $\boldsymbol{\xi}_k \to \boldsymbol{\xi}_0$, then $|\boldsymbol{\psi}_k(t)| \leq M$ for some M, for all k, and for $0 \leq t \leq T$. Let $B_M = \{\boldsymbol{\xi} : |\boldsymbol{\xi}| \leq M\}$, and on B_M we have $\mathbf{g}_k(\boldsymbol{\xi}) \rightrightarrows \mathbf{g}(\boldsymbol{\xi})$. Because $\mathbf{g}$ and the $\mathbf{g}_k$ are continuous, there is a Q with $|\mathbf{g}_k(\boldsymbol{\xi})| \leq Q$ on B_M for all k, and because $\boldsymbol{\psi}_k(s) \in B_M$ on $[0, T]$, if t and $t_1 \in [0, T]$, then

$$|\boldsymbol{\psi}_k(t) - \boldsymbol{\psi}_k(t_1)| = \left| \int_{t_1}^t \mathbf{g}_k(\boldsymbol{\psi}_k(s)) \, ds \right| \leq Q|t - t_1|.$$

Thus, $\{\boldsymbol{\psi}_k\}$ is a uniformly bounded and equicontinuous sequence on $[0, T]$. Hence, there is a subsequence $\boldsymbol{\psi}_{k_j} \rightrightarrows \boldsymbol{\psi}(t)$, for some $\boldsymbol{\psi}$. Therefore

$$\mathbf{g}_{k_j}(\boldsymbol{\psi}_{k_j}(t)) \rightrightarrows \mathbf{g}(\boldsymbol{\psi}(t)) \quad \text{on} \quad [0, T].$$

As

$$\boldsymbol{\psi}_{k_j}(t) = \boldsymbol{\xi}_{k_j} + \int_0^t \mathbf{g}_{k_j}(\boldsymbol{\psi}_{k_j}(s)) \, ds,$$

letting $j \to \infty$ we obtain

$$\boldsymbol{\psi}(t) = \boldsymbol{\xi}_0 + \int_0^t \mathbf{g}(\boldsymbol{\psi}(s)) \, ds,$$

so that $\boldsymbol{\psi}$ is some solution of (3.4.7), as required.

Theorem 3.4.3. *Let* $\mathbf{g} : R^n \to R^n$ *be continuous with* $|\mathbf{g}(\boldsymbol{\xi})| \leq K(1+|\boldsymbol{\xi}|)$ *on* R^n. *If for each* $\boldsymbol{\xi}_0 \in R^n$ *the problem*

$$\boldsymbol{\xi}' = \mathbf{g}(\boldsymbol{\xi})\,, \quad \boldsymbol{\xi}(0) = \boldsymbol{\xi}_0$$

has a unique solution $\boldsymbol{\psi}(t, \boldsymbol{\xi}_0)$, *if* $\boldsymbol{\xi}_k \to \boldsymbol{\xi}_0$ *and* $t_k \to t_0$, $t_i \geq 0$, *then we have*

$$\boldsymbol{\psi}(t_k, \boldsymbol{\xi}_k) \to \boldsymbol{\psi}(t_0, \boldsymbol{\xi}_0)\,.$$

That is, uniqueness implies continuity of the solution in all variables.

Proof. Let $T > 0$ be such that $0 \leq t_k \leq T$, and identify $\mathbf{g}_k(\xi)$ with $\mathbf{g}(\xi)$ and $\boldsymbol{\psi}_k(t)$ with $\boldsymbol{\psi}(t, \boldsymbol{\xi}_k)$. By Theorem 3.4.2, there is a subsequence $k_j \to \infty$ such that

$$\boldsymbol{\psi}(t, \boldsymbol{\xi}_{k_j}) \rightrightarrows \widetilde{\boldsymbol{\psi}}(t) \quad \text{on} \quad [0, T] \quad \text{as} \quad j \to \infty\,.$$

Because $\widetilde{\boldsymbol{\psi}}(t)$ is a solution of

$$\boldsymbol{\xi}' = \mathbf{g}(\boldsymbol{\xi})\,, \quad \widetilde{\boldsymbol{\psi}}(0) = \boldsymbol{\xi}_0\,,$$

then $\widetilde{\boldsymbol{\psi}}(t) = \boldsymbol{\psi}(t, \boldsymbol{\xi}_0)$ by uniqueness. Thus $\boldsymbol{\psi}(t, \boldsymbol{\xi}_{k_j}) \rightrightarrows \boldsymbol{\psi}(t, \boldsymbol{\xi}_0)$ on $[0, T]$. For, by way of contradiction, suppose there were a subsequence for which this were not true. Then by Theorem 3.4.2 there would be a subsequence of that one tending to a solution $\boldsymbol{\psi}^*$ of

$$\boldsymbol{\xi}' = \mathbf{f}(\boldsymbol{\xi})\,, \quad \boldsymbol{\psi}^*(0) = \boldsymbol{\xi}_0\,,$$

with $\boldsymbol{\psi}^*(t) \not\equiv \boldsymbol{\psi}(t, \boldsymbol{\xi}_0)$. This contradicts uniqueness. Thus $\boldsymbol{\psi}(t, \boldsymbol{\xi}_k) \rightrightarrows \boldsymbol{\psi}(t, \boldsymbol{\xi}_0)$ on $[0, T]$, so $\boldsymbol{\psi}(t_k, \boldsymbol{\xi}_k) \to \boldsymbol{\psi}(t_0, \boldsymbol{\xi}_0)$ because $\boldsymbol{\psi}(t_k, \boldsymbol{\xi}_0) \to \boldsymbol{\psi}(t_0, \boldsymbol{\xi}_0)$ and $\boldsymbol{\psi}(t, \boldsymbol{\xi}_0)$ is continuous in t. This completes the proof.

When we set out to formulate a counterpart to Theorem 3.4.2 for

$$\mathbf{x}(t) = \mathbf{f}(t) + \int_0^t \mathbf{g}(t, s, \mathbf{x}(s))\,ds\,,$$

it is clear that we want a sequence $\mathbf{g}_k(t, s, \mathbf{x}) \rightrightarrows \mathbf{g}(t, s, \mathbf{x})$ on compact subsets of $[0, \infty) \times [0, \infty) \times R^n$. But $\mathbf{f}(t)$ contains the initial conditions, and we therefore desire a sequence of functions $\mathbf{f}_k(t) \to \mathbf{f}(t)$. However, the type of convergence needed is not very clear. The fact that $\boldsymbol{\xi}_k \to \boldsymbol{\xi}_0$ in Theorem 3.4.2 is of little help for functions $\mathbf{f}_k(t)$. A simple solution is to ask for equicontinuity of $\{\mathbf{f}_k\}$ and a form of equicontinuity of $\{\mathbf{g}_k(t, s, \mathbf{x})\}$ in t. In particular, if there is a $P > 0$ with $|\mathbf{g}_k(t, s, \mathbf{x}) - \mathbf{g}_k(t_1, s, \mathbf{x})| \leq P|t - t_1|$ on compact sets, this works very well.

Theorem 3.4.4. *Let $\{\mathbf{g}_k\}$ be a sequence of continuous functions with $\mathbf{g}_k :$ $[0, a] \times [0, a] \times R^n \to R^n$ satisfying $|\mathbf{g}_k(t, s, \mathbf{x})| \le K(1 + |\mathbf{x}|)$ on its domain. Suppose that $\{\mathbf{f}_k\}$ is a sequence of uniformly bounded and equicontinuous functions with $\mathbf{f}_k : [0, a] \to R^n$ and $\mathbf{f}_k(t) \rightrightarrows \mathbf{f}(t)$ on $[0, a]$. Suppose also*

(a) *for each compact subset $B \subset R^n$, then $\mathbf{g}_k(t, s, \mathbf{x}) \rightrightarrows g(t, s, \mathbf{x})$ on $[0, a] \times [0, a] \times B$;*

(b) *for each $k, \boldsymbol{\psi}_k(t)$ is a solution of*

$$\boldsymbol{\psi}_k(t) = \mathbf{f}_k(t) + \int_0^t \mathbf{g}_k(t, s, \boldsymbol{\psi}_k(s))\, ds\,,$$

$0 \le t \le a;$

(c) *for each $\varepsilon > 0$ and $M > 0$, there exists $\delta > 0$ such that [k an integer, $s \in [0, a]$, $|t - t_1| < \delta$, $t, t_1 \in [0, a]$, $|x| \le M$] imply $|\mathbf{g}_k(t, s, \mathbf{x}) - \mathbf{g}_k(t_1, s, \mathbf{x})| \le \varepsilon|t - t_1|$.*

Then there is a subsequence $k_j \to \infty$ with j such that $\boldsymbol{\psi}_{k_j}(t) \rightrightarrows \boldsymbol{\psi}(t)$ on $[0, a]$ as $j \to \infty$, and $\boldsymbol{\psi}$ satisfies

$$\boldsymbol{\psi}(t) = \mathbf{f}(t) + \int_0^t \mathbf{g}(t, s, \boldsymbol{\psi}(s))\, ds$$

on $[0, a]$.

Proof. If $|\mathbf{f}_k(t)| \le J$, then from (b) we have

$$|\boldsymbol{\psi}_k(t)| \le J + \int_0^t K\big(1 + |\boldsymbol{\psi}_k(s)|\big)\, ds$$

$$\le J + aK + K \int_0^t |\boldsymbol{\psi}_k(s)|\, ds\,,$$

so that

$$|\boldsymbol{\psi}_k(t)| \le (J + aK)e^{aK} \overset{\text{def}}{=} M \quad \text{on} \quad [0, a]\,.$$

Let $B_M = \{(t, s, \mathbf{x}) : 0 \le s \le a,\ 0 \le t \le a,\ |\mathbf{x}| \le M\}$. Now on B_M we have $\mathbf{g}_k(t, s, x) \rightrightarrows \mathbf{g}(t, s, \mathbf{x})$ Also, by assumption, there is a $Q > 0$ with

$|\mathbf{g}_k(t, s, \mathbf{x})| \leq Q$ on B_M for all k. If $t, t_1 \in [0, a]$ with $t > t_1$, as $|\boldsymbol{\psi}_k(s)| \leq M$ we have

$$
\begin{aligned}
&|\boldsymbol{\psi}_k(t) - \boldsymbol{\psi}_k(t_1)| \\
&= \left| \mathbf{f}_k(t) - \mathbf{f}_k(t_1) + \int_0^t \mathbf{g}_k(t, s, \boldsymbol{\psi}_k(s))\, ds - \int_0^{t_1} \mathbf{g}_k(t_1, s, \boldsymbol{\psi}_k(s))\, ds \right| \\
&\leq |\mathbf{f}_k(t) - \mathbf{f}_k(t_1)| + \left| \int_0^{t_1} \left[\mathbf{g}_k(t, s, \boldsymbol{\psi}_k(s)) - \mathbf{g}_k(t_1, s, \boldsymbol{\psi}_k(s)) \right] ds \right| \\
&\quad + \left| \int_{t_1}^t \mathbf{g}_k(t, s, \boldsymbol{\psi}_k(s))\, ds \right| \\
&\leq |\mathbf{f}_k(t) - \mathbf{f}_k(t_1)| + \varepsilon a |t - t_1| + Q|t - t_1| \,.
\end{aligned}
$$

Because the $\mathbf{f}_k$ are equicontinuous, so is $\{\boldsymbol{\psi}_k\}$. Hence, there is a subsequence of the $\boldsymbol{\psi}_k$, say $\boldsymbol{\psi}_k$ again, with $\boldsymbol{\psi}_k(t) \rightrightarrows \boldsymbol{\psi}(t)$ on $[0, a]$.

We have

$$
\boldsymbol{\psi}_k(t) = \mathbf{f}_k(t) + \int_0^t \mathbf{g}_k(t, s, \boldsymbol{\psi}(s))\, ds \,,
$$

and as $k \to \infty$ we obtain

$$
\boldsymbol{\psi}(t) = \mathbf{f}(t) + \int_0^t \mathbf{g}(t, s, \boldsymbol{\psi}(s))\, ds \,,
$$

as required.

Notice that if $\mathbf{g}_k(t, s, \mathbf{x}) \equiv \mathbf{g}(t, s, \mathbf{x})$ and if solutions are unique, then the result states that as the initial functions $\mathbf{f}_k(t)$ converges to $\mathbf{f}(t)$, then the solutions converge. That is, solutions depend continuously on initial functions. In short, uniqueness implies continual dependence of solutions on initial conditions.

Quite obviously, continual dependence of solutions on initial conditions implies uniqueness.

Chapter 4

History, Examples, and Motivation

4.0   Introduction

This chapter is devoted to a selection of problems and historical events that have affected the development of the subject. Many of the formulations are quite different from the traditional derivations seen in mathematical physics, which proceed from first principles. At least in the early development of the subject, problems were formulated from the descriptive point of view; a physical situation was observed and a mathematical model was constructed that described the observations. The aim was to discover properties from the mathematical model that had not been observed in the physical situation, which could assist the observer in better understanding the outside object. Much of mathematical biology has proceeded in this fashion. And though its critics abound, the successes have been marked and important. An authoritative case for proceeding in this way is made by the eminent biologist J. Maynard Smith (1974, p. 19) in a modern monograph on mathematical biology.

In this chapter we briefly discuss numerous problems related, in at least some way, to Volterra equations. In some cases we present substantial results; in other cases we formulate the problems so they may be solved using methods of later chapters; and finally, some problems are briefly introduced as examples to which the general theory applies. In all cases we attempt to provide references, so that the interested reader may pursue the topic in some depth.

4.1   Volterrra and Mathematical Biology

In this section we study the work that went into the formulation of a pair
of predator-prey equations

$$x' = x[a - by - dx],$$

$$y' = y\left[-c + kx + \int_0^t K(t - s)x(s)\,ds \right] \qquad (4.1.1)$$

and then transform these equations into the form treated in the general
theory discussed in Chapters 5 and 6.

The study of Volterra's work on competing species is a fundamental example of the progressive improvement of a model to explain a description of
a physical process. It shows, in particular, how a description of observable
facts can lead to the suggestion of new information.

Fairly accurate records had been kept by Italian port authorities of
the ratio of food fish to trash fish (rays, sharks, skates, etc.) netted by
Italian fisherman from 1914 to 1923. The period spanned World War I and
displayed a very curious and unexpected phenomenon. The proportion of
food fish markedly decreased during the war years and then increased to
the prewar levels.

Fishing was much less intense during the war; it was hazardous, and
many fisherman were otherwise occupied. Intuitively, one would think
fishing would be much improved after the slow period. Indeed, rare is
that person who has not dreamed of the glorious fishing to be had in some
virgin mountain lake or stream.

The Italian biologist Umberto D'Ancona considered several possible
explanations, rejected all of them, and in 1925 consulted the distinguished
Italian mathematician Vito Volterra in search of a mathematical model
explaining this fishing phenomenon.

The problem interested Volterra for the remainder of his life and provided a new setting for his functionals. Moreover, his initial work inspired
such widespread interest that by 1940 the literature on the problem was
positively enormous. A brief description of his concern with it is quite
worthwhile. It was, to begin with, quite clearly a problem of predator
and prey. The trash fish fed on the food fish. Moreover, the literature on
descriptive growth of species was not at all empty.

In 1798 Thomas Robert Malthus, an English economist and historian,
published a work (of inordinate title length) contending that a population
increases geometrically (e.g., 3, 9, 27, 81, ...) whereas food production
increases arithmetically (e.g., 3, 6, 9, 12, ...). He contended that population will always tend to a limit of subsistence at which it will be held by

famine, disease, and war. (See Encyclopaedia Britannica, 1970, vol. 14 for a synopsis.) This contention, although far from ludicrous, has been under attack since its publication.

One proceeds as follows to formulate a mathematical model of *Malthusian growth*. Let $p(t)$ denote the population size (or density) at a given time t. If there is unlimited space and food, while the environment does not change, then it seems plausible that the population will increase at a rate proportional to the number of individuals present at a given time.

If $p(t)$ is quite large, it may be fruitful to conceive of $p(t)$ as being continuous or even differentiable. (Indeed, the science philosopher Charles S. Peirce (1957, pp. 57–60) contends that the "application of continuity to cases where it does not really exist ... illustrates ... the great utility which fictions sometimes have in science." He seems to consider it a cornerstone of scientific progress.) In that case we would say

$$dp(t)/dt = kp(t), \quad p(t_0) = p_0, \tag{4.1.2}$$

where k is the constant of proportionality. As it is assumed that the population increases, $k > 0$. This problem has the unique solution

$$p(t) = p_0 \exp[k(t - t_0)]. \tag{4.1.3}$$

Notice that when time is divided into equal intervals, say, years, this does yield a geometric increase.

Obviously, no environment could sustain such growth, and by about 1842 the Belgian statistician L. A. J. Quetelet had noticed that a population able to reproduce freely with abundant space and resources tends to increase geometrically, while obstacles tend to slow the growth, causing the population to approach the upper limit, resulting in an S-shaped population curve with a limiting population L (see Fig. 4.1). Such curves had been observed by Edward Wright in 1599 and were called *logistic curves*, a term still in use.

The history of the problem of modeling such a curve mathematically is an interesting one. A colleague of Quetelet, P. F. Verhulst, assumed that the population growth was inhibited by a factor proportional to the square of the population. Thus, the equation for Malthusian growth was modified to

$$p'(t) = kp(t) - rp^2(t) \tag{4.1.4}$$

for k and r positive. This has become known as the *logistic equation* and $rp^2(t)$ the *logistic load*. It is a simple Riccati equation, which is equivalent to a second-order, linear differential equation. Its solution is

$$p(t) = m/[1 + M \exp(-kt)], \tag{4.1.5}$$

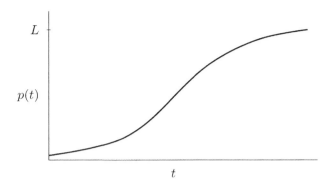

Figure 4.1: S-shaped population curve with limit L.

which may be obtained by separation of variables and partial fractions. Its limiting population is

$$m = k/r \,,$$

called the *carrying capacity* of the environment. It describes the curve of Fig. 4.1 and, moreover, if $p(t_0) > k/r$, it describes a curve of negative slope approaching k/r as $t \to \infty$. Thus, for example, if a fishpond is initially over stocked, the population declines to k/r.

With the proper choice of k and r, (4.1.5) describes the growth of many simple populations, such as yeast [see Maynard Smith (1974, p. 18)]. Although the logistic equation is a descriptive statement, it has received several pseudo derivations.

The *law of mass action* states, roughly, that if m molecules of a substance x combine with n molecules of a substance y to form a new substance z, then the rate of reaction is proportional to $[x]^m [y]^n$, where $[x]$ and $[y]$ denote the concentrations of substances x and y, respectively.

Thus, one might argue that for population $x(t)$ with density $p(t)$, the members compete with one another for space and resources, and the rate of encounter is proportional to $p(t)p(t)$. So, one postulates that population increases at a rate proportional to the density and decreases at a rate proportional to the square of the density

$$p'(t) = kp(t) - rp^2(t) \,.$$

Derivations based on the Taylor series may be found in Pielou (1969, p. 20). One may ask: What is the simplest series representation for

$$p'(t) = f(p) \,,$$

where f is some function of the population? To answer this question, write

$$f(p) = a + bp + cp^2 + \cdots .$$

First, we must agree that $f(0) = 0$, as a zero population does not change; hence, $a = 0$. Next, if the population is to grow for small p, then b must be positive. But if the population is to be self-limiting and if we wish to work with no more than a quadratic, then c must be negative. This yields (4.1.4).

Detractors have always argued that the growth of certain populations are S-shaped, and hence, *any* differential equation having S-shaped solutions with parameters that can be fitted to the situation could be advanced as an authoritative description.

Enter Volterra: Let $x(t)$ denote the population of the prey (food fish) and $y(t)$ the population of the predator (trash fish). Because the Mediterranean Sea (actually the upper Adriatic) is large, let us imagine unlimited resources, so that in the absence of predators,

$$x' = ax \quad a > 0 , \tag{4.1.6}$$

which is Malthusian growth. But $x(t)$ should decrease at a rate proportional to the encounter of prey with predator, yielding

$$x' = ax - bxy , \quad a > 0 \text{ and } b > 0 . \tag{4.1.7}$$

Now imagine that, in the absence of prey, the predator population would decrease at a rate proportional to its size

$$y' = -cy , \quad c > 0 .$$

But y should increase at a rate proportional to its encounters with the prey, yielding

$$y' = -cy + kxy , \quad c > 0 \text{ and } k > 0 . \tag{4.1.8}$$

We now have the simplest predator-prey system

$$\begin{aligned} x' &= ax - bxy , \\ y' &= -cy + kxy , \end{aligned} \tag{4.1.9}$$

and we readily reason that a, b, c, and k are positive, with $b > k$, because y does not have 100% efficiency in converting prey.

Incidentally, (4.1.9) had been independently derived and studied by Lotka (1924) and, hence, is usually called the Lotka-Volterra system. The system may be solved for a first integral as follows. We have

$$dy/dx = (-c + kx)y/(a - by)x \,, \tag{4.1.10}$$

so that separation of variables and an integration yields

$$(y^a/e^{by})(x^c/e^{kx}) = K \,, \tag{4.1.11}$$

K a constant. The solution curves are difficult to plot, but Volterra (1931; p. 29) [See Davis (1962, p. 103)] devised an ingenious graphical scheme for displaying them.

The predator-prey system makes sense only for $x \geq 0$ and $y \geq 0$. Also, there is an equilibrium point ($x' = y' = 0$) in the open first quadrant at ($x = c/k$, $y = a/b$), which means that populations at that level remain there. May we say that the predator and prey would "live happily ever after" at that level?

The entire open first quadrant is then filled with (noncrossing) simple closed curves (corresponding to periodic solutions), all of which encircle the equilibrium point (c/k, a/b) (see Fig. 4.2). We will not go into the details of this complex graph now but a simplifying transformation presented later will enable the reader easily to see the form.

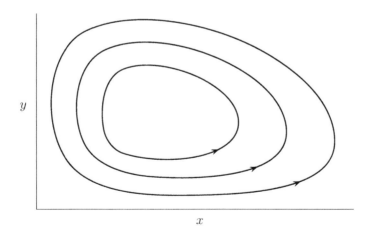

Figure 4.2: Periodic solutions of predator-prey systems.

We are unable to solve for $x(t)$ and $y(t)$ explicitly, but we may learn much from the paths of the solutions, called orbits, displayed in Fig. 4.2. An orbit that is closed and free of equilibrium points represents a periodic solution. Each of those in Fig. 4.2 may have a different period T.

Let us interpret the action taking place during one period. We trace out a solution once in the counterclockwise direction starting near the point $(0, 0)$. Because there are few predators, the prey population begins to increase rapidly. This is good for the predators, which now find ample food and begin to multiply, but as the predators increase, they devour the prey, which therefore diminish in number. As the prey decrease, the predators find themselves short on food and lose population rapidly. The cycle continues.

Although we cannot find $x(t)$, $y(t)$, nor T, surprisingly, we can find the average value of the population densities over any cycle. The average of a periodic function f over a period T is

$$\bar{f} = (1/T) \int_0^T f(t) \, dt \,.$$

From (4.1.9) we have

$$(1/T) \int_0^T [x'(t)/x(t)] \, dt = (1/T) \int_0^T [a - by(t)] \, dt$$
$$= (1/T)aT - (b/T) \int_0^T y(t) \, dt \,;$$

furthermore

$$(1/T) \int_0^T [x'(t)/x(t)] \, dt = (1/T) \ln [x(T)/x(0)] = 0 \,,$$

because $x(T) = x(0)$. This yields

$$\bar{y} = (1/T) \int_0^T y(t) \, dt = a/b \,.$$

A symmetric calculation shows $\bar{x} = c/k$. Thus, the coordinates of the equilibrium point $(c/k, a/b)$ are the average populations of any cycle. Notice that statistics on catches would be averages, and those averages are the equilibrium populations.

To solve the problem presented to Volterra (in this simple model), we must take into account the effects of fishing. The fishing was by net, so the densities of x and y are decreased by the same proportional factor, namely, $-\varepsilon x$ and $-\varepsilon y$, respectively. The predator-prey fishing equations become

$$x' = ax - bxy - \varepsilon x \,,$$
$$y' = -cy + kxy - \varepsilon y \,. \tag{4.1.12}$$

[The reader should consider and understand why $b \neq k$, but ε is the same in both directions.] The new equilibrium point (or average catch) is

$$\left(\frac{c + \varepsilon}{k} , \frac{a - \varepsilon}{b} \right) .$$

In other words, a moderate amount of fishing ($a > \varepsilon$) actually diminishes the proportion of predator and increases the proportion of prey.

If one believes the model (and not even Volterra did, he continued to refine it), there are far-reaching implications. For example, spraying poison on insects tends to kill many kinds, in the way the net catches many kinds of fish. Would spraying fruit trees increase the average prey density and decrease the average predator density? Here, the prey are leaf and fruit eaters and the predators are the friends of the orchard. The controversy rages, and we will, of course, settle nothing here. Let it be said, however, that elderly orchardists in southern Illinois claim that prior to 1940 they raised highly acceptable fruit crops without spraying. Chemical companies showed them that a little spraying would correct even their small problems. Now they are forced to spray every 3 to 10 days during the growing season to obtain marketable fruit.

In a more scientific vein, there is hard evidence that the feared outcome of spraying did occur in the apple orchards of the Wenatchee area of Washington state. There, DDT was used to control the McDaniel spider mite, which attacked leaves, but the spraying more effectively controlled its predator [see Burton (1980b, p. 257)].

We return now to Volterra's problem and consider the effect of logistic loads. Thus we examine

$$x' = ax - dx^2 - bxy \,,$$
$$y' = -cy + kxy \,, \tag{4.1.13}$$

with equilibrium at

$$\left(\frac{c}{k} , \frac{ak - dc}{bk} \right) \overset{\text{def}}{=} (\bar{x}, \bar{y}) \,, \tag{4.1.14}$$

requiring $ak > dc$, so that it is in the first quadrant.

It is easy to see that any solution $(x(t), y(t))$ in the open first quadrant is bounded, because

$$kx' + by' = kax - kdx^2 - bcy$$

is negative for $x^2 + y^2$ large, x and y positive. Thus, an integration yields $kx(t) + by(t)$ bounded.

In fact, one may show that all of these solutions approach the equilibrium point of (4.1.14). To that end, define

$$u = \ln\left[x/\bar{x}\right] \quad \text{and} \quad v = \ln\left[y/\bar{y}\right], \tag{4.1.15}$$

so that

$$x = \bar{x}e^u \quad \text{and} \quad y = \bar{y}e^v. \tag{4.1.16}$$

Then using (4.1.13)–(4.1.16) we obtain

$$u' = d\bar{x}(1 - e^u) + b\bar{y}(1 - e^v),$$

and   (4.1.17)

$$v' = k\bar{x}(e^u - 1).$$

If we multiply the first equation in (4.1.17) by $k\bar{x}(e^u - 1)$ and the second by $b\bar{y}(e^v - 1)$, then adding we obtain

$$k\bar{x}(e^u - 1)u' + b\bar{y}(e^v - 1)v' = -dk\bar{x}^2(e^u - 1)^2$$

or

$$(d/dt)\left[k\bar{x}(e^u - u) + b\bar{y}(e^v - v)\right] \leq 0.$$

Thus the function

$$V(u, v) = k\bar{x}(e^u - u) + b\bar{y}(e^v - v)$$

is a Liapunov function. It has a minimum at $(0,0)$ (by the usual derivative tests), and $V(u, v) \to \infty$ as $u^2 + v^2 \to \infty$. As $V'_{(4.1.17)}(u, v) \leq 0$, all solutions are bounded. Moreover, if we examine the set in which $V'(u, v) = 0$, we have

$$e^u - 1 = 0,$$

or $u = 0$. Now, if $u = 0$, then $v' = 0$ and $u' = -b\bar{y}(e^v - 1)$, which is nonzero unless $v = 0$. Thus, a solution intersecting $u = 0$ will leave $u = 0$

unless $v = 0$ also. This situation is covered in the work of Barbashin and Krasovskii (see our Section 6.1, Theorems 6.1.4 and 6.1.5). We may conclude that all solutions of (4.1.17) tend to $(0,0)$. But, in view of (4.1.16), all solutions of (4.1.13) approach the equilibrium $(\bar{x}, \bar{y})$ of (4.1.14). [Incidentally, transforming (4.1.9) by (4.1.15) will simplify the graphing problem.]

It seems appropriate now to summarize much of this work in the following result.

Theorem 4.1.1. *Consider (4.1.13) and (4.1.14) with a, b, c, and k positive, $ak > dc$, and $d \geq 0$.*

(a) *If $d > 0$, then all solutions in Quadrant I approach $(\bar{x}, \bar{y})$.*

(b) *If $d = 0$, all solutions in Quadrant I are periodic. The mean value of any solution $(x(t), y(t))$ is $(c/k, a/b)$.*

The predator-prey-fishing equations become

$$x' = ax - dx^2 - bxy - \varepsilon x \,,$$
$$y' = -cy + kxy - \varepsilon y \,, \tag{4.1.18}$$

so that the new equilibrium point is

$$\left(\frac{c + \varepsilon}{k} \,, \; \frac{(a - \varepsilon)k - d(c + \varepsilon)}{bk} \right) .$$

Thus, the asymptotic population of prey increases with moderate fishing and the predator population decreases.

Much solid scientific work has gone into experimental verification of Volterra's model, with mixed results. A critical summary is given in Goel *et al.* (1971, pp. 121–124).

The next observation is that, although the prey population immediately decreases upon contact with predator, denoted by $-bxy$, it is clear that the predator population does not immediately increase upon contact with the prey. There is surely a time delay, say, T, required for the predator to utilize the prey. This suggests the system

$$x' = ax - bxy \,,$$
$$y' = -cy + kx(t - T)y(t - T) \,, \tag{4.1.19}$$

which is a system of delay differential equations. Actually, (4.1.19) does not seem to have been studied by Volterra, but rather by Wangersky and Cunningham (1957). Yet, the system seems logically to belong here in the successive refinement of the problem given Volterra.

The initial condition for (4.1.19) needs to be a pair of continuous initial functions $x(t) = \phi(t)$, $y(t) = \psi(t)$ for $-T \leq t \leq 0$. Notice that equilibrium populations still occur. For if $x' = y' = 0$, then $x(t) = x(t - T)$ and $y(t) = y(t - T)$, so we need $x = c/k$ and $y = a/b$.

Although (4.1.19) must be regarded as far more sophisticated than the Lotka-Volterra system, which we did not explicitly solve, interestingly enough (4.1.19) can be solved by the method of steps. For if $0 \leq t \leq T$, then

$$y' = -cy + k\phi(t - T)\psi(t - T) \,, \tag{4.1.20}$$

which is a linear, first-order equation with the initial condition $y(0) = \psi(0)$ having a unique solution, say, $\eta(t)$, on $[0, T]$. Then on $[0, T]$ we have

$$x' = ax - bx\eta(t) \,, \quad x(0) = \phi(0) \,. \tag{4.1.21}$$

An integration yields $x(t) = \gamma(t)$ on $[0, T]$.

Now the new initial function is $x(t) = \gamma(t)$ and $y(t) = \eta(t)$ on $[0, T]$, so system (4.1.19) can be solved on $[T, 2T]$. One may continue by quadratures alone and obtain a solution on any interval $[0, t]$.

The more general system

$$\begin{aligned} x' &= ax - dx^2 - bxy \,, \\ y' &= -cy + kx(t - T)y(t - T) \end{aligned} \tag{4.1.22}$$

may be solved in the same way when we note that

$$x' = ax - dx^2 - bx\eta(t) \tag{4.1.23}$$

is simply a Bernoulli equation.

Obviously, the integrations quickly become fierce, and one longs for a qualitative approach. Wangersky and Cunningham (1957) used computer simulation on (4.1.22) with somewhat inconclusive results. (Careful consideration of (4.1.22) under the transformation (4.1.15) may well lead to a Liapunov functional.) The computer simulations indicated that for $d = 0$ there are large amplitude oscillations. When $d > 0$, then increasing d tends to stabilize the behavior, whereas increasing T tends to create instability.

Before returning to Volterra's own formulation, we point out that several alternative forms of the predator-prey system have been seriously studied. Leslie (1948) suggested the system

$$\begin{aligned} x' &= ax - dx^2 - bxy \,, \\ y' &= -ey - (ky^2/x) \,. \end{aligned} \tag{4.1.24}$$

Rosenzweig and MacArthur (1963) studied the general system

$$x' = f(x) - \phi(x, y),$$
$$y' = -ey + k\phi(x, y) \tag{4.1.25}$$

and conjectured various forms for f and ϕ. Later, Rosenzweig (1969) analyzed, in some detail, the biological significance of the shape of the curve for $\dot{x} = 0$,

$$f(x) = \phi(x, y). \tag{4.1.26}$$

In the fourth chapter of his book, Volterra (1931) formulates the object of that section of this book. We describe it here in his own notation and with his explanation. He begins with the Lotka-Volterra system written in differential form

$$dN_1 = \varepsilon_1 N_1 \, dt - \gamma_1 N_1 N_2 \, dt,$$
$$dN_2 = -\varepsilon_2 N_2 \, dt + \gamma_2 N_1 N_2 \, dt, \tag{4.1.27}$$

with $\varepsilon_i > 0$ and $\gamma_i > 0$.

Let us assume that, in the second species at least, the distributions by age of the individuals remains constant, and let $\phi(\xi)d\xi$ be the ratio of the number of ages lying between ξ and $\xi + d\xi$ to the total number of individuals of the predatory species. Then the number of predatory individuals, being at time t of an age at least equal to θ, is

$$N_2(t) \int_\theta^\infty \phi(\xi) \, d\xi = N_2(t) f(\theta), \quad f \geq 0, \; f \not\equiv 0.$$

Among $N_2(t)$ individuals existing at time t, there are then $N_2 f$ which existed already at the previous τ. By the law of mass action, one can see then that the quantity of nourishment in individuals of the first species absorbed during the interval $(\tau, \tau + d\tau)$ by the individuals of the second species who existed at both times τ and t, is

$$\gamma f(t - \tau) N_1(\tau) N_2(t) d\tau.$$

We can assume that this nourishment creates an increment of

$$\psi(t - \tau) dt \, \gamma f(t - \tau) N_1(\tau) N_2(t) d\tau, \quad \psi \geq 0, \; \psi \not\equiv 0,$$

individuals of the second species during the interval $(t, t + dt)$, so that, in adding these supposedly independent effects, one obtains

$$N_2(t) dt \int_{-\infty}^t \gamma \psi(t - \tau) f(t - \tau) N_1(\tau) \, d\tau.$$

We then replace the second equation in (4.1.27) by

$$dN_2 = -\varepsilon_2 N_2 dt + N_2 dt \int_{-\infty}^t F(t - \tau) N_1(\tau) \, d\tau \,, \quad F \geq 0 \ F \not\equiv 0 \,.$$

We then have the system

$$N_1' = N_1(t)\big[\varepsilon_1 - \gamma_1 N_2(t)\big]$$

$$N_2' = N_2(t)\bigg[-\varepsilon_2 + \int_{-\infty}^t F(t - \tau)N_1(\tau) \, d\tau \bigg] \tag{4.1.28}$$

or the more symmetric system

$$N_1' = \bigg[\varepsilon_1 - \gamma_1 N_2(t) - \int_{-\infty}^t F_1(t - \tau)N_2(\tau) \, d\tau \bigg] N_1(t) \,,$$

$$N_2' = \bigg[-\varepsilon_2 + \gamma_2 N_1(t) + \int_{-\infty}^t F_2(t - \tau)N_1(\tau) \, d\tau \bigg] N_2(t) \tag{4.1.29}$$

with $\varepsilon_1, \varepsilon_2, \gamma_1, \gamma_2 \geq 0$, $F_1 \geq 0$, $F_2 \geq 0$, and especially $\gamma_1 > 0$ and $F_2 \not\equiv 0$.
Volterra emphasizes that these integrals may take the form of

$$\int_{t_0}^t \quad \text{or} \quad \int_{t-T}^t$$

depending on the duration of the heredity.

Although the complete stability analysis of the problems formulated by Volterra has not been given, it is enlightening to view some properties of equations of that general type. We might call

$$x' = x\bigg[a - bx + \int_{t_0}^t K(t, s)x(s) \, ds \bigg] \,, \tag{4.1.30}$$

with a and b positive constants and K continuous for $t_0 \leq s \leq t < \infty$ a scalar Verhulst-Volterra equation. Here, t_0 may be $-\infty$, in which case we would, of course, write $t_0 < s \leq t < \infty$. Thus we are taking into account the entire past history of x. Frequently, $K(t, s)$ is discontinuous and the integral is taken in the sense of Stieltjes as described, for example, by Cushing (1976).

We shall suppose for the next two results that $t_0 \leq 0$ and that we have an initial function on $[t_0, 0]$. Thus, we shall be discussing the solutions for $t > 0$. We could let t_0 be any value and let the initial function be given on any interval $[t_0, t_1]$, but the historical setting of such problems tends to be of the type chosen here.

Theorem 4.1.2. *Let r, R, and m be positive constants with*

$$\int_{t_0}^{t} |K(t,s)|\, ds \le m < b$$

for all $t \ge t_0$, and suppose that for $t_0 \le s \le 0$ we have $0 < r < x(s) < R$ with $a - bR + mR < 0$ and $a - br - Rm > 0$. Then $r < x(t) < R$ for $0 \le t < \infty$.

Proof. Suppose that the result is false and that t_1 is the first value of t for which $r < x(t) < R$ is violated; to be definite, let $x(t_1) = R$. Then from the differential equation we have

$$x'(t_1)/x(t_1) \le a - bR + R \int_{t_0}^{t_1} |K(t_1,s)|\, ds$$
$$< a - bR + Rm < 0\,,$$

a contradiction because we must have $x'(t_1) \ge 0$.

Now suppose $x(t_1) = r$. Then $x'(t_1)/r \ge a - br - Rm > 0$, a contradiction because we must have $x'(t_1) \le 0$. This completes the proof.

Roughly speaking, if m is quite small, then solutions are bounded and extinction does not occur. It would be very interesting to learn precisely how such solutions behave. For example, can carrying capacity be defined, and do solutions oscillate around that carrying capacity?

Next, we consider the system

$$x' = x\left[a - bx - cy - \int_{t_0}^{t} K_1(t,s)y(s)\, ds\right],$$

$$y' = y\left[-\alpha + \beta x + \int_{t_0}^{t} K_2(t,s)x(s)\, ds\right]. \tag{4.1.31}$$

Theorem 4.1.3. *If $K_i(t,s) \ge 0$ and continuous for $t_0 \le s \le t < \infty$, if a, b, c, α, and β are positive constants, and if there is an $\varepsilon > 0$ with*

$$\alpha > (a/b) \int_{t_0}^{t} K_2(t,s)\, ds + \varepsilon \quad \text{for} \quad t \ge t_0\,,$$

then all solutions remaining in the first quadrant and satisfying $x(t) \le a/b$ on the initial interval $[t_0, 0]$ are bounded.

Proof. First notice that $x' < 0$ if $x > a/b$, hence, $x(t) \leq a/b$ if $t \geq t_0$. Next,

$$\beta x' + cy' \leq \beta x(a - bx - cy) + cy\left(-\alpha + \beta x + \int_{t_0}^t K_2(t,s)x(s)\,ds\right)$$

$$\leq \beta x(a - bx) + cy\left[-\alpha + (a/b)\int_{t_0}^t K_2(t,s)\,ds\right]$$

$$\leq \beta x(a - bx) - c\varepsilon y\,,$$

which is negative if

$$c\varepsilon y > a\beta x - b\beta x^2\,.$$

Hence, if the line $\beta x + cy = \text{constant}$ lies above the parabola

$$c\varepsilon y = a\beta x - b\beta x^2\,,$$

then

$$\left[\beta x(t) + cy(t)\right]' \leq 0\,,$$

so that $y(t)$ is bounded.

It would be very interesting to obtain information about the qualitative behavior of these bounded solutions. It is our view that one of the real deficiencies of the attempts to analyze (4.1.31) is the absence of a true equilibrium of any type. For example, Cushing (1976) considers the system

$$x_1' = x_1(a_1 - c_1 x_2)$$

$$x_2' = x_2\left[-a_2 + \int_{t_0}^t k_2(t-s)x_1(s)\,ds\right]$$

and speaks of the equilibrium point $\left(a_2/\left[b_1 + \int_0^\infty k_2(s)\,ds\right],\, a_1/c_1\right)$, where b_1 comes from another equation. Clearly, $x_2 = a_1/c_1$, but that value in x_2' does not yield $x_2' \equiv 0$ for any constant x_1. Similarly, (4.1.31) does not have a constant equilibrium solution. It seems that one needs to locate an asymptotic equilibrium and then work on perturbation problems.

Volterra's derivation suggests that we consider $K_1 \equiv 0$. Thus, let us consider

$$x' = x[a - dx - by]\,,$$

$$y' = y\left[-c + kx + \int_0^t K(t-s)x(s)\,ds\right]\,,$$

(4.1.32)

in which a, b, c, d, and k are positive constants, K is continuous with $K(t) \geq 0$, and $\int_0^\infty K(s)\,ds = r < \infty$.

To locate an asymptotic equilibrium we write

$$\int_0^t K(t-s)\,ds = \int_0^t K(s)\,ds$$
$$= \int_0^\infty K(s)\,ds - \int_t^\infty K(s)\,ds$$
$$\overset{\text{def}}{=} r - \gamma(t)\,.$$

Then we write (4.1.32) as

$$x' = x[a - dx - by]$$
$$y' = y\left\{ -c + kx + r\bar{x} - \bar{x}\gamma(t) + \int_0^t K(t-s)[x(s) - \bar{x}]\,ds \right\} \tag{4.1.33}$$

where $\bar{x}$ is defined by

$$-c + k\bar{x} - r\bar{x} = 0 \quad \text{or} \quad \bar{x} = c/(k+r)\,.$$

Then

$$a - d\bar{x} - b\bar{y} = 0$$

yields

$$\bar{y} = (a - d\bar{x})/b\,.$$

Let

$$u = \ln[x/\bar{x}] \quad \text{and} \quad v = \ln[y/\bar{y}]\,,$$

so that (4.1.33) becomes

$$u' = d\bar{x}(1 - e^u) + b\bar{y}(1 - e^v)$$
$$v' = k\bar{x}(e^u - 1) - \bar{x}\gamma(t) + \int_0^t \bar{x}K(t-s)(e^{u(s)} - 1)\,ds\,. \tag{4.1.34}$$

Now the linear approximation of this is

$$u' = -d\bar{x}u - b\bar{y}v\,,$$
$$v' = k\bar{x}u + \int_0^t \bar{x}K(t-s)u(s)\,ds - \bar{x}\gamma(t)\,, \tag{4.1.35}$$

which, in matrix form, is

$$
\begin{pmatrix} u \\ v \end{pmatrix}' = \begin{pmatrix} -d\bar{x} & -b\bar{y} \\ k\bar{x} & 0 \end{pmatrix} \begin{pmatrix} u \\ v \end{pmatrix}
$$
$$
+ \int_0^t \begin{pmatrix} 0 & 0 \\ \bar{x}K(t-s) & 0 \end{pmatrix} \begin{pmatrix} u(s) \\ v(s) \end{pmatrix} ds + \begin{pmatrix} 0 \\ -\bar{x}\gamma(t) \end{pmatrix} \qquad (4.1.36)
$$

or

$$
\mathbf{X}' = A\mathbf{X} + \int_0^t C(t-s)\mathbf{X}(s)\, ds + \mathbf{\Gamma}(t)\,, \qquad (4.1.37)
$$

where all characteristic roots of A have negative real parts, $\int_0^\infty |C(s)|\, ds < \infty$, and $\mathbf{\Gamma}(t) \to 0$ as $t \to \infty$. Moreover, it is consistent with the problem to ask that

$$
\int_0^\infty |\gamma(t)|\, dt < \infty\,. \qquad (4.1.38)
$$

We then find a matrix $B = B^T$ satisfying $A^T B + BA = -I$ and form a Liapunov functional for (4.1.37) in the form

$$
V(t, \mathbf{X}(\cdot)) = \left\{ [\mathbf{X}^T B\mathbf{X}]^{1/2} + \bar{K}\int_0^t \int_t^\infty |C(u-s)|\, du\, |\mathbf{X}(s)|\, ds + 1 \right\}
$$
$$
\times \exp\left[-L\int_0^t |\gamma(s)|\, ds \right].
$$

These forms are precisely the ones considered in Chapter 6. See the perturbation result Theorem 6.4.5 for both (4.1.37) and (4.1.34).

Now return to the nonlinear form (4.1.34) and consider

$$
\begin{aligned}
u' &= d\bar{x}(1 - e^u) + b\bar{y}(1 - e^v)\,, \\
v' &= k\bar{x}(e^u - 1)\,.
\end{aligned} \qquad (4.1.39)
$$

The work leading to Theorem 4.1.1 yields uniform asymptotic stability in the large. Under these conditions we shall see in Chapter 6 (Theorem 6.1.6) that there is a Liapunov function W for (4.1.39) that is positive definite and satisfies

$$
W'_{(4.1.39)}(u, v) \le -cW(u, v)\,, \qquad (4.1.40)
$$

where $c > 0$. We then show in Chapter 6 (see Theorems 6.4.1–6.4.3 and 6.4.5) how W may be used to show global stability for the nonlinear system (4.1.34).

In addition, Brauer (1978) has an interesting discussion of such equilibrium questions as raised here. He applies certain linearization techniques of Grossman and Miller to systems of the form of (4.1.34) with (4.1.38) holding.

4.2   Renewal Theory

The renewal equation is an example of an integral equation attracting interest in many areas over a long period of time. An excellent account, given by Bellman and Cooke (1963), contains 41 problems (solved and unsolved) of historical interest.

Consider the scalar equation

$$u(t) = g(t) + \int_0^t u(t-s)f(s)\, ds \,, \tag{4.2.1}$$

where f and $g : [0, \infty) \to [0, \infty)$ are continuous. Note that we are assuming that f and g are nonnegative.

Our discussion here is brief and is taken from the excellent classical paper by Feller (1941), which appeared at an interesting time historically. The work of Volterra in Section 4.1 had been well circulated and had received much attention. Moreover, just two years earlier Lotka (1939) had published a paper containing 74 references to the general questions considered in Section 4.1.

The work by Feller represents an attempt to synthesize, simplify, and correct much of the then current investigation. He gives very exact results concerning the behavior of solutions of (4.2.1). This is in contrast to the stability objective of this book. His work is important here in revealing the kinds of behavior one might attempt to prove in qualitative terms. Moreover, he provides two excellent formulations of concrete problems. The entire paper is strongly recommended to the interested reader.

Although we make no use of it here, Feller points out that (4.2.1) can be put into another form of particular interest when f is not continuous. We have frequently differentiated an integral equation in order to use the techniques of integro-differential equations. By contrast, one can integrate (4.2.1) and obtain a new integral equation.

Define U, F, and G by

$$U(t) = \int_0^t u(s)\,ds\,,$$

$$F(t) = \int_0^t f(s)\,ds\,,$$

and   (4.2.2)

$$G(t) = \int_0^t g(s)\,ds\,,$$

so that we may write

$$U(t) = G(t) + \int_0^t U(t - s)\,dF(s)\,.$$ (4.2.3)

The main objective is to study the mean value of $u(t)$, namely,

$$u^*(t) = (1/t) \int_0^t u(s)\,ds\,.$$

Equation (4.2.1) has at least two practical applications. The first is Lotka's formulation.

In the abstract theory of industrial replacement, each time an individual drops out that individual is replaced by a new one of age zero. (One may think, at times, of light bulbs, for example.) Here $f(t)$ denotes the probability density at the moment of replacement that an individual of age t will drop out. Now let $\eta(s)$ denote the age distribution of the population at time $t = 0$. Thus the number of individuals between ages s and $s + (\delta s)$ is $\eta(s)(\delta s) + O(\delta s)$. Then $g(t)$ defined by

$$g(t) = \int_0^t \eta(s)f(t - s)\,ds$$ (4.2.4)

represents the rate of removal at time t of individuals belonging to the parent population. The function $u(t)$ gives the removal rate at time t of individuals of the total population. Note that each individual dropping out at time t either belonged to the parent population or entered the population by the process of replacement at some time $t - s$ for $0 < s < t$. Hence, $u(t)$ satisfies (4.2.1). Because f is a probability density function, we have

$$\int_0^\infty f(t)\,dt = 1\,.$$ (4.2.5)

The next formulation is for a single species and is akin to Volterra's own derivation of the predator-prey system of Section 4.1.

Let $f(t)$ denote the reproduction rate of females of a certain species at age t. In particular, the average number of females born during a time interval $(t, t+(\delta t))$ from a female of age t is $f(t)(\delta t)+O(\delta t)$. If $\eta(s)$ denotes the age distribution of the parent population at $t = 0$, then Eq. (4.2.4) yields the rate of production of females at time t by members of the parent population. Then $u(t)$ in (4.2.1) measures the rate of female births at time $t > 0$. This time f is not a probability density function, and we have

$$\int_0^\infty f(t)\,dt \quad \text{being any nonnegative number.} \tag{4.2.6}$$

This integral is a measure of the population's tendency to increase or decrease.

We list without proof some sample results by Feller.

Theorem 4.2.1. *Suppose that $\int_0^\infty f(t)\,dt = a$ and $\int_0^\infty g(t)\,dt = b$, both are finite, $f \geq 0$, and $g \geq 0$.*

(a) *In order that*

$$u^*(t) = (1/t) \int_0^t u(s)\,ds \to c$$

as $t \to \infty$, where c is a positive constant, it is necessary and sufficient that $a = 1$ and that $\int_0^\infty tf(t)\,dt = m$, a finite number. In this case, $c = b/m$.

(b) *If $a < 1$, then $\int_0^\infty u(t)\,dt = b/(1-a)$.*

Notice that according to Theorem 2.6.1, this result concerns uniform asymptotic stability.

Theorem 4.2.2. *Let $\int_0^\infty f(t)\,dt = 1$ and $\int_0^\infty g(t)\,dt = b < \infty$. Suppose there is an integer $n \geq 2$ with*

$$m_k = \int_0^\infty t^k f(t)\,dt \quad \text{for} \quad k = 1, 2, \ldots, n$$

all being finite and that the functions

$$f(t), tf(t), \ldots, t^{n-2} f(t)$$

are of bounded total variation over $(0, \infty)$. Suppose also that

$$\lim_{t\to\infty} t^{n-2} g(t) = 0 \quad \text{and} \quad \lim_{t\to\infty} t^{n-2} \int_t^\infty g(s)\,ds = 0.$$

Then $\lim_{t\to\infty} u(t) = b/m_1$ and

$$\lim_{t\to 0} t^{n-2}\big[u(t) - (b/m_1)\big] = 0\,.$$

4.3 Examples

In this section, we give a number of examples of physical processes that give rise to integro-differential or integral equations. In most cases the examples are very brief and are accompanied by references, so that the interested reader may pursue them in depth. The main point here is that applications of the general theory are everywhere.

If $\mathbf{f}(t, \mathbf{x})$ is smooth, then the problem

$$\mathbf{x}' = \mathbf{f}(t, \mathbf{x})\,, \quad \mathbf{x}(t_0) = \mathbf{x}_0$$

has one and only one solution. If that problem is thought to model a given physical situation, then we are postulating that the future behavior depends only on the object's position at time t_0. Frequently this position is extreme. Physical processes tend to depend very strongly on their past history.

The point was made by Picard (1907), in his study of celestial mechanics, that the future of a body in motion depends on its present state (including velocity) and the previous state taken back in time infinitely far. He calls this *heredity* and points out that students of classical mechanics claim that this is only apparent because too few variables are being considered.

A. Torsion of a Wire

In the same vein, Volterra (1913, pp. 138–139, 150) considered the first approximation relation between the couple of torsion P and the angle of torsion W as $W = kP$. He claimed that the elastic body had inherited characteristics from the past because of fatigue from previously experienced distortions. His argument was that hereditary effects could be represented by an integral summing the contributions from some t_0 to t, so that the approximation $W = kP$ could be replaced by

$$W(t) = kP(t) + \int_{t_0}^{t} K(t, s)P(s)\,ds\,. \tag{4.3.1}$$

He called $K(t, s)$ the *coefficient of heredity*.

The expression of W is a function of a function, and Volterra had named such expressions "functions of lines." Hadamard suggested the name "functionals," and that name persists. [This problem is also discussed by Davis (1962, p. 112) and by Volterra (1959, p. 147).]

B.   Dynamics

Lagrange's form for the general equations of dynamics is

$$\frac{d}{dt}\frac{\partial T}{\partial q_i} - \frac{\partial(T - \boldsymbol{\Omega})}{\partial q_i} = Q_i\,, \tag{4.3.2}$$

where $q_i, \ldots, q_n$ are the independent coordinates,

$$T = \frac{1}{2}\sum_i\sum_s a_{is}q_i'q_s'$$

the kinetic energy,

$$-\boldsymbol{\Omega} = -\frac{1}{2}\sum_i\sum_s b_{is}q_iq_s$$

the potential energy, and $Q_1, \ldots, Q_n$ the external forces. See Rutherford (1960) or Volterra (1959, pp. 191–192) for details.

When a_{is} and b_{is} are constants, then the equations take the linear form

$$\sum_s a_{is}q_s'' + \sum_s b_{is}q_s = Q_i\,. \tag{4.3.3}$$

Volterra (1928) shows that in the case of hereditary effects (4.3.3) becomes

$$\sum_s a_{is}q_s'' + \sum_s b_{is}q_s + \sum_s \int_{-\infty}^{t}\boldsymbol{\Phi}_{is}(t,r)q_s(r)\,dr = Q_i\,. \tag{4.3.4}$$

If the system has only one degree of freedom, if $\boldsymbol{\Phi}$ is of convolution type, and if the duration of heredity is T, then the system becomes the single equation

$$q'' + bq + \int_0^T \boldsymbol{\Phi}(s)q(t-s)\,ds = Q\,. \tag{4.3.5}$$

If we suppose that $\boldsymbol{\Phi}(s)$ is continuous, nonpositive, increasing, and zero for $s \geq T$ and if $b > 0$, then $b + \int_0^T \boldsymbol{\Phi}(s)\,ds = m > 0$. In this way we may write (4.3.5) as

$$q'' + mq - \int_0^T \boldsymbol{\Phi}(s)\big[q(t) - q(t-s)\big]\,ds = Q\,. \tag{4.3.6}$$

Then

$$\frac{1}{2}mq^2 - \frac{1}{2}\int_0^T \Phi(s)\left[q(t) - q(t-s)\right]^2 ds \qquad (4.3.7)$$

is called the *potential* of all forces.

Potentials are always important in studying the stability of motion. Frequently a potential function can be used directly as a Liapunov function, an idea going back to Lagrange (well before Liapunov). See Chapter 6 and the discussion surrounding Eq. (6.2.4) for such construction. A suggestion by Volterra (1928) concerning energy enabled Levin (1963) to construct a very superior Liapunov functional.

C. Viscoelasticity

We consider a one-dimensional viscoelastic problem in which the material lies on the interval $0 \le x \le L$ and is subjected to a displacement given by

$$u(t, x) = f(t, x) - x, \qquad (4.3.8)$$

where $f : [0, \infty) \times [0, L] \to R$. If $\rho_0 : [0, L] \to [0, \infty)$ is the initial density function, then, from Newton's law of $F = ma$, we have

$$\sigma_x(t, x) = [\rho_0(x)]\left[f_{tt}(t, x)\right], \qquad (4.3.9)$$

where σ is the stress.

For linear viscoelasticity the stress is given by

$$\sigma(t, x) = \int_0^t G(t - s, x)u_{xt}(s, x)\, ds, \qquad (4.3.10)$$

where $G : [0, \infty) \times [0, L] \to [0, \infty)$ is the relaxation function and satisfies $G_t < 0$, $G_{tt} \ge 0$.

If we integrate (4.3.10) by parts we obtain

$$\sigma(t, x) = G(0, x)u_x(t, x) - G(t, x)u_x(0, x)$$

$$+ \int_0^t G_t(t - s, x)u_x(s, x)\, ds. \qquad (4.3.11)$$

Because $\rho_0(x)f_{tt}(t, x) = \sigma_x(t, x)$ it follows that

$$\rho_0(x)u_{tt} = \left[G(0, x)u_x(t, x) - \int_0^t G_t(t - s, x)u_x(s, x)\, ds\right]_x. \qquad (4.3.12)$$

If the material is homogeneous in a certain sense, then we take $\rho_0(x) \equiv 1$ and G to be independent of x, say, $G(t, x) = G(t)$. This yields

$$u_{tt} = G(0)u_{xx}(t, x) + \int_0^t G_t(t - s)u_{xx}(s, x)\,ds. \qquad (4.3.13)$$

With well-founded trepidation, one separates variables

$$u(t, x) = g(t)h(x)$$

and obtains (where the overdot indicates d/dt and the prime indicates d/dx *for this section only*)

$$\ddot{g}(t)h(x) = G(0)g(t)h''(x) + \int_0^t \dot{G}(t - s)g(s)h''(x)\,ds, \qquad (4.3.14)$$

so that

$$h(x)/h''(x) = \left[G(0)g(t) + \int_0^t \dot{G}(t - s)g(s)\,ds\right]\bigg/\ddot{g}(t) \qquad (4.3.15)$$

K a constant (which may need to be negative to satisfy boundary conditions). This yields

$$\ddot{g}(t) = KG(0)g(t) + K\int_0^t \dot{G}(t - s)g(s)\,ds \qquad (4.3.16)$$

Let $g = y$, $\dot{g} = z$, and obtain

$$\begin{pmatrix} \dot{y} \\ \dot{z} \end{pmatrix} = \begin{pmatrix} 0 & 1 \\ KG(0) & 0 \end{pmatrix}\begin{pmatrix} y \\ z \end{pmatrix} + \int_0^t \begin{pmatrix} 0 & 0 \\ K\dot{G}(t - s) & 0 \end{pmatrix}\begin{pmatrix} y(s) \\ z(s) \end{pmatrix}\,ds,$$

which we write as

$$\mathbf{X}' = A\mathbf{X} + \int_0^t C(t - s)\mathbf{X}(s)\,ds.$$

If $K < 0$, then the characteristic roots of A have zero real parts and the stability theory developed in Chapter 2 fails to apply. A detailed discussion of the problem may be found in Bloom (1981, Chapter II, especially pp. 29–31, 73–75). Stability analysis was performed by MacCamy (1977b).

D. Electricity

Even the very simplest RLC circuits lead to integro-differential equations. For if a single-loop circuit contains resistance R, capacitance C, and inductance L with impressed voltage $E(t)$, then Kirchhoff's second law yields

$$LI' + RI + (1/C)Q = E(t), \qquad (4.3.17)$$

with $Q = \int_{t_0}^{t} I(s)\, ds$. Although this is usually thought to be a trivial integro-differential equation, if E is too rough to be differentiated, then the equation must be treated in its present form, perhaps by Laplace transforms.

At the other end of the spectrum, Hopkinson (1877) considers an electromagnetic field in a nonconducting material, where $E = (E_1, E_2, E_3)$ is the electric field and $D = (D_1, D_2, D_3)$ the electric displacement. He uses Maxwell's equations (indeed, the problem was suggested by Maxwell) to write

$$\mathbf{D}(t) = \varepsilon \mathbf{E}(t) + \int_{-\infty}^{t} \phi(t-s)\mathbf{E}(s)\, ds, \qquad (4.3.18)$$

where ε is constant and ϕ is continuous.

Interestingly enough, the problem is as current today as it was in 1877. A recent discussion may be found in Davis (1975) as well as Bloom (1981, Chapter III).

Another interesting example of an integro-differential equation in electrical theory arises in the construction of a field-theoretical model for electron-beam devices. Snyder (1975) develops a one-dimensional transmission-line theory, which is too complicated to develop here. The resulting integro-differential equation may be written as

$$F''(x) + G(x)F(x) = [c/h(x)] \int_{0}^{x} K(x-s)h(s)F(s)\, ds, \qquad (4.3.19)$$

with an exponential kernel and

$$F'' + G(x)F = 0, \qquad (4.3.20)$$

a Hill equation with bounded solutions. This problem does not fit into the theory of Chapter 2 because (4.3.20) is never asymptotically stable. However, in certain cases (4.3.19) can be stabilized by the methods of Chapter 5, especially Theorem 5.3.2.

E.   Reactor Dynamics

Levin and Nohel (1960) consider a continuous-medium nuclear reactor with the model

$$du/dt = -\int_{-\infty}^{\infty} \alpha(x)T(t,x)\,dx\,, \tag{4.3.21}$$

$$aT_t = bT_{xx} + \eta(x)u\,, \tag{4.3.22}$$

for $0 \le t < \infty$, and satisfying the initial condition

$$u(0) = u_0\,, \quad T(0,x) = f(x)\,, \quad -\infty < x < \infty\,.$$

Here $u(t)$ and $T(t,x)$ are the unknown functions, α, η, and f given, real-valued functions, u_0 a real constant, and a, b given, positive constants. The various quantities are interpreted as follows:

$$t = \text{time,}$$

$$x = \text{position along the reactor, regarded as a doubly infinite rod,}$$

$$u(t) = \text{logarithm of the total reactor power,}$$

$$T(t,x) = \text{deviation of the temperature from equilibrium,}$$

$$-\alpha(x) = \text{ratio of the temperature coefficient of reactivity to the}$$

$$\text{mean life of neutrons,}$$

$$\eta(x) = \text{fraction of the power generated at } x,$$

$$a = \text{heat capacity, and}$$

$$b = \text{thermal conductivity.}$$

When f, α, and η are $L^2[0,\infty)$, then an application of Fourier transform theory [see Miller (1968)] shows that $u(t)$ satisfies

$$u'(t) = -\int_0^t m_1(t-s)u(s)\,ds - m_2(t)\,, \quad u(0) = u_0\,, \tag{4.3.23}$$

where

$$m_j(t) = (1/\pi)\int_0^\infty \exp[-s^2 t]h_j(s)\,ds$$

with

$$h_1(s) = \operatorname{Re} \eta^*(s)\alpha^*(-s) \,,$$

$$h_2(s) = \operatorname{Re} f^*(s)\alpha^*(-s) \,,$$

and the asterisk being the L^2 Fourier transform.

Notice that when we can differentiate $m_1(t)$ through the integral then $m_1'(t) < 0$, $m_1''(t) > 0$, and $m_1'''(t) < 0$. Also, (4.3.23) is linear, so that we can consider the homogeneous form and then use the variation of parameters theorem.

F. Heat Flow

In many of the applications we begin with a partial differential equation and, through simplifying assumptions, arrive at an integral or integro-differential equation. If one casts the problem in a Hilbert space with unbounded nonlinear operators, then these problems appear to merge into one and the same thing.

A particularly pleasing example of the merging of many problems and concepts occurs in the work of MacCamy (1977b) who considers the problem of one-dimensional heat flow in materials with "memory" modeled by

$$u_t(t,x) = \int_0^t a(t-s)\sigma_x(u_x(s,x))\,ds \, + f(t,x)\,,$$
$$0 < x < 1\,,\ t > 0\,, \tag{4.3.24}$$
$$u(t,0) = u(t,1) = 0\,,$$

and

$$u(0,x) = u_0(x)\,.$$

Now (4.3.24) is an example of an integro-differential equation

$$u'(t) = -\int_0^t a(t-s)g(u(s))\,ds \, + f(t)\,, \tag{4.3.25}$$
$$u(0) = u_0$$

on a Hilbert space with g a nonlinear unbounded operator. Moreover, (4.3.25) is equivalent to

$$u''(t) + k(0)u'(t) + g(u(t)) + \int_0^t k'(t-s)u'(s)\,ds \, = \phi(t) \tag{4.3.26}$$

for some kernel k, and the damped wave equation

$$u_{tt} + \alpha u_t - \sigma_x(u_x) = 0, \quad -\infty < x < \infty, \ t > 0,$$
$$u(0,x) = u_0(x), \quad u_t(0,x) = u_1(x), \quad \alpha > 0$$

(4.3.27)

is a special case of (4.3.26). Finally, the problem of nonlinear viscoelasticity is formally the same as (4.3.26).

Thus, we see a merging of the wave equation, the heat equation, viscoelasticity, partial differential equations, and integro-differential equations. The literature is replete with such merging. In Burton (1991) there is a lengthy, detailed, and elementary presentation of the damped wave equation as a Liénard equation. The classical Liapunov functionals for the Liénard equation are parlayed into Liapunov functions for the damped wave equation and corresponding stability results are obtained.

G.   Chemical Oscillations

The Lotka-Volterra equations of Section 4.1 are closely related to certain problems in chemical kinetics. The problem discussed here was also discussed by Prigogine (1967), who gives a linear stability analysis of the resulting equations.

Consider an autocatalytic reaction represented by

$$A + X \underset{h}{\overset{1}{\rightleftarrows}} 2X,$$

$$X + Y \underset{h}{\overset{1}{\rightleftarrows}} 2Y,$$

(4.3.28)

$$Y \underset{h}{\overset{1}{\rightleftarrows}} B,$$

where the concentrations of A and B do not vary with time. Here, all kinetic constants for the forward reactions are taken as unity and the reverse as h. The reaction rates,

$$v_1 = AX - hX^2,$$

$$v_2 = XY - hY^2,$$

(4.3.29)

$$v_3 = Y - hB$$

are based on the law of mass action [see the material in Section 4.1 following Eq. (4.1.5)]. Thus, the differential equations are

$$X' = AX - XY - hX^2 + hY^2 \,,$$
$$Y' = XY - Y - hY^2 + hRA \,. \qquad (4.3.30)$$

Note that as $h \to 0$ we obtain the Lotka-Volterra equations (4.1.9) with $a = b = c = k = 1$. The total affinity of the reaction is

$$A = -\log h^3 R \quad \text{with} \quad R = B/A \,.$$

Although it is difficult to find even the equilibrium point in the open first quadrant for (4.3.30), much can be said about the system.

Solutions starting in the open Quadrant I remain there, according to our uniqueness theory. Also,

$$X' + Y' = AX - Y - hX^2 + hRA < 0 \qquad (4.3.31)$$

if $X^2 + Y^2$ is large. Hence, all solutions starting in the open Quadrant I are bounded.

Moreover, if we write (4.3.31) as $X' = P$ and $Y' = Q$, then

$$(\partial P/\partial X) + (\partial Q/\partial Y) = A - Y - 2hX - 1 - 2hY < 0 \qquad (4.3.32)$$

in Quadrant I provided that

$$h > \frac{1}{2} \quad \text{and} \quad A < 1 \,. \qquad (4.3.33)$$

By Bendixson's criterion [see Lefschetz (1957, p. 238)], there are no closed orbits in Quadrant I. Because solutions are bounded, one may argue from the Poincaré-Bendixson theorem [see Lefschetz (1957, p. 230)] that all solutions in the first quadrant approach the equilibrium point.

Prigogine (1967, pp. 122–124) gives a linear discussion of some interest and discusses the case $h \to 0$, yielding the Lotka-Volterra system

$$X' = AX - XY \,,$$
$$Y' = XY - Y \,, \qquad (4.3.34)$$

which we recall, has all periodic solutions in Quadrant I. One wonders if for very small $h > 0$, Eq. (4.3.30) may also possess at least one nontrivial periodic solution in Quadrant I.

Also recall that such a solution follows a simple closed curve, along which $X(t)$ and $Y(t)$ alternate between large and small values. Indeed, even if a solution $(X(t), Y(t))$ spirals toward the equilibrium point, then $X(t)$ and $Y(t)$ alternate between larger and smaller values. That is, the concentrations of the products are changing. This may be called chemical oscillation.

Some chemical oscillators have attracted enormous interest. The best known such oscillator was discovered by Belousov (1959). It is now referred to as the Belousov-Zhabotinski chemical reaction and consists of a cerium-ion-catalyzed oxidation by bromate ion of brominated organic material. The medium is kept well stirred, so that oscillations occur in the ratio of the oxidized and reduced forms of the metal ion catalyst. The oscillations are visible as sharp color changes caused by a redox indicator. A nice discussion, together with differential equations involved, is given by Troy (1980).

Chapter 5

Instability, Stability, and Perturbations

5.1 The Matrix $A^T B + BA$

The following result was obtained in Section 2.4. It is listed here for handy reference.

Theorem 5.1.1. *Let A be a real $n \times n$ constant matrix. The equation*

$$A^T B + BA = -I \tag{5.1.1}$$

can be solved for a unique, positive definite symmetric matrix B if and only if all characteristic roots of A have negative real parts..

Thus, given the system of ordinary differential equations

$$\mathbf{x}' = A\mathbf{x}, \tag{5.1.2}$$

if the characteristic roots of A all have negative real parts, then one forms the Liapunov function

$$V(\mathbf{x}) = \mathbf{x}^T B \mathbf{x}, \tag{5.1.3}$$

finds that

$$V'_{(5.1.2)}(\mathbf{x}) = -\mathbf{x}^T \mathbf{x}, \tag{5.1.4}$$

and easily concludes uniform asymptotic stability as seen in Chapter 2.

133

Now consider the possibly nonlinear system

$$\mathbf{x}' = \mathbf{G}(t, \mathbf{x}),\tag{5.1.5}$$

in which $\mathbf{G} : [0, \infty) \times D \to R^n$, where D is an open set in R^n and $\mathbf{0} \in D$. We suppose that $\mathbf{G}$ is continuous and $\mathbf{G}(t, \mathbf{0}) = \mathbf{0}$, so that $\mathbf{x}(t) = \mathbf{0}$ is a solution. As in Chapter 2, we define stability.

Definition 5.1.1. *The zero solution of* (5.1.5) *is stable if for each $\varepsilon > 0$ and $t_0 \geq 0$, there is a $\delta > 0$ such that*

$$|\mathbf{x}_0| < 0 \quad and \quad t \geq t_0$$

imply $|\mathbf{x}(t, t_0, \mathbf{x}_0)| < \varepsilon$.

We have seen substantial work with stability and Liapunov's direct method in Chapter 2, but the basic stability result for (5.1.5) may be stated as follows.

Theorem 5.1.2. *Suppose that $V : [0, \infty) \times D \to [0, \infty)$ has continuous first partial derivatives, $W : D \to [0, \infty)$ is continuous with $W(\mathbf{0}) = 0$, $W(\mathbf{x}) > 0$ if $\mathbf{x} \neq \mathbf{0}$, $V(t, \mathbf{x}) \geq W(\mathbf{x})$ on D, and $V(t, \mathbf{0}) \equiv 0$. If $V'_{(5.1.5)}(t, \mathbf{x}) \leq 0$, then the zero solution of* (5.1.5) *is stable.*

Proof. Let $\varepsilon > 0$ and $t_0 \geq 0$ be given. Assume ε so small that $|\mathbf{x}| \leq \varepsilon$ implies $\mathbf{x} \in D$. Because W is continuous on the compact set $L = \{\mathbf{x} \in R^n : |\mathbf{x}| = \varepsilon\}$, W has a positive minimum, say, α, on L. Because V is continuous and $V(t, \mathbf{0}) = 0$, there is a $\delta > 0$ such that $|\mathbf{x}_0| < \delta$ implies $V(t_0, \mathbf{x}_0) < \alpha$.

Now for $|\mathbf{x}_0| < \delta$, we note that $V'(t, \mathbf{x}(t, t_0, \mathbf{x}_0)) \leq 0$, so that $t \geq t_0$ implies

$$W(\mathbf{x}(t, t_0, \mathbf{x}_0)) \leq V(t, \mathbf{x}(t, t_0, \mathbf{x}_0))$$
$$\leq V(t_0, \mathbf{x}_0) < \alpha.$$

This implies that $|\mathbf{x}(t, t_0, \mathbf{x}_0)| < \varepsilon$, completing the proof.

As previously noted, when $\mathbf{G}(t, \mathbf{x}) = A\mathbf{x}$ with all characteristic roots of A having negative real parts, then one chooses $B = B^T$ with $A^T B + BA = -I$, and

$$V(t, \mathbf{x}) = \mathbf{x}^T B \mathbf{x} = W(\mathbf{x}).$$

This condition on A actually implies uniform asymptotic stability.

We negate Definition 5.1.1 and obtain the definition of instability.

Definition 5.1.2. *The zero solution of* (5.1.5) *is unstable if there exists* $\varepsilon > 0$ *and there exists* $t_0 \geq 0$ *such that for any* $\delta > 0$ *there is an* $\mathbf{x}_0$ *with* $|\mathbf{x}_0| < \delta$ *and a* $t_1 > t_0$ *such that* $|\mathbf{x}(t_1, t_0, \mathbf{x}_0)| \geq \varepsilon$.

Notice that stability requires all solutions starting near zero to stay near zero, but instability calls for the existence of some solutions starting near zero to move well away from zero. Ordinary differential equations frequently have a property that is not seen in most types of functional differential equations.

Definition 5.1.3. *The zero solution of* (5.1.5) *is completely unstable if there exists a* $t_0 \geq 0$ *and an* $\varepsilon > 0$ *such that for any* $\delta > 0$ *if* $0 < |\mathbf{x}_0| < \delta$, *then there is a* $t_1 > t_0$ *with* $|\mathbf{x}(t_1, t_0, \mathbf{x}_0)| \geq \varepsilon$.

Here, t_1 depends on ε, t_0, and $\mathbf{x}_0$; however, every solution starting near zero (but not at zero) at t_0 moves well away from zero.

Theorem 5.1.3. *Consider* (5.1.5) *and suppose there is a function* $V :$ $D \to (-\infty, \infty)$ *having continuous first partial derivatives and a sequence* $\{\mathbf{x}_n\}$ *converging to zero with* $V(\mathbf{x}_n) < 0$ *and* $V(\mathbf{0}) = 0$. *Suppose also that there is a continuous function* $U : D \to [0, \infty)$ *with* $U(\mathbf{x}) > 0$ *if* $\mathbf{x} \neq 0$ *and*

$$V'_{(5.1.5)}(t, \mathbf{x}) \leq -U(\mathbf{x}) \quad on \quad [0, \infty) \times D \,.$$

Then the zero solution of (5.1.5) *is unstable.*

Proof. Assume, by way of contradiction, that $\mathbf{x} = 0$ is stable. Choose $\varepsilon > 0$ such that $|\mathbf{x}| \leq \varepsilon$ implies $\mathbf{x} \in D$. Choose $t_0 = 0$ and fix δ using the definition of stability. Now, select $\mathbf{x}_0$ with $|\mathbf{x}_0| < \delta$ and $V(\mathbf{x}_0) < 0$. Let $\mathbf{x}(t) = \mathbf{x}(t, 0, \mathbf{x}_0)$. We shall show that $|\mathbf{x}(t)|$ cannot be bounded by ε.
Because $V'(\mathbf{x}(t)) \leq 0$, we have

$$V(\mathbf{x}(t)) \leq V(\mathbf{x}_0) < 0 \,,$$

so that $|\mathbf{x}(t)| \geq \alpha$ for some $\alpha > 0$. Also, there is a $\beta > 0$ with $U(\mathbf{x}) \geq \beta$ if $\alpha \leq |\mathbf{x}| \leq \varepsilon$. Because

$$V'(\mathbf{x}(t)) \leq -U(\mathbf{x}(t)) \leq -\beta \,,$$

we have

$$V(\mathbf{x}(t)) \leq V(\mathbf{x}_0) - \beta t \to -\infty$$

as $t \to \infty$, so that $|\mathbf{x}(t)| < \varepsilon$ is impossible. This completes the proof.

One may generalize the result to allow a $V(t, \mathbf{x})$ with $V(t, \mathbf{x}) \geq -H(\mathbf{x})$, where H is continuous and positive definite.

If the characteristic roots of A in (5.1.2) all have positive real parts, then the V of Theorem 5.1.3 is easily selected.

Theorem 5.1.4.   *Suppose that the characteristic roots of A all have positive real parts. Then there is a unique, symmetric, positive definite matrix B with $A^T B + BA = I$.*

Proof. All characteristic roots of $-A$ have negative real parts, so that

$$(-A)^T B + B(-A) = -I$$

has a unique, positive definite, symmetric solution matrix B. Thus, $A^T B + BA = I$, as required.

The function V for (5.1.2) that satisfies Theorem 5.1.3 may be chosen as

$$V(\mathbf{x}) = -\mathbf{x}^T B \mathbf{x},$$

so that

$$V'_{(5.1.2)}(x) = -\mathbf{x}^T \mathbf{x} = -U(\mathbf{x}).$$

Clearly, it is excessive to require that all characteristic roots of A have positive real parts to prove instability.

With Theorem 5.1.3 in mind one might attempt to find a matrix $B = B^T$ with $A^T B + BA = -I$ whenever A has at least one characteristic root with a positive real part. However, such a matrix B will never exist if A has a characteristic root with a zero real part.

Theorem 5.1.5.   *If A has a characteristic root with a zero real part, then $A^T B + BA = -I$ has no solution.*

Proof. Suppose there is a solution B. If λ is a characteristic root of A with a zero real part, and if $\mathbf{X}$ is a characteristic vector belonging to λ, then $\mathbf{x}(t) = \mathbf{X} e^{\lambda t}$ is a solution of (5.1.2). Because A is real, the real part of $\mathbf{x}(t)$ and the imaginary part of $x(t)$ are both solutions of (5.1.2); moreover, at least one of the solutions is nonzero. This means (5.1.2) has at least one solution, say, $\mathbf{z}(t)$, that is bounded and bounded away from zero, say, $|\mathbf{z}(t)| \geq \alpha$, for $\alpha > 0$.

If we write $V(\mathbf{x}) = \mathbf{x}^T B \mathbf{x}$, then $V(\mathbf{z}(t)) = \mathbf{z}^T(t) B \mathbf{z}(t)$, and because $\mathbf{z}(t)$ is a solution of (5.1.2), we have

$$V'(\mathbf{z}(t)) = -\mathbf{z}^T(t)\mathbf{z}(t) \le -\alpha^2$$

for $0 \le t < \infty$. Thus,

$$V(\mathbf{z}(t)) \le V(\mathbf{z}(0)) - \alpha^2 t \to -\infty$$

as $t \to \infty$ is a contradiction to $V(\mathbf{z}(t))$ being bounded. This completes the proof.

Remark 5.1.1. Barbashin(1968) notes that for any given matrix $C = C^T$, the equation

$$A^T B + BA = C$$

may be uniquely solved for $B = B^T$ provided that the characteristic roots of A are such that $\lambda_i + \lambda_k$ does not vanish for any i and k.

To give a unified exposition, we have consistently taken $C = -I$. However, any negative definite C would suit our purpose, and when

$$\lambda_i + \lambda_k = 0\,,$$

we sometimes *must* take $C \ne -I$. That is, if $\lambda_i + \lambda_k = 0$, then we may still be able to solve

$$A^T B + BA = C$$

with C being negative definite; but the solution may not be unique.

The Barbashin result is an interesting one. We give three examples and a general construction idea.

Example 5.1.1. Let

$$A = \begin{pmatrix} 1 & 0 \\ 0 & -1 \end{pmatrix}, \quad B = \begin{pmatrix} b_1 & b_2 \\ b_2 & b_3 \end{pmatrix},$$

and solve $A^T B + BA = -I$ for B. We have

$$A^T B + BA = \begin{pmatrix} 2b_1 & 0 \\ 0 & -2b_3 \end{pmatrix} = \begin{pmatrix} -1 & 0 \\ 0 & -1 \end{pmatrix},$$

so that $b_1 = -\frac{1}{2}$, $b_3 = \frac{1}{2}$, and b_2 is not determined. Any choice for b_2 will produce a matrix B such that $V(\mathbf{x}) = \mathbf{x}^T B \mathbf{x}$ will satisfy Theorem 5.1.3 for $\mathbf{x}' = A\mathbf{x}$. Thus, B exists for $C = -I$, but B is not unique.

When B can be determined, but not uniquely, it usually best serves our purpose to make $|B|$ a minimum. See, for example, Theorems 5.3.1 and 5.3.3.

The next example has B unique, and A an unstable matrix.

Example 5.1.2. Let

$$A = \begin{pmatrix} -1 & 1 \\ -1 & 2 \end{pmatrix}, \quad B = \begin{pmatrix} b_1 & b_2 \\ b_2 & b_3 \end{pmatrix},$$

and solve for B in $A^T B + BA = -I$. We have

$$\begin{pmatrix} -2(b_1 + b_2) & b_1 + b_2 - b_3 \\ b_1 + b_2 - b_3 & 2b_2 + 4b_3 \end{pmatrix} = \begin{pmatrix} -1 & 0 \\ 0 & -1 \end{pmatrix},$$

so that

$$b_1 + b_2 = \tfrac{1}{2},$$
$$b_1 + b_2 = b_3,$$

and

$$2b_2 + 4b_3 = -1.$$

The determinant of coefficients is

$$\begin{vmatrix} 1 & 1 & 0 \\ 1 & 1 & -1 \\ 0 & 2 & 4 \end{vmatrix} = 4 - [-2 + 4] \neq 0,$$

so the solution

$$b_3 = \tfrac{1}{2}, \quad b_2 = -\tfrac{3}{2}, \quad \text{and} \quad b_1 = 2$$

is unique. Because $\lambda_i + \lambda_j \neq 0$, this was predicted by the Barbashin result. If we write $\mathbf{x} = (x_1, x_2)^T$, we find

$$V(\mathbf{x}) = \mathbf{x}^T B \mathbf{x}$$
$$= 2x_1^2 - 3x_1 x_2 + \tfrac{1}{2} x_2^2.$$

Along $x_1 = x_2$ we have $V(\mathbf{x}) = -\tfrac{1}{2} x_1^2$, so that the conditions of Theorem 5.1.3 are satisfied.

Our next example shows that there are cases when we must select C other then $-I$.

Example 5.1.3. Consider

$$A = \begin{pmatrix} 1 & 1 \\ 0 & -1 \end{pmatrix}, \quad B = \begin{pmatrix} b_1 & b_2 \\ b_2 & b_3 \end{pmatrix},$$

and try to solve $A^T B + BA = -I$. We obtain

$$A^T B + BA = \begin{pmatrix} 2b_1 & b_1 \\ b_1 & 2(b_2 - b_3) \end{pmatrix},$$

which cannot be $-I$; however, if $b_1 = -1$, $b_2 = 0$, and $b_3 = 1$, then

$$V = (x_1, x_2) \begin{pmatrix} -1 & 0 \\ 0 & 1 \end{pmatrix} \begin{pmatrix} x_1 \\ x_2 \end{pmatrix}$$
$$= -x_1^2 + x_2^2,$$

and

$$V' = -2(x_1^2 + x_1 x_2 + x_2^2)$$
$$\leq -(x_1^2 + x_2^2).$$

The conditions of Theorem 5.1.3 are satisfied when $x_2 = (1/n, 0)$.

The Barbashin result, stated in Remark 5.1.1, takes care of all matrices A with $\lambda_i + \lambda_k \neq 0$. We know that there is no possible B when A has a characteristic root with a zero real part. However, we do wish to find some V satisfying Theorem 5.1.3 when no root of A has a zero real part. The following procedure shows that it is possible to do so. Some of the details were provided by Prof. L. Hatvani.

Let $x' = Ax$ with A real and having no characteristic root with zero real part. Transform A into its Jordan form as follows. Let $x = Qy$ so that $x' = Qy' = AQy$ or $y' = Q^{-1}AQy$. Now

$$Q^{-1}AQ = \begin{pmatrix} P_1 & 0 \\ 0 & P_2 \end{pmatrix}$$

where P_1 and P_2 are blocks in which all characteristic roots of P_1 have positive real parts and all roots of P_2 have negative real parts.

Let M^* denote the conjugate-transpose of a matrix M and define

$$B_1 = -\int_{-\infty}^{0} (\exp P_1^* t)(\exp P_1 t)\, dt$$

and

$$B_2 = +\int_{0}^{\infty} (\exp P_2^* t)(\exp P_2 t)\, dt.$$

Notice that

$$-I = (\exp P_1^* t)(\exp P_1 t)\big|_{-\infty}^{0}$$

$$= -\int_{-\infty}^{0} (d/dt)\big[(\exp P_1^* t)(\exp P_1 t)\big]\, dt$$

$$= \int_{-\infty}^{0} -P_1^*(\exp P_1^* t)(\exp P_1 t)\, dt$$

$$\quad - \int_{-\infty}^{0} (\exp P_1^* t)(\exp P_1 t)P_1\, dt$$

$$= P_1^* B_1 + B_1 P_1 \,.$$

In the same way,

$$-I = +(\exp P_2^* t)(\exp P_2 t)\big|_{0}^{\infty}$$

$$= \int_{0}^{\infty} (d/dt)\big[(\exp P_2^* t)(\exp P_2 t)\big]\, dt$$

$$= \int_{0}^{\infty} P_2^*(\exp P_2^* t)(\exp P_2 t)\, dt$$

$$\quad + \int_{0}^{\infty} (\exp P_2^* t)(\exp P_2 t)P_2\, dt$$

$$= P_2^* B_2 + B_2 P_2 \,.$$

Next, form the matrices

$$P = \begin{pmatrix} P_1 & 0 \\ 0 & P_2 \end{pmatrix} \quad \text{and} \quad B = \begin{pmatrix} B_1 & 0 \\ 0 & B_2 \end{pmatrix},$$

and notice that

$$P^* B + BP = \begin{pmatrix} P_1^* & 0 \\ 0 & P_2^* \end{pmatrix}\begin{pmatrix} B_1 & 0 \\ 0 & B_2 \end{pmatrix} + \begin{pmatrix} B_1 & 0 \\ 0 & B_2 \end{pmatrix}\begin{pmatrix} P_1 & 0 \\ 0 & P_2 \end{pmatrix}$$

$$= \begin{pmatrix} P_1^* B_1 & 0 \\ 0 & P_2^* B_2 \end{pmatrix} + \begin{pmatrix} B_1 P_1 & 0 \\ 0 & B_2 P_2 \end{pmatrix}$$

$$= \begin{pmatrix} P_1^* B_1 + B_1 P_1 & 0 \\ 0 & P_2^* B_2 + B_2 P_2 \end{pmatrix} = -I \,.$$

Thus $y' = Q^{-1}AQy = Py$, and so $V(y) = y^T By$ yields $V'(y) = -y^T y$. We then have

$$(Q^{-1}AQ)^* B + B(Q^{-1}AQ) = -I$$

or

$$Q^* A^T (Q^{-1})^* B + BQ^{-1}AQ = -I \, .$$

Now, left multiply by $(Q^*)^{-1}$ and right multiply by Q^{-1} obtaining

$$A^T (Q^{-1})^* BQ^{-1} + (Q^*)^{-1} BQ^{-1}A = -(Q^*)^{-1} Q^{-1}$$

and, because $(Q^{-1})^* = (Q^*)^{-1}$, we have

$$A^T (Q^{-1})^* BQ^{-1} + \left((Q^{-1})^* BQ^{-1} \right)^* A = -(Q^{-1})^* Q^{-1} \stackrel{\text{def}}{=} -C \, .$$

Now Q^{-1} is nonsingular and

$$\mathbf{x}^T (Q^{-1})^* Q^{-1} \mathbf{x} = (Q^{-1}\mathbf{x})^* (Q^{-1}\mathbf{x}) \, .$$

Because $(Q^{-1})^* Q^{-1}$ is Hermite-symmetric, this quadratic form is real and we have

$$V = \mathbf{x}^T (Q^{-1})^* BQ^{-1} \mathbf{x}$$

with

$$V' = -\mathbf{x}^T C\mathbf{x} = -\mathbf{x}^T (Q^{-1})^* Q^{-1} \mathbf{x}$$

being negative definite. Thus when A has no characteristic root with a zero real part, then a positive definite matrix C can be found so that

$$A^T B + BA = -C$$

can be solved.

Remark 5.1.2. We will continue to write $A^T B + BA = -I$ for the sake of definiteness, even though we understand that when

$$\lambda_i + \lambda_k = 0$$

then $-I$ may have to be replaced with a different negative definite matrix.

Naturally, if we wished only to determine stability or instability properties of $\mathbf{x}' = A\mathbf{x}$, we would not bother with a Liapunov function. We have in mind a perturbed system

$$\mathbf{x}' = A\mathbf{x} + \mathbf{G}(t, \mathbf{x})$$

that we expect to inherit the stability properties of $\mathbf{x}' = A\mathbf{x}$. Thus, we construct a Liapunov function for $\mathbf{x}' = A\mathbf{x}$, say $V(\mathbf{x}) = \mathbf{x}^T B\mathbf{x}$, and apply it to $\mathbf{x}' = A\mathbf{x} + \mathbf{G}(t, \mathbf{x})$ for $|\mathbf{x}|$ small. This is developed in Chapter 6, Eq. (6.2.27) (Liapunov's Theorem).

In the same way, we present this material for $\mathbf{x}' = A\mathbf{x}$, but we shall apply it to

$$\mathbf{x}' = A\mathbf{x} + \int_0^t C(t, s)\mathbf{x}(s)\, ds$$

in later sections.

Theorem 5.1.6.   *Let A be a real matrix, $C = C^T$ a negative definite matrix, and $B = B^T$ any solution of*

$$A^T B + BA = C \,.$$

Then the zero solution of

$$\mathbf{x}' = A\mathbf{x}$$

is stable if and only if $\mathbf{x}^T B\mathbf{x} > 0$ for each $x \neq 0$.

Proof. If $V(\mathbf{x}) = \mathbf{x}^T B\mathbf{x} > 0$ for each $\mathbf{x} \neq 0$, then V is positive definite and $V'(\mathbf{x}) = \mathbf{x}^T C\mathbf{x} \leq 0$, so that stability readily follows. Indeed, $\mathbf{x} = 0$ is uniformly asymptotically stable.

Suppose there is some $\mathbf{x}_0 \neq \mathbf{0}$ with $\mathbf{x}_0^T B\mathbf{x}_0 \leq 0$. If $\mathbf{x}_0^T B\mathbf{x}_0 = 0$, let $\mathbf{x}(t) = \mathbf{x}(t, 0, \mathbf{x}_0)$ and consider $V(\mathbf{x}(t)) = \mathbf{x}^T(t) B\mathbf{x}(t)$ with $V'(\mathbf{x}(t)) = x^T(t) C\mathbf{x}(t) < 0$ at $t = 0$. Thus, V decreases, so if $t_1 > 0$, $V(\mathbf{x}(t_1)) = \mathbf{x}^T(t_1) B\mathbf{x}(t_1) < 0$. Thus, we may suppose $\mathbf{x}_0^T B\mathbf{x}_0 < 0$. If $\mathbf{x}_0 = n\mathbf{y}_n$ defines $\mathbf{y}_n$, then $\mathbf{x}_0^T B\mathbf{x}_0 = n^2 \mathbf{y}_n^T B\mathbf{y}_n < 0$, so $\{\mathbf{y}_n\}$ converges to zero and $V(\mathbf{y}_n) < 0$. All parts of Theorem 5.1.3 are satisfied and $\mathbf{x} = 0$ is unstable. This completes the proof.

5.2   The Scalar Equation

The concept in Theorem 5.1.6 is a key one, and it will be extended to systems of Volterra equations after we lay some groundwork with scalar equations.

Consider the scalar equation

$$x' = A(t)x + \int_0^t C(t,s)x(s)\,ds \tag{5.2.1}$$

with $A(t)$ continuous on $[0, \infty)$ and $C(t,s)$ continuous for $0 \le s \le t < \infty$. Select a continuous function $G(t,s)$ with $\partial G/\partial t = C(t,s)$, so that (5.2.1) may also be written as

$$x' = Q(t)x + (d/dt)\int_0^t G(t,s)x(s)\,ds \tag{5.2.2}$$

with

$$Q(t) + G(t,t) = A(t)\,.$$

Note that (5.2.1) and (5.2.2) are, in fact, the same equation.

For reference we repeat the definition of stability of the zero solution of (5.2.1) and then negate it.

Definition 5.2.1. *The zero solution (5.2.1) is stable if for each $\varepsilon > 0$ and each $t_0 \ge 0$ there exists a $\delta > 0$ such that*

$$\phi : [0, t_0] \to R, \ \phi \text{ continuous}, \ |\phi(t)| < \delta \text{ on } [0, t_0], \text{ and } t \ge t_0$$

imply $x(t, t_0, \phi) < \varepsilon$.

Definition 5.2.2. *The zero solution of (5.2.1) is unstable if there exists an $\varepsilon > 0$ and there exists a $t_0 \ge 0$ such that, for each $\delta > 0$, there is a continuous function $\phi : [0, t_0] \to R$ with $|\phi(t)| < \delta$ on $[0, t_0]$ and with $|x(t_1, t_0, \phi)| \ge \varepsilon$ for some $t_1 > t_0$.*

Theorem 5.2.1. *Suppose there are constants M_1 and M_2, such that for $0 \le t < \infty$ we have*

$$\int_0^t |C(t,s)|\,ds + \int_t^\infty |C(u,t)|\,du \le M_1 < M_2 \le 2|A(t)|\,. \tag{5.2.4}$$

Then the zero solution of (5.2.1) is stable if and only if $A(t) < 0$.

Proof. First suppose that $A(t) < 0$. We shall show that $x = 0$ is stable. Define

$$V_1(t, x(\cdot)) = x^2 + \int_0^t \int_t^\infty |C(u, s)| \, du \, x^2(s) \, ds \,,$$

so that the derivative of V_1 along a solution of (5.2.1) satisfies

$$V_{1(5.2.1)}'(t, x(\cdot)) \leq 2Ax^2 + 2\int_0^t |C(t, s)| \, |x(s)x(t)| \, ds$$

$$+ \int_t^\infty |C(u, t)| \, du \, x^2 - \int_0^t |C(t, s)| x^2(s) \, ds$$

$$\leq 2Ax^2 + \int_0^t |C(t, s)| \left[x^2(s) + x^2(t) \right] ds$$

$$+ \int_t^\infty |C(u, t)| \, ds \, x^2 - \int_0^t |C(t, s)| x^2(s) \, ds$$

$$= \left[2A + \int_0^t |C(t, s)| \, ds + \int_t^\infty |C(u, t)| \, du \right] x^2$$

$$\leq [2A + M_1] x^2 \leq [-M_2 + M_1] x^2$$

$$\overset{\text{def}}{=} -\alpha x^2 \,, \quad \alpha > 0 \,.$$

As V_1 is positive definite and $V_1' \leq 0$, it readily follows that $x = 0$ is stable.

Now suppose that $A(t) > 0$. (Note that $M_1 < M_2 \leq 2|A(t)|$ implies that $A(t) \neq 0$). Define

$$V_2(t, x(\cdot)) = x^2 - \int_0^t \int_t^\infty |C(u, s)| \, du \, x^2(s) \, ds \,,$$

so that

$$V_{2(5.2.1)}'(t, x(\cdot)) \geq 2A(t)x^2 - 2\int_0^t |C(t, s)| \, |x(s)x(t)| \, ds$$

$$- \int_t^\infty |C(u, t)| \, du \, x^2 + \int_0^t |C(t, s)| x^2(s) \, ds$$

$$\geq 2A(t)x^2 - \int_0^t |C(t, s)| \left[x^2(t) + x^2(s) \right] ds$$

$$- \int_t^\infty |C(u, t)| \, du \, x^2 + \int_0^t |C(t, s)| x^2(s) \, ds$$

$$= \left[2A(t) - \int_0^t |C(t,s)|\,ds - \int_t^\infty |C(u,t)|\,du \right] x^2$$

$$\geq [2A(t) - M_1]x^2 \geq [M_2 - M_1]x^2$$
$$\overset{\text{def}}{=} \alpha x^2, \quad \alpha > 0.$$

Now, if $x(t) = x(t, t_0, \phi)$ is a solution of (5.2.1), then we have

$$x^2(t) \geq V_2(t, x(\cdot)) \geq V_2(t_0, \phi(\cdot)) + \alpha \int_{t_0}^t x^2(s)\,ds.$$

Given any $t_0 \geq 0$ and any $\delta > 0$ we can pick a continuous initial function $\phi : [0, t_0] \to R$ with $V_2(t_0, \phi(\cdot)) > 0$. Thus, $x^2(t) \geq V_2(t_0, \phi(\cdot))$, so that

$$x^2(t) \geq V_2(t, x(\cdot))$$
$$\geq V_2(t_0, \phi(\cdot)) + \alpha \int_{t_0}^t V_2(t_0, \phi(\cdot))\,ds$$
$$= V_2(t_0, \phi(\cdot)) + \alpha V_2(t_0, \phi(\cdot))(t - t_0),$$

and so $|x(t)| \to \infty$ as $t \to \infty$. This completes the proof.

Corollary 1. *Let (5.2.4) hold, let $A(t)$ be bounded, and let $A(t) < 0$. Then $x = 0$ is asymptotically stable.*

Proof. We showed that $V'_{1(5.2.1)}(t, x(\cdot)) \leq -\alpha x^2$, so we have $x^2(t)$ in $L^1[0, \infty)$ and $x^2(t)$ bounded. Note also that $x'(t)$ is bounded. Thus, $x(t) \to 0$ as $t \to \infty$.

Exercise 5.2.1. Try to eliminate the requirement that $A(t)$ be bounded from the corollary. In Chapter 6 we develop three ways of doing this.

Corollary 2. *Let (5.2.4) hold and let $A(t) > 0$. Then the unbounded solution $x(t)$ produced in the proof of Theorem 5.2.1 satisfies $|x(t)| \geq c_1 + c_2(t - t_0)$ for c_1 and c_2 positive.*

Theorem 5.2.2. *Suppose there are constants J, Q_1, Q_2, and R with $R < 2$, and*

$$0 < Q_1 \leq |Q(t)| \leq Q_2 , \tag{5.2.5}$$

$$2|A| + \int_0^t |C(t,s)|\, ds + \int_t^\infty |C(u,t)|\, du \leq J , \tag{5.2.6}$$

and

$$\int_t^\infty |G(u,t)|\, du + \int_0^t |G(t,s)|\, ds \leq RQ_1/Q_2 \tag{5.2.7}$$

for $0 \leq t < \infty$. Suppose also that there is a continuous function $h : [0,\infty) \to [0,\infty)$ with

$$|G(t,s)| \leq h(t-s) \quad \text{and} \quad h(u) \to 0 \text{ as } u \to \infty . \tag{5.2.8}$$

Then the solution of (5.2.2) is stable if and only if $Q(t) < 0$.

Proof. First suppose that $Q(t) < 0$ and define

$$V_3(t, x(\cdot)) = \left[x - \int_0^t G(t,s)x(s)\, ds \right]^2 + Q_2 \int_0^t \int_t^\infty |G(u,s)|\, du\, x^2(s)\, ds ,$$

so that along a solution $x(t)$ of (5.2.2) we have

$$V'_{3(5.2.2)}(t, x(\cdot)) = 2\left(x - \int_0^t G(t,s)x(x)\, ds \right) Q(t)x$$

$$+ Q_2 \int_t^\infty |G(u,t)|\, du\, x^2 - Q_2 \int_0^t |G(t,s)| x^2(s)\, ds$$

$$\leq 2Q(t)x^2 + Q_2 \int_0^t |G(t,s)| \left[x^2(s) + x^2(t) \right] ds$$

$$+ Q_2 \int_t^\infty |G(u,t)|\, du\, x^2 - Q_2 \int_0^t |G(t,s)| x^2(s)\, ds$$

$$= \left[2Q(t) + Q_2 \left(\int_0^t |G(t,s)|\, ds + \int_t^\infty |G(u,t)|\, du \right) \right] x^2$$

$$\leq [-2Q_1 + RQ_1] x^2$$

$$\overset{\text{def}}{=} -\beta x^2 , \quad \beta > 0 .$$

Recall that (5.2.1) and (5.2.2) are the same and consider $V_1(t, x(\cdot))$ once more. It is certainly true that $V'_{1(5.2.1)}(t, x(\cdot)) = V'_{1(5.2.2)}(t, x(\cdot))$. Hence, $V'_{1(5.2.2)}$ may be obtained by taking

$$V'_{1(5.2.1)}(t, x(\cdot)) \leq 2|A(t)|x^2$$
$$+ \int_0^t |C(t,s)| \left[x^2(s) + x^2(t)\right] ds$$
$$+ \int_t^\infty |C(u,t)| \, du \, x^2 - \int_0^t |C(t,s)|x^2(s) \, ds$$
$$= \left[2|A(t)| + \int_0^t |C(t,s)| \, ds + \int_t^\infty |C(u,t)| \, du\right] x^2$$
$$\leq Jx^2 .$$

Hence, if we let

$$W_1(t, x(\cdot)) = (\beta/2J)V_1(t, x(\cdot)) + V_3(t, x(\cdot)),$$

then we have

$$W'_{1(5.2.2)}(t, x(\cdot)) \leq (\beta/2)x^2 - \beta x^2 = -\beta x^2/2 .$$

Because W_1 is positive definite, it follows that $x = 0$ is stable.

Now, let $Q > 0$ and define

$$V_4(t, x(\cdot)) = \left(x - \int_0^t G(t,s)x(s) \, ds\right)^2 - Q_2 \int_0^t \int_t^\infty |G(u,s)| \, du \, x^2(s) \, ds ,$$

so that

$$V'_{4(5.2.2)}(t, x(\cdot)) = 2\left(x - \int_0^t G(t,s)x(s) \, ds\right)Q(t)x$$
$$+ Q_2 \int_0^t |G(t,s)|x^2(s) \, ds - Q_2 \int_t^\infty |G(u,t)| \, du \, x^2$$
$$\geq 2Q(t)x^2 - Q_2 \int_0^t |G(t,s)| \left[x^2(s) + x^2(t)\right] ds$$
$$+ Q_2 \int_0^t |G(t,s)|x^2(s) \, ds - Q_2 \int_t^\infty |G(u,t)| \, du \, x^2$$
$$= \left[2Q(t) - Q_2\left(\int_0^t |G(t,s)| \, ds + \int_t^\infty |G(u,t)| \, du\right)\right] x^2$$
$$\geq [2Q(t) - RQ_1]x^2 \geq [2Q_1 - RQ_1]x^2$$
$$\overset{\text{def}}{=} \gamma x^2, \quad \gamma > 0 .$$

Now, by way of contradiction, we suppose $x = 0$ is stable. Thus, given $\varepsilon > 0$ and $t_0 \geq 0$, there is a $\delta > 0$ such that for any continuous $\phi : [0, t_0] \to R$ with $|\phi(t)| < \delta$ on $[0, t_0]$, we have $|x(t, t_0, \phi)| < \varepsilon$ for $t \geq t_0$. For this t_0 and this δ, we may choose such a ϕ with

$$V_4(t_0, \phi(\cdot)) > 0$$

and let

$$x(t) = x(t, t_0, \phi) \,.$$

As $\int_0^t |G(t, s)| \, ds$ is bounded and $x(t)$ is bounded, so is $\int_0^t G(t, s) x(s) \, ds$. If $x^2(t)$ is not in $L^1[0, \infty)$, then $x(t)$ is unbounded, and we have

$$\left[x(t) - \int_0^t G(t, s) x(s) \, ds \right]^2 \geq V_4(t, x(\cdot))$$

$$\geq V_4(t_0, \phi(\cdot)) + \gamma \int_{t_0}^t x^2(s) \, ds \,.$$

Hence, we suppose $x^2(t)$ is in $L^1[0, \infty)$.

Next, note that

$$\left(\int_0^t |G(t, s)| \, |x(s)| \, ds \right)^2 = \left(\int_0^t |G(t, s)|^{1/2} |G(t, s)|^{1/2} |x(s)| \, ds \right)^2$$

$$\leq \int_0^t |G(t, s)| \, ds \int_0^t |G(t, s)| x^2(s) \, ds$$

by the Schwartz inequality. Moreover, $|G(t, s)| \leq h(t - s)$ and $h(u) \to 0$ as $u \to \infty$. Thus,

$$\int_0^t |G(t, s)| x^2(s) \, ds \leq \int_0^t h(t - s) x^2(s) \, ds \,,$$

which is the convolution of an L^1 function with a function tending to zero, and hence, the integral tends to zero.

We then have

$$\left| x - \int_0^t G(t, s) x(s) \, ds \right| \geq \left[V_4(t_0, \phi(\cdot)) \right]^{1/2}.$$

As the integral tends to zero, it follows that $|x(t)| \geq \alpha$ for large t and some $\alpha > 0$. This contradicts $x^2(t)$ in L^1, thereby completing the proof.

Corollary 1.   *Suppose that A is constant, $C(t, s) = C(t - s)$, $G(t, s) = G(t - s)$, $\int_0^\infty |C(t)|\, dt < \infty$, and $G(t) = -\int_t^\infty C(s)\, ds$. Also suppose that*

$$\int_0^\infty |G(u)|\, du < 1. \tag{5.2.9}$$

Then the zero solution of (5.2.2) is stable if and only if $Q < 0$.

Proof. We verify the conditions of Theorem 5.2.2. Because A is constant and G is of convolution type, it follows that Q is constant. As $Q \neq 0$, (5.2.5) holds with $Q_1 = Q_2 = |Q|$.

Now

$$\int_0^t |C(t - s)|\, ds = \int_0^t |C(s)|\, ds < \infty$$

and

$$\int_t^\infty |C(u - t)|\, du = \int_0^\infty |C(v)|\, dv < \infty,$$

so that (5.2.6) holds.

Finally,

$$\int_t^\infty |G(u - t)|\, du = \int_0^\infty |G(s)|\, ds < 1$$

and

$$\int_0^t |G(t - s)|\, ds = \int_0^t |G(s)|\, ds \leq \int_0^\infty |G(s)|\, ds < 1,$$

so that (5.2.7) holds. This completes the proof.

Exercise 5.2.2.   Formulate a condition from Theorem 5.2.2 ensuring asymptotic stability.

Exercise 5.2.3.   Theorem 5.2.1 says that if $C(t, s)$ is "small", then the stability of (5.2.1) depends on the sign of A. Theorem 5.2.2 says that if $G(t, s)$ is "small", then the stability of (5.2.2) depends on the sign of $Q(t)$. Can other fundamental quantities be found besides A and Q?

Theorems 5.2.1 and 5.2.2 are very closely related. The covering assumption in Theorem 5.2.1 relates to the conclusion in Theorem 5.2.2.

A necessary and sufficient condition is strong or weak depending on the mildness or severity of the covering assumptions. One would call Theorem 5.2.1 a strong result, for example, if we could say the following: Suppose $A(t) < 0$; then the zero solution of (5.2.1) is stable if and only if the covering assumption (5.2.4) holds. In the present generality, that seems too complicated to check. However, when we check for it under simplifying assumptions, then we see the relation between the two results.

To that end, we suppose that A is constant, $C(t,s) = C(t-s)$, $C(t) \geq 0$, $G(t) = -\int_t^\infty C(v)\,dv$, and

$$\int_0^\infty |G(v)|\,dv < 1\,. \tag{5.2.10}$$

This last condition requires the "memory" of the initial conditions to fade, a concept discussed in some detail in Chapters 6 and 8.

Then Eq. (5.2.1) becomes

$$x' = Ax + \int_0^t C(t-s)x(s)\,ds\,; \tag{5.2.11}$$

(5.2.2) becomes

$$x' = Qx + (d/dt)\int_0^t G(t-s)x(s)\,ds\,; \tag{5.2.12}$$

(5.2.3) becomes

$$Q = A + \int_0^\infty C(v)\,dv\,; \tag{5.2.13}$$

(5.2.4) becomes

$$\int_0^\infty C(v)\,dv < |A|\,; \tag{5.2.14}$$

and (5.2.7) becomes (5.2.10).

We negate (5.2.14) in two steps.

Theorem 5.2.3.   *Let (5.2.10) hold, let $C(t) \geq 0$, and let $A < 0$. If*

$$\int_0^\infty C(v)\,dv = |A|\,,$$

then the zero solution of (5.2.11) is stable.

Proof. Write (5.2.11) as (5.2.12). Because $A < 0$, $\int_0^\infty C(v)\,dv = -A$ or $0 = A + \int_0^\infty C(v)\,dv = Q$. Let $\varepsilon > 0$ and $t_0 \geq 0$ be given. Let $\phi : [0,t_0] \to R$

be a continuous initial function with $|\phi(t)| < \delta$ on $[0, t_0]$, where $\delta \leq \varepsilon$ is yet to be determined. Integrate (5.2.12) from t_0 to $t > t_0$ and obtain

$$x(t) = x(t_0) + \int_0^t G(t - s)x(s)\,ds \; - \int_0^{t_0} G(t_0 - s)\phi(s)\,ds\,.$$

If we let $M = \left| x(t_0) - \int_0^{t_0} G(t_0 - s)\phi(s)\,ds \right|$, then $M \leq 2\delta$ and

$$|x(t)| \leq 2\delta + \int_0^t |G(t - s)|\,|x(s)|\,ds\,.$$

Now, if $|x(t)| < \delta$ on $[0, \infty)$, there is nothing to prove, because we may take $\delta = \varepsilon$. So suppose that there is a $t_1 > t_0$ with $|x(t_1)|$ the maximum of $|x(t)|$ on $[0, t_1]$. Then

$$
\begin{aligned}
|x(t_1)| &\leq 2\delta + |x(t_1)| \int_0^{t_1} |G(t_1 - s)|\,ds \\
&\leq 2\delta + |x(t_1)| \int_0^{\infty} |G(v)|\,dv \\
&= 2\delta + |x(t_1)|P, \quad 0 < P < 1\,.
\end{aligned}
$$

Hence, $|x(t_1)|\,[1 - P] \leq 2\delta$ or $|x(t_1)| \leq 2\delta/[1 - P] < \varepsilon$ if $\delta < \varepsilon[1 - P]/2$. This proves stability.

We continue and negate (5.2.14).

Theorem 5.2.4. *Let (5.2.10) hold, let $C(t) \geq 0$, and let $A < 0$. If*

$$\int_0^{\infty} C(v)\,dv > |A|\,,$$

then $Q > 0$, so the zero solution of (5.2.11) is unstable.

Proof. Because $A < 0$, $\int_0^{\infty} C(v)\,dv > -A$ or $Q = A + \int_0^{\infty} C(v)\,dv > 0$. Clearly, both (5.2.5) and (5.2.6) are satisfied. Because (5.2.7) reduces to (5.2.10), the conditions of Theorem 5.2.2 are satisfied with $Q > 0$, so the zero solution is unstable. This completes the proof.

Theorem 5.2.5. *Let A be constant, $A < 0$, $C(t) \geq 0$, $\int_0^{\infty} C(s)\,ds < \infty$, $G(t) = -\int_t^{\infty} C(s)\,ds$, and let*

$$\int_0^{\infty} |G(v)|\,dv < 1\,. \tag{5.2.10}$$

The zero solution is asymptotically stable if and only if Eq. (5.2.14) holds.

Proof. If (5.2.14) holds, then we do have asymptotic stability by Corollary 1 to Theorem 5.2.1.

If $\int_0^\infty C(v)\,dv = |A| = -A$ then $A + \int_0^\infty C(v)\,dv = Q = 0$. Thus, by (5.2.12), we have upon integration

$$x(t) = x_0 + \int_0^t G(t-s)x(s)\,ds\,, \quad x_0 > 0\,.$$

Because $C \geq 0$, $G \leq 0$, then as long as $x(t) > 0$, we have $x(t)$ decreasing and

$$x(t) \geq x_0 + x_0 \int_0^t G(t-s)\,ds$$

$$\geq x_0 \left(1 + \int_0^\infty G(s)\,ds \right) \stackrel{\text{def}}{=} x_0(1-\alpha)\,, \quad \alpha > 0\,.$$

Hence, $x(t)$ is bounded strictly away from zero.

If $\int_0^\infty C(v)\,dv > |A| = -A$, then $Q = A + \int_0^\infty C(v)\,dv > 0$, so $x = 0$ is unstable by Theorem 5.2.2. This completes the proof.

Exercise 5.2.4. Obtain the counterparts of Theorems 5.2.3 and 5.2.4 under the assumption that $A > 0$.

Exercise 5.2.5. Give an example of a function $C(t) > 0$ with $\int_0^\infty C(t)\,dt > 10$ but $\int_0^\infty |G(v)|\,dv < 1$.

We pass from (5.2.1) to (5.2.2) in an effort to make A and C more tractable; thus, we integrate C to obtain G. We have seen before that the same end may sometimes be accomplished by differentiation. For example, if $C'(t)$ is continuous, then differentiation of (5.2.11) yields

$$x'' = Ax' + C(0)x + \int_0^t C'(t-s)x(s)\,ds$$

or the system

$$\mathbf{X}' = \begin{pmatrix} 0 & 1 \\ C(0) & A \end{pmatrix} \mathbf{X} + \int_0^t \begin{pmatrix} 0 & 0 \\ C'(t-s) & 0 \end{pmatrix} \mathbf{X}(s)\,ds\,,$$

which can be treated using the results of the next section.

The theorems here concerning A and Q are boundary results and an infinite number of results lie in between. Write $C(t,s) = C_1(t,s) + C_2(t,s)$ and consider

$$x' = A(t)x + \int_0^t C_1(t,s)x(s)\,ds \; + \int_0^t C_2(t,s)x(s)\,ds\,, \qquad (5.2.15)$$

so that if $G(t,s)$ now satisfies

$$\partial G(t,s)/\partial t = C_2(t,s) \qquad (5.2.16)$$

and if

$$Q(t) = A(t) - G(t,t)\,, \qquad (5.2.17)$$

then

$$x' = Q(t)x + \int_0^t C_1(t,s)x(s)\,ds \; + (d/dt)\int_0^t G(t,s)x(s)\,ds \qquad (5.2.18)$$

is the same as (5.2.15). Thus, we consider the scalar equation

$$x' = L(t)x + \int_0^t C_1(t,s)x(s)\,ds \; + (d/dt)\int_0^t H(t,s)x(s)\,ds\,, \qquad (5.2.19)$$

in which $L(t)$ is continuous for $0 \le t < \infty$, and $C_1(t,s)$ and $H(t,s)$ are continuous for $0 \le s \le t < \infty$. Here, L, C_1, H, and x are all scalars.

We assume that $\int_t^\infty |C_1(u,t)|\,du$ is defined for $t \ge 0$, and let

$$P = \sup_{t \ge 0} \int_0^t |C_1(t,s)|\,ds\,, \qquad (5.2.20)$$

$$J = \sup_{t \ge 0} \int_0^t |H(t,s)|\,ds\,, \qquad (5.2.21)$$

and agree that $0 \times P = 0$.

Theorem 5.2.6. *Suppose that $J < 1$, $J|L(t)| \le JQ$, for some $Q > 0$, and*

$$\int_0^t \left[|C_1(t,s)| + Q|H(t,s)| \right] ds$$

$$+ \int_t^\infty \left[(1+J)\,|C_1(u,t)| + (Q+P)\,|H(u,t)| \right] du$$

$$- 2|L(t)| \le -\alpha\,, \quad \text{for some } \alpha > 0\,. \qquad (5.2.22)$$

In addition, suppose there is a continuous function $h : [0,\infty) \to [0,\infty)$ such that $|H(t,s)| \le h(t-s)$ and $h(u) \to 0$ as $u \to \infty$. Then the zero solution of (5.2.19) is stable if and only if $L(t) < 0$.

The proof is left as an exercise. It is very similar to earlier ones, except that when using the Schwartz inequality one needs to shift certain functions from one integral to the other. Details may be found in Burton and Mahfoud (1983, 1985). Numerous examples and more exact qualitative behavior are also found in those papers.

5.3   The Vector Equation

We now extend the results of Section 5.2 to systems of Volterra equations and present certain perturbation results. Owing to the greater complexity of systems over scalars, it seems preferable to reduce the generality of A and G.

Consider the system

$$\mathbf{x}' = A\mathbf{x} + \int_0^t C(t,s)\mathbf{x}(s)\, ds \,, \tag{5.3.1}$$

in which A is a constant $n \times n$ matrix and C an $n \times n$ matrix of functions continuous for $0 \le s \le t < \infty$. We suppose there is a symmetric matrix B with

$$A^T B = BA = -I \,. \tag{5.3.2}$$

Moreover, we refer the reader to Remark 5.1.2 and to Theorem 5.1.6, which show that (5.3.2) can be replaced by the more general condition that

$$A^T B + BA = -C$$

has a solution B for some positive definite matrix C.

Theorem 5.3.1. *Let (5.3.2) hold and suppose there is a constant $M > 0$ with*

$$|B|\left(\int_0^t |C(t,s)|\, ds + \int_t^\infty |C(u,t)|\, du \right) \le M < 1 \,. \tag{5.3.3}$$

Then the zero solution of (5.3.1) is stable if and only if $\mathbf{x}^T B\mathbf{x} > 0$ for each $x \ne 0$.

Proof. We define

$$V_1(t, \mathbf{x}(\cdot)) = \mathbf{x}^T B \mathbf{x} + |B| \int_0^t \int_t^\infty |C(u, s)| \, du \, \mathbf{x}^2(s) \, ds \,,$$

where $\mathbf{x}^2 = \mathbf{x}^T \mathbf{x}$. Then

$$V_{1(5.3.1)}'(t, \mathbf{x}(\cdot)) = \left[\mathbf{x}^T A^T + \int_0^t \mathbf{x}^T(s) C^T(t, s) \, ds \right] B \mathbf{x}$$

$$+ \mathbf{x}^T B \left[A\mathbf{x} + \int_0^t C(t, s) \mathbf{x}(s) \, ds \right]$$

$$+ |B| \int_t^\infty |C(u, t)| \, du \, \mathbf{x}^2 - \int_0^t |B| \, |C(t, s)| \mathbf{x}^2(s) \, ds$$

$$= -\mathbf{x}^2 + 2\mathbf{x}^T B \int_0^t C(t, s) \mathbf{x}(s) \, ds$$

$$+ |B| \int_t^\infty |C(u, t)| \, du \, \mathbf{x}^2 - |B| \int_0^t |C(t, s)| \mathbf{x}^2 \, ds$$

$$\leq -\mathbf{x}^2 + |B| \int_0^t |C(t, s)| \left[\mathbf{x}^2(s) + \mathbf{x}^2(t) \right] ds$$

$$+ |B| \int_t^\infty |C(u, t)| \, du \, \mathbf{x}^2 - |B| \int_0^t |C(t, s)| \mathbf{x}^2(s) \, ds$$

$$= \left[-1 + |B| \left(\int_0^t |C(t, s)| \, ds + \int_t^\infty |C(u, t)| \, du \right) \right] \mathbf{x}^2$$

$$\leq [-1 + M] \mathbf{x}^2 \stackrel{\text{def}}{=} -\alpha \mathbf{x}^2 \,, \quad \alpha > 0 \,.$$

Now, if $\mathbf{x}^T B \mathbf{x} > 0$ for all $\mathbf{x} \neq 0$, then V_1 is positive definite and V_1' is negative definite, so $\mathbf{x} = 0$ is stable.

Suppose there is an $\mathbf{x}_0 \neq 0$ with $\mathbf{x}_0^T B \mathbf{x}_0 \leq 0$. Argue as in the proof of Theorem 5.1.6 that there is also an $\mathbf{x}_0$ with $\mathbf{x}_0^T B \mathbf{x}_0 < 0$. By way of contradiction, we suppose that $\mathbf{x} = 0$ is stable. Thus, for $\varepsilon = 1$ and $t_0 = 0$, there is a $\delta > 0$ such that $|\mathbf{x}_0| < \delta$ and $t \geq 0$ implies $|\mathbf{x}(t, 0, \mathbf{x}_0)| < 1$. We may choose $\mathbf{x}_0$ with $|\mathbf{x}_0| < \delta$ and $\mathbf{x}_0^T B \mathbf{x}_0 < 0$. Let $\mathbf{x}(t) = \mathbf{x}(t, 0, \mathbf{x}_0)$ and have $V_1(0, \mathbf{x}_0) < 0$ and $V_1'(t, \mathbf{x}(\cdot)) \leq -\alpha x^2$, so that

$$\mathbf{x}^T(t) B \mathbf{x}(t) \leq V_1(t, \mathbf{x}(\cdot))$$

$$\leq V_1(0, \mathbf{x}_0) - \alpha \int_0^t \mathbf{x}^2(s) \, ds$$

$$= \mathbf{x}_0^T B \mathbf{x}_0 - \alpha \int_0^t \mathbf{x}^2(s) \, ds \,.$$

If there is a sequence $\{t_n\}$ tending to infinity monotonically such that $\mathbf{x}(t_n) \to 0$, then $\mathbf{x}^T(t_n)B\mathbf{x}(t_n) \to 0$, and this would contradict $\mathbf{x}^T(t)B\mathbf{x}(t) \leq \mathbf{x}_0^T B\mathbf{x}_0 < 0$. Thus, there is a $\gamma > 0$ with $\mathbf{x}^2(t) \geq \gamma$, so that

$$\mathbf{x}^T(t)B\mathbf{x}(t) \leq \mathbf{x}_0^T B\mathbf{x}_0 - \alpha\gamma t,$$

which implies that $|\mathbf{x}(t)| \to \infty$ as $t \to \infty$. This contradicts $|\mathbf{x}(t)| < 1$ and completes the proof.

In Eq. (5.3.1) we suppose C is of convolution type, $C(t,s) = C(t-s)$, and select a matrix G with $G'(t) = C(t)$. Then write (5.3.1) as

$$\mathbf{x}' = Q\mathbf{x} + (d/dt)\int_0^t G(t-s)\mathbf{x}(s)\,ds\,, \tag{5.3.4}$$

where

$$Q + G(0) = A\,.$$

Note that (5.3.1) and (5.3.4) are the same under the convolution assumption.

We suppose there is a constant matrix D with

$$D^T = D \quad \text{and} \quad Q^T D + DQ = -I\,. \tag{5.3.5}$$

Refer also to Remark 5.1.2.

Theorem 5.3.2. *Let (5.3.5) hold and suppose $G(t) \to 0$ as $t \to \infty$. Suppose also that there are constants N and P with*

$$2|DQ|\int_0^\infty |G(t)|\,dt \leq N < 1 \tag{5.3.6}$$

and

$$2\left[|A| + \int_0^\infty |C(t)|\,dt\right] \leq P\,. \tag{5.3.7}$$

Then the zero solution of (5.3.4) is stable if and only if $\mathbf{x}^T D\mathbf{x} > 0$ for all $\mathbf{x} \neq 0$.

Proof. Define

$$V_2(t, \mathbf{x}(\cdot)) = \left[\mathbf{x} - \int_0^t G(t-s)\mathbf{x}(s)\,ds\right]^T D\left[\mathbf{x} - \int_0^t G(t-s)\mathbf{x}(s)\,ds\right]$$

$$+ |DQ|\int_0^t \int_t^\infty |G(u-s)|\,du\,\mathbf{x}^2(s)\,ds\,,$$

so that

$$V'_{2(5.3.4)}(t, \mathbf{x}(\cdot))$$

$$= \mathbf{x}^T Q^T D \left[\mathbf{x} - \int_0^t G(t-s)\mathbf{x}(s)\, ds \right]$$

$$+ \left[\mathbf{x} - \int_0^t G(t-s)\mathbf{x}(s)\, ds \right]^T DQ\mathbf{x}$$

$$+ |DQ| \int_t^\infty |G(u-t)|\, du\, \mathbf{x}^2 - |DQ| \int_0^t |G(t-s)|\mathbf{x}^2(s)\, ds$$

$$\leq -\mathbf{x}^2 + |DQ| \int_0^t |G(t-s)| \left[\mathbf{x}^2(t) + \mathbf{x}^2(s) \right] ds$$

$$+ |DQ| \int_t^\infty |G(u-t)|\, du\, \mathbf{x}^2 - |DQ| \int_0^t |G(t-s)|\mathbf{x}^2(s)\, ds$$

$$= \left[-1 + |DQ| \left(\int_0^t |G(t-s)|\, ds + \int_t^\infty |G(u-t)|\, du \right) \right] \mathbf{x}^2$$

$$\leq \left[-1 + 2|DQ| \int_0^\infty |G(t)|\, dt \right] \mathbf{x}^2$$

$$\leq [-1 + N]x^2 \overset{\text{def}}{=} -\mu x^2, \quad \mu > 0.$$

Now consider

$$V_3(t, \mathbf{x}(\cdot)) = \mathbf{x}^2 + \int_0^t \int_t^\infty |C(u-s)|\, du\, \mathbf{x}^2(s)\, ds$$

and find $V'_{3(5.3.4)}$ by computing

$$V'_{3(5.3.1)}(t, \mathbf{x}(\cdot)) \leq \left[\mathbf{x}^T A^T + \int_0^t \mathbf{x}^T(s) C^T(t-s)\, ds \right] \mathbf{x}$$

$$+ \mathbf{x}^T \left[A\mathbf{x} + \int_0^t C(t-s)\mathbf{x}(s)\, ds \right]$$

$$+ \int_t^\infty |C(u-t)|\, du\, \mathbf{x}^2 - \int_0^t |C(t-s)|\mathbf{x}^2(s)\, ds$$

$$\leq 2|A|\mathbf{x}^2 + \int_0^t |C(t-s)| \left[\mathbf{x}^2(s) + \mathbf{x}^2(t) \right] ds$$

$$+ \int_t^\infty |C(u-t)|\, du\, \mathbf{x}^2 - \int_0^t |C(t-s)|\mathbf{x}^2(s)\, ds$$

$$\leq 2 \left[|A| + \int_0^\infty |C(t)|\, dt \right] \mathbf{x}^2 \leq P\mathbf{x}^2.$$

Define
$$W(t, \mathbf{x}(\cdot)) = (\mu/2P)V_3(t, \mathbf{x}(\cdot)) + V_2(t, \mathbf{x}(\cdot))$$
and obtain
$$W'_{(5.3.4)}(t, \mathbf{x}(\cdot)) \leq -(\mu/2)\mathbf{x}^2 \,.$$

We now suppose that $\mathbf{x}^T D\mathbf{x} > 0$ for all $\mathbf{x} \neq 0$, so that
$$(\mu/2P)\mathbf{x}^2 \leq W(t, \mathbf{x}(\cdot))$$
$$\leq W(t_0, \boldsymbol{\phi}(\cdot)) - (\mu/2)\int_{t_0}^{t} \mathbf{x}^2(s)\, ds$$

and conclude that $x = 0$ is stable.

Next, we suppose there is an $\mathbf{x}_0 \neq 0$ with $\mathbf{x}_0^T D\mathbf{x}_0 \leq 0$, arguing as before that we may suppose $\mathbf{x}_0^T D\mathbf{x}_0 < 0$. Hence, $(k\mathbf{x}_0)^T D(k\mathbf{x}_0) < 0$ for each $k > 0$, so we may suppose $|\mathbf{x}_0|$ to be as small as we please.

By way of contradiction, we suppose $\mathbf{x} = 0$ is stable. Thus, for $\varepsilon = 1$ and $t_0 = 0$, there is a $\delta > 0$ such that
$$|\mathbf{x}_0| < \delta \quad \text{and} \quad t \geq 0$$

imply $|\mathbf{x}(t, 0, \mathbf{x}_0)| < 1$. Choose $|\mathbf{x}_0| < \delta$ and $\mathbf{x}_0^T D\mathbf{x}_0 < 0$. Then let $\mathbf{x}(t) = \mathbf{x}(t, 0, \mathbf{x}_0)$, so that $V'_{2(5.3.4)}(t, \mathbf{x}(\cdot)) \leq -\mu\mathbf{x}^2(t)$. We then conclude that
$$V_2(t, \mathbf{x}(\cdot)) \leq V_2(0, \mathbf{x}_0) - \mu \int_0^t \mathbf{x}^2(s)\, ds$$
$$= \mathbf{x}_0^T D\mathbf{x}_0 - \mu \int_0^t \mathbf{x}^2(s)\, ds$$
$$\overset{\text{def}}{=} -\eta - \mu \int_0^t \mathbf{x}^2(s)\, ds \,, \quad \eta > 0 \,.$$

Thus,
$$\left[\mathbf{x}(t) - \int_0^t G(t - s)\mathbf{x}(s)\, ds\right]^T D\left[\mathbf{x}(t) - \int_0^t G(t - s)\mathbf{x}(s)\, ds\right]$$
$$\leq V_2(t, \mathbf{x}(\cdot)) \leq -\eta - \mu \int_0^t \mathbf{x}^2(s)\, ds \,. \tag{5.3.8}$$

Use the Schwartz inequality to conclude
$$\left[\int_0^t |G(t - s)|\, |\mathbf{x}(s)|\, ds\right]^2$$
$$\leq \int_0^t |G(t - s)|\, ds \int_0^t |G(t - s)|\mathbf{x}^2(s)\, ds \,. \tag{5.3.9}$$

As $\int_0^t |G(t-s)|\,ds$ is bounded and $|\mathbf{x}(t)| < 1$, we have $\int_0^t G(t-s)\mathbf{x}(s)\,ds$ bounded.

Thus, in (5.3.8) if $\mathbf{x}^2(t)$ is not in $L^1[0, \infty)$, then $\mathbf{x}(t)$ is unbounded. Hence, we suppose $\mathbf{x}^2(t)$ in L^1. Now $G(t) \to 0$ as $t \to \infty$ and $\mathbf{x}^2$ in L^1 implies $\int_0^t G(t-s)\mathbf{x}(s)\,ds \to 0$ as $t \to \infty$. In (5.3.8) we see that for large t, then $\mathbf{x}^T(t)D\mathbf{x}(t) \leq -\eta/2$. Moreover, as $\mathbf{x} \to \mathbf{0}$ it follows that $\mathbf{x}^T D\mathbf{x} \to 0$; hence, we conclude that $\mathbf{x}^2(t) \geq \gamma$ for some $\gamma > 0$ and all large t. This contradicts $\mathbf{x}^2(t)$ in L^1. The proof is now complete.

Our conditions on $C(t, s)$ have been fairly exacting and one would want the results to hold for equations similar to (5.3.1). Thus, we consider a perturbed form of (5.3.1).

Let U be an open set in R^n with $\mathbf{0} \in U$, let H_1, H_2 be $n \times n$ matrices of continuous functions of $\mathbf{x}$, and consider the system

$$\mathbf{x}' = A\mathbf{x} + A_1(t)\mathbf{x} + H_1(\mathbf{x})\mathbf{x}$$
$$+ \int_0^t \big[C(t, s) + C_1(t, s) + C_2(t, s)H_2(\mathbf{x}(s)) \big]\mathbf{x}(s)\,ds, \qquad (5.3.10)$$

in which A is a constant $n \times n$ matrix, A_1 an $n \times n$ matrix of functions continuous on $[0, \infty)$, and C_1 and C_2 $n \times n$ matrices continuous for $0 \leq s \leq t < \infty$. We suppose there are constants m_1, m_2, and J with

$$|C_1(t, s)| \leq m_1 |C(t, s)| \quad \text{and} \quad |A_1(t)| \leq m_2 \qquad (5.3.11)$$

and

$$|C_2(t, s)| \leq J|C(t, s)| \qquad (5.3.12)$$

for $0 \leq s \leq t < \infty$ and

$$H_i(\mathbf{0}) = 0, \quad i = 1, 2. \qquad (5.3.13)$$

Finally, we suppose there is a symmetric matrix B with

$$A^T B + BA = -I. \qquad (5.3.2)$$

Relative to (5.3.2), the reader should review Remark 5.1.2.

Theorem 5.3.3. *Let (5.3.2) and (5.3.10)–(5.3.13) hold. Suppose there is a constant $M > 0$ with*

$$|B|\left[\int_0^t |C(t, s)|\,ds + \int_t^\infty |C(u, t)|\,du \right] \leq M < 1. \qquad (5.3.3)$$

Then for the m_i sufficiently small, the zero solution of (5.3.10) is stable if and only if $\mathbf{x}^T B\mathbf{x} > 0$ for each $\mathbf{x} \neq \mathbf{0}$.

Note. The details on the size of the m_i are as follows. Because the $H_i(\mathbf{x})$ are continuous and $H_i(\mathbf{0}) = 0$, for each $\eta > 0$ there is a $\gamma > 0$ such that $|\mathbf{x}| < \gamma$ implies $|H_i(\mathbf{x})| < \eta$. Let $r = (m_1 + J\eta)|B|$ and pick r, m_i, and η small enough that, for some $\bar{M} < 1$, we have

$$|B| \left\{ 2m_2 + 2\eta + (1 + m_1 + J\eta) \int_0^t |C(t, s)| \, ds \right.$$

$$\left. + [1 + (r/|B|)] \int_t^\infty |C(u, t)| \, du \right\} \leq \bar{M}.$$

Proof of Theorem 5.3.3. Solutions of (5.3.10) with continuous initial functions in U will exist as long as they do not approach the boundary of U. Let $D = |B| + r$, $r > 0$ and to be determined, and define

$$V(t, \mathbf{x}(\cdot)) = \mathbf{x}^T B \mathbf{x} + D \int_0^t \int_t^\infty |C(u, s)| \, du \, \mathbf{x}^2(s) \, ds$$

$$V'_{(5.3.10)}(t, \mathbf{x}(\cdot))$$

$$= - \left\{ \mathbf{x}^T A^T + \mathbf{x}^T A_1^T + \mathbf{x}^T H_1^T + \int_0^t \mathbf{x}^T(s) \big[C^T(t, s) + C_1^T(t, s) \right.$$

$$+ H_2^T(\mathbf{x}(s)) C_2^T(t, s) \big] \, ds \left. \right\} B \mathbf{x} + \mathbf{x}^T B \left\{ A \mathbf{x} + A_1(t) \mathbf{x} + H_1(\mathbf{x}) \mathbf{x} \right.$$

$$+ \int_0^t \big[C(t, s) + C_1(t, s) + C_2(t, s) H_2(\mathbf{x}(s)) \big] \mathbf{x}^2(s) \, ds \left. \right\}$$

$$+ D \int_t^\infty |C(u, t)| \, du \, \mathbf{x}^2 - D \int_0^t |C(t, s)| \mathbf{x}^2(s) \, ds$$

$$\leq -\mathbf{x}^2 + 2|A_1(t)| \, |B| \mathbf{x}^2 + 2|H_1(\mathbf{x})| \, |B| \mathbf{x}^2$$

$$+ |B| \int_0^t \big[|C(t, s)| + |C_1(t, s)|$$

$$+ |C_2(t, s)| \, |H_2(\mathbf{x}(s))| \, \big[\mathbf{x}^2(s) + \mathbf{x}^2(t) \big] \big] \, ds$$

$$+ D \int_t^\infty |C(u, t)| \, du \, \mathbf{x}^2 - D \int_0^t |C(t, s)| \mathbf{x}^2(s) \, ds$$

$$\leq -\mathbf{x}^2 + 2|B| m_2 \mathbf{x}^2 + 2|B| \, |H_1(\mathbf{x})| \mathbf{x}^2$$

$$+ |B| \int_0^t \big(|C(t, s)| + m_1 |C(t, s)|$$

$$+ J|C(t, s)| \, |H_2(\mathbf{x}(s))| \big) \, \big[\mathbf{x}^2(s) + \mathbf{x}^2(t) \big] \, ds$$

$$+ D \int_t^\infty |C(u, t)| \, du \, \mathbf{x}^2 - D \int_0^t |C(t, s)| \mathbf{x}^2(s) \, ds$$

$$= \Bigg[-1 + |B| \bigg(2m_2 + 2|H_1(\mathbf{x})| $$

$$+ \int_0^t |C(t,s)| \left(1 + m_1 + J|H_2(\mathbf{x}(s))|\right) ds \bigg) \Bigg] \mathbf{x}^2$$

$$+ |B| \int_0^t |C(t,s)| \left[1 + m_1 + J|H_2(\mathbf{x}(s))|\right] \mathbf{x}^2(s) \, ds$$

$$+ D \int_t^\infty |C(u,t)| \, du \, \mathbf{x}^2 - D \int_0^t |C(t,s)| \mathbf{x}^2(s) \, ds .$$

Now, $H_i(\mathbf{x})$ are continuous and $H_i(\mathbf{0}) = 0$, so for each $\eta > 0$ there is a $\gamma > 0$ such that $|\mathbf{x}| \leq \gamma$ implies $|H_i(\mathbf{x})| < \eta$. Let $\eta > 0$ be given and find γ such that as long as $|\mathbf{x}(t)| < \gamma$ we have

$$V'(t, \mathbf{x}(\cdot)) \leq \bigg\{ -1 + |B| \bigg[2m_2 + 2\eta + \int_0^t |C(t,s)| \left(1 + m_1 + J\eta\right) ds$$

$$+ (D/|B|) \int_t^\infty |C(u,t)| \, du \bigg] \bigg\} \mathbf{x}^2$$

$$+ |B| \int_0^t |C(t,s)| \left(1 + m_1 + J\eta\right) \mathbf{x}^2(s) \, ds$$

$$- D \int_0^t |C(t,s)| \mathbf{x}^2(s) \, ds .$$

If $r = |B|(m_1 + J\eta)$, then

$$V'(t, \mathbf{x}(\cdot)) \leq \bigg\{ -1 + |B| \bigg[2m_2 + 2\eta + \int_0^t |C(t,s)| \left(1 + m_1 + J\eta\right) ds$$

$$+ (D/|B|) \int_t^\infty |C(u,t)| \, du \bigg] \bigg\} \mathbf{x}^2$$

$$\leq \{-1 + \bar{M}\} \mathbf{x}^2$$

as long as $|\mathbf{x}(s)| \leq \gamma$ for $0 \leq s \leq t$.

Suppose that $\mathbf{x}^T B \mathbf{x} > 0$ for all $\mathbf{x} \neq 0$, and let $\varepsilon > 0$ and $t_0 \geq 0$ be given. Assume $\varepsilon < \gamma$. Because B is positive definite, we may pick $d > 0$ with $d\mathbf{x}^2 \leq \mathbf{x}^T B \mathbf{x} \leq V(t, \mathbf{x}(\cdot))$. Then, for $\varepsilon > 0$ and $t_0 \geq 0$, we can find $\delta > 0$ such that $|\boldsymbol{\phi}(t)| < \delta$ on $[0, t_0]$ implies $V(t_0, \boldsymbol{\phi}(\cdot)) < \varepsilon^2 / d$. Thus, as long as $|\mathbf{x}(t)| \leq \gamma$, then $V'(t, \mathbf{x}(\cdot)) \leq 0$. And, as long as $V'(t, \mathbf{x}(\cdot)) \leq 0$, then

$$d\mathbf{x}^2(t) \leq \mathbf{x}^T(t) B \mathbf{x}(t) \leq V(t, \mathbf{x}(\cdot))$$

$$\leq V(t_0, \boldsymbol{\phi}(\cdot)) < d\varepsilon^2 ,$$

so that $|\mathbf{x}(t)| < \varepsilon$. That is, for $|\mathbf{x}(t)| \leq \gamma$, we have $V' \leq 0$ and for $V' \leq 0$, we have $|\mathbf{x}(t)| < \varepsilon$. Because $\varepsilon < \gamma$, it follows that $|\mathbf{x}(t)|$ will always remain smaller than γ. Thus, $\mathbf{x} = 0$ is stable.

Now, suppose there is an $\mathbf{x}_0 \neq 0$ with $\mathbf{x}_0^T B \mathbf{x}_0 \leq 0$. Argue as before that there is an $\mathbf{x}_0$ with $\mathbf{x}_0^T B \mathbf{x}_0 < 0$. By way of contradiction, we suppose $\mathbf{x} = \mathbf{0}$ is stable and pick $\varepsilon = \gamma$. Then take $t_0 = 0$ and select $\delta > 0$ for stability. Let $\mathbf{x}_0$ be chosen with $|\mathbf{x}_0| < \delta$ and $\mathbf{x}_0^T B \mathbf{x}_0 < 0$. Argue as in the proof of Theorem 5.3.1 that $|\mathbf{x}(t, 0, \mathbf{x}_0)| = \gamma$ for some $t_1 > 0$. That will complete the proof.

There is another interesting perturbation that we wish to study. Consider the system

$$\mathbf{x}' = A\mathbf{x} + \mathbf{f}(t, \mathbf{x}) + \int_0^t C(t, s)\mathbf{x}(s)\,ds\,, \tag{5.3.14}$$

with A and C as in (5.3.1),

$$A^T B + BA = -I \tag{5.3.2}$$

(taking into account Remark 5.1.2), $\mathbf{f} : [0, \infty) \times R^n \to R^n$ continuous, and

$$|\mathbf{f}(t, \mathbf{x})| \leq \lambda(t)(|\mathbf{x}| + 1)\,, \tag{5.3.15}$$

where $\lambda : [0, \infty) \to [0, \infty)$ is continuous,

$$\int_0^\infty \lambda(s)\,ds < \infty \quad \text{and} \quad \lambda(t) \to 0 \text{ as } t \to \infty\,. \tag{5.3.16}$$

Theorem 5.3.4.   *Suppose that (5.3.2), (5.3.3), (5.3.15), and (5.3.16) hold. All solutions of (5.3.14) are bounded if and only if $\mathbf{x}^T B \mathbf{x} > 0$ for each $\mathbf{x} \neq 0$.*

Proof. In the proof of Theorem 5.3.1 we found $V'_{1(5.3.1)}(t, \mathbf{x}(\cdot)) \leq -\alpha \mathbf{x}^2$, $\alpha > 0$. Select $L > 0$ so that

$$-\alpha \mathbf{x}^2 + 2|B|\,|\mathbf{x}|\,(|\mathbf{x}| + 1)\lambda(t) - L\lambda(t) \leq -\bar{\alpha}\mathbf{x}^2\,,$$

$\bar{\alpha} > 0$, for all $\mathbf{x}$ when t is large enough, say, $t \geq S$. Next, define

$V(t, \mathbf{x}(\cdot))$

$$= \left[\mathbf{x}^T B \mathbf{x} + 1 + |B| \int_0^t \int_t^\infty |C(u, s)|\,du\,\mathbf{x}^2(s)\,ds\right] \exp\left[-L \int_0^t \lambda(s)\,ds\right],$$

so that

$$V'_{(5.3.14)}(t, \mathbf{x}(\cdot))$$

$$\leq -L\lambda(t)V + \exp\left[-\int_0^t L\lambda(s)\,ds\right]\left[V'_{1(5.3.1)}(t, \mathbf{x}(\cdot)) + 2|B|\,|\mathbf{x}|\,|\mathbf{f}(t,\mathbf{x})|\right]$$

$$\leq -L\lambda(t)V + \exp\left[-L\int_0^t \lambda(s)\,ds\right]\left[-\alpha\mathbf{x}^2 + 2|B|\,|\mathbf{x}|\lambda(t)(|\mathbf{x}|+1)\right]$$

$$\leq \exp\left[-L\int_0^t \lambda(s)\,ds\right]\left[-\alpha\mathbf{x}^2 + 2|B|\,|\mathbf{x}|\lambda(t)(|\mathbf{x}|+1) - L\lambda(t)\right]$$

$$\leq -\beta\mathbf{x}^2\,, \quad \beta > 0 \text{ if } t \geq S\,.$$

Suppose that $\mathbf{x}_0^T B\mathbf{x}_0 > 0$ for all $\mathbf{x}_0 \neq \mathbf{0}$. If $\mathbf{x}(t)$ is any solution of (5.3.14), then, by the growth condition on $\mathbf{f}$, it can be continued for all future time. Hence, for $t \geq S$ we have $V(t, \mathbf{x}(t)) \leq V(S, \mathbf{x}(\cdot))$, so that $\mathbf{x}(t)$ is bounded.

Suppose that $\mathbf{x}_0^T B\mathbf{x}_0 \leq 0$ for some $\mathbf{x}_0 \neq \mathbf{0}$. Because B is independent of $\mathbf{f}$, one may argue that $\mathbf{x}_0^T B\mathbf{x}_0 < 0$ for some $\mathbf{x}_0$. Pick $t_0 = S$ and select ϕ on $[0, t_0]$ with $V(t_0, \phi) < 0$. Then $V'(t, \mathbf{x}(\cdot)) \leq 0$ implies $V(t, \mathbf{x}(\cdot)) \leq V(t_0, \phi(\cdot))$. One may argue that $|\mathbf{x}(t)|$ is bounded strictly away from zero, say, $|\mathbf{x}(t)| \geq \mu$, for some $\mu > 0$. As $\lambda(t) \to 0$, if $t \geq S$, then $V'(t, \mathbf{x}(\cdot)) \leq -\beta\mathbf{x}^2$, so $V'(t, \mathbf{x}(\cdot)) \leq -\beta\mu^2$ for large t. Thus, $V(t, \mathbf{x}(\cdot)) \to -\infty$ and $\mathbf{x}(t)$ is unbounded. This completes the proof.

Exercise 5.3.1. Review the proof and notice that when B is positive definite one may take L so large that the condition $\lambda(t) \to 0$ is not needed.

Exercise 5.3.2. Formulate and prove Theorems 5.3.3 and 5.3.4 for Eq. (5.3.4)

5.4 Complete Instability

In this section we focus on three facts relative to instability of Volterra equations.

First, when all characteristic roots of A have positive real parts, then an instability result analogous to Theorem 5.3.1 may be obtained by integrating only one coordinate of $C(t, s)$, as opposed to integrating both coordinates in (5.3.3).

Next, we point out the existence of complete instability in this case. Indeed, this also could have been done in Section 5.2 or 5.3.

Finally, we note that Volterra equations have a solution space that is far simpler than one might expect. Generally, complete instability is impossible for functional differential equations.

Consider the system

$$\mathbf{x}' = D\mathbf{x} + \int_0^t C(t,s)\mathbf{x}(s)\,ds\,, \tag{5.4.1}$$

where D is an $n \times n$ constant matrix whose characteristic roots all have positive real parts. Find the unique matrix $L = L^T$ that is positive definite and satisfies

$$D^T L + LD = I\,, \tag{5.4.2}$$

and find positive constants m and M with

$$|\mathbf{x}| \geq 2m(\mathbf{x}^T L\mathbf{x})^{1/2} \tag{5.4.3}$$

and

$$|L\mathbf{x}| \leq M(\mathbf{x}^T L\mathbf{x})^{1/2}\,. \tag{5.4.4}$$

Theorem 5.4.1. *Let (5.4.2)–(5.4.4) hold and let $\int_t^\infty |C(u,s)|\,du$ be continuous. Suppose there is an $H > M$ and $\gamma > 0$ with $m - H \int_t^\infty |C(u,t)|\,du \geq \gamma$. Then each solution $\mathbf{x}(t)$ of (5.4.1) on $[0,\infty)$ with $\mathbf{x}(0) \neq \mathbf{0}$ satisfies $|\mathbf{x}(t)| \geq c_1 + c_2 t$ for $0 \leq t < \infty$, where c_1 and c_2 are positive constants depending on $\mathbf{x}(0)$. Also, if $t_0 > 0$, then, for each $\delta > 0$, there is a continuous initial function $\boldsymbol{\phi} : [0,t_0] \to R^n$ with $|\boldsymbol{\phi}(t)| < \delta$ and $|\mathbf{x}(t,t_0,\boldsymbol{\phi})| \geq c_1 + c_2(t - t_0)$ for $t_0 \leq t < \infty$, where c_1 and c_2 are positive constants depending on $\boldsymbol{\phi}$ and t_0.*

Proof. Let $H > 0$, define

$$V(t,\mathbf{x}(\cdot)) = (\mathbf{x}^T L\mathbf{x})^{1/2} - H \int_0^t \int_t^\infty |C(u,s)|\,du\,|\mathbf{x}(s)|\,ds\,,$$

and for $\mathbf{x} \neq 0$, obtain

$$V'_{(5.4.1)}(t,\mathbf{x}(\cdot))$$

$$= \left\{ \left[\mathbf{x}^T D^T + \int_0^t \mathbf{x}^T(s)C^T(t,s)\,ds \right] L\mathbf{x} \right.$$

$$\left. + \mathbf{x}^T L \left[D\mathbf{x} + \int_0^t C(t,s)\mathbf{x}(s)\,ds \right] \right\} \Big/ \left\{ 2(\mathbf{x}^T L\mathbf{x})^{1/2} \right\}$$

$$- H \int_t^\infty |C(u,t)|\,du\,|\mathbf{x}| + H \int_0^t |C(t,s)|\,|\mathbf{x}(s)|\,ds$$

$$\geq \{\mathbf{x}^T \mathbf{x}/2(\mathbf{x}^T L \mathbf{x})^{1/2}\}$$

$$+ \int_0^t \{\mathbf{x}^T(t) L C(t,s) \mathbf{x}(s)/[\mathbf{x}^T(t) L \mathbf{x}(t)]^{1/2}\} \, ds$$

$$- H \int_t^\infty |C(u,t)| \, du \, |\mathbf{x}| + H \int_0^t |C(t,s)| \, |\mathbf{x}(s)| \, ds$$

$$\geq m|\mathbf{x}| - H \int_t^\infty |C(u,t)| \, du \, |\mathbf{x}|$$

$$+ H \int_0^t |C(t,s)| \, |\mathbf{x}(s)| \, ds \; - M \int_0^t |C(t,s)| \, |\mathbf{x}(s)| \, ds$$

$$= \left[m - H \int_t^\infty |C(u,t)| \, du \right] |\mathbf{x}| + (H - M) \int_0^t |C(t,s)| \, |\mathbf{x}(s)| \, ds$$

$$\geq \gamma|\mathbf{x}| + (H - M) \int_0^t |C(t,s)| \, |\mathbf{x}(s)| \, ds \,.$$

Hence, there is a $\mu > 0$ with

$$V'_{(5.4.1)}(t, \mathbf{x}(\cdot)) \geq \mu \big[|\mathbf{x}(t)| + |\mathbf{x}'(t)| \big] \,. \tag{5.4.5}$$

From the form of V and an integration of (5.4.5), for some $\alpha > 0$, we have

$$\alpha|\mathbf{x}(t)| \geq [\mathbf{x}^T(t) L \mathbf{x}(t)]^{1/2} \geq V(t, \mathbf{x}(\cdot))$$

$$\geq V(t_0, \boldsymbol{\phi}(\cdot)) + \int_{t_0}^t \mu|\mathbf{x}(s)| \, ds \,,$$

where $\mathbf{x}(t)$ is any solution of (5.4.1) on $[t_0, t)$ with $t_0 \geq 0$.
If $t_0 = 0$, then

$$|\mathbf{x}(t)| \geq \left\{ [\mathbf{x}^T(0) L \mathbf{x}(0)]^{1/2} + \int_0^t \mu|\mathbf{x}(s)| \, ds \right\} \Big/ \alpha$$

$$\geq [\mathbf{x}^T(0) L \mathbf{x}(0)]^{1/2}/\alpha \,,$$

so that

$$|\mathbf{x}(t)| \geq \left\{ [\mathbf{x}^T(0) L \mathbf{x}(0)]^{1/2} + t\mu[\mathbf{x}^T(0) L \mathbf{x}(0)]^{1/2}/\alpha \right\} / \alpha$$

$$\overset{\text{def}}{=} c_1 + c_2 t \,.$$

If $t_0 > 0$, select ϕ on $[0, t_0]$ with

$$[\boldsymbol{\phi}^T(t_0) L \boldsymbol{\phi}(t_0)]^{1/2} > \int_0^{t_0} \int_{t_0}^\infty |C(u,s)| \, du \, |\phi(s)| \, ds$$

and draw a conclusion, as before, to complete the proof.

Roughly speaking, a functional differential equation is one in which $\mathbf{x}'(t)$ depends explicitly on part or all of the past history of $\mathbf{x}(t)$. Such dependence is clear in (5.4.1). Explicit dependence is absent in

$$\mathbf{x}'(t) = \mathbf{f}(t, \mathbf{x}(t)) \,,$$

although it may become implicit through continual dependence of solutions on initial conditions.

Conceptually, one of the most elementary functional differential equations is a scalar delay equation of the form

$$x'(t) = ax(t) + bx(t-1) \,, \tag{5.4.6}$$

where a and b are constants with $b \neq 0$. Recall that we encountered a system of such equations in Section 4.1. To specify a solution we need a continuous initial function $\phi : [t_0 - 1, t_0] \to R$. We may then integrate

$$x'(t) = ax(t) + b\phi(t-1) \,, \quad x(t_0) = \phi(t_0)$$

on the interval $[t_0, t_0 + 1]$ to obtain a solution, say, $\psi(t)$. Now on the interval $[t_0, t_0 + 1]$ the function ψ becomes the initial function. We then solve

$$x'(t) = ax(t) + b\psi(t-1) \,, \quad x(t_0 + 1) = \psi(t_0 + 1) \,,$$

on the interval $[t_0 + 1, t_0 + 2]$. We may, in theory, continue this process to any point $T > t_0$. This is called the *method of steps* and it immediately yields existence and continuation of solutions. We can say, with much justice, that (5.4.6) is a completely elementary problem whose solution is within the reach of a college sophomore. Indeed, letting a and b be functions of t does not put the problem beyond our grasp.

By contrast, (5.4.1) is exceedingly complicated. Unless $C(t, s)$ is of such a special type that (5.4.1) can be reduced to an ordinary differential equation, there is virtually no hope of displaying a solution in terms of integrals, even on the interval $[0, 1]$.

Yet, it turns out that the solution space of (5.4.6) is enormously complicated. With a and b constant, try for a solution of the form $x = e^{rt}$ with r constant. Thus, $x' = re^{rt}$, so

$$re^{rt} = ae^{rt} + be^{r(t-1)}$$

or

$$r = a + be^{-r} \,, \tag{5.4.7}$$

which is called the *characteristic quasi-polynomial*. It is known that there is an infinite sequence $\{r_n\}$ of solutions of (5.4.7) [see El'sgol'ts(1966)].

Moreover, $\operatorname{Re} r_n \to -\infty$ as $n \to \infty$. Each function $x(t) = c e^{r_n t}$ is a solution for each constant c. Because we may let c be arbitrarily small, the zero solution cannot be completely unstable.

As simple as (5.4.6) may be, its solution space on $[0, \infty)$ is infinite-dimensional, whereas that of (5.4.1) on $[0, \infty)$ is finite-dimensional. This contributes to the contrast in degree of instability. The infinite-dimensionality would appear to have a stabilizing effect.

Roughly speaking, any n-dimensional linear and homogeneous, functional differential equation whose delay at some t_0 reduces to a single point and that enjoys unique solutions will have a finite-dimensional solution space starting at t_0. For example, the delay equation

$$x'(t) = ax(t) + bx[t - r(t)] \tag{5.4.8}$$

with $r(t)$ continuous, $r(t) \geq 0$, and $r(t_0) = 0$ for some t_0, should have exactly one linearly independent solution starting at t_0.

5.5 Non-exponential Decay

In this section we discuss work of J. Appleby and D. Reynolds on a linear scalar equation

$$z'(t) = -az(t) + \int_0^t k(t - s)z(s)\,ds\,, \quad t > 0\,, \quad z(0) = 1\,, \tag{5.5.1}$$

whose solutions decay slower than exponential. We make the assumption that k is a continuously differentiable, integrable function with $k(t) > 0$ for all $t \geq 0$. Then (5.5.1) has a unique continuous solution on $[0, \infty)$. It is known that $z \in L^1(0, \infty)$ if and only if $a > \int_0^\infty k(s)\,ds$, and that in this case $z(t) \to 0$ as $t \to \infty$. On the other hand, if $z(t) \to 0$ as $t \to \infty$, then $a \geq \int_0^\infty k(s)\,ds > 0$.

We ask that the kernel further satisfy

$$\lim_{t \to \infty} \frac{k'(t)}{k(t)} = 0\,, \tag{5.5.2}$$

which forces $k(t) \to 0$ as $t \to \infty$ more slowly than any decaying exponential. To see this, put $p(t) = k'(t)/k(t)$, and let $\varepsilon > 0$. Then there is a $T > 0$ such that $p(t) \geq -\varepsilon/2$ for all $t \geq T$. Since

$$k(t) = k(T)e^{\int_T^t p(s)\,ds}\,,$$

it follows by multiplying both sides by $e^{\varepsilon t}$, that $e^{\varepsilon t}k(t) \geq k(T)e^{\varepsilon(t-T)/2} \to \infty$ as $t \to \infty$. Hence (5.5.2) implies that

$$\lim_{t \to \infty} e^{\varepsilon t}k(t) = \infty\,, \quad \text{for every} \quad \varepsilon > 0\,. \tag{5.5.3}$$

Suppose that the solution of (5.5.1) obeys $z(t) \to 0$ as $t \to \infty$. Then it satisfies the ordinary differential equation

$$z'(t) = -az(t) + f(t), \quad t > 0,$$

with the forcing term given by

$$f(t) = \int_0^t k(t-s)z(s)\,ds \to 0 \quad \text{as} \quad t \to \infty.$$

Since $a > 0$ and the solution can be represented using the variation of parameters formula,

$$z(t) = e^{-at} + \int_0^t e^{-a(t-s)} f(s)\,ds, \quad t \geq 0, \qquad (5.5.4)$$

the asymptotic behaviour of $f(t)$ as $t \to \infty$ influences the rate at which $z(t) \to 0$ as $t \to \infty$. This is brought out in the proof of the following result.

Theorem 5.5.1.   *Suppose that k is an integrable and continuously differentiable function on $[0,\infty)$, with $k(t) > 0$ as $t \to \infty$. Moreover assume that $k'(t)/k(t) \to 0$ as $t \to \infty$. If the solution of (5.5.1) obeys $z(t) \to 0$ as $t \to \infty$, then*

$$\liminf_{t\to\infty} \frac{z(t)}{k(t)} \geq \frac{1}{a^2}. \qquad (5.5.5)$$

Consequently $\lim_{t\to\infty} e^{\varepsilon t} z(t) = \infty$ for every $\varepsilon > 0$.

Proof. Firstly note that $z(t) \geq e^{-at}$ for all $t \geq 0$. Since $z(0) = 1$,

$$t_0 = \inf\left\{t \in [0,\infty) : z(t) = 0\right\}.$$

Since $k(t) > 0$ and $z(t) > 0$ for all $0 \leq t < t_0$, $f(t) \geq 0$ for all $0 \leq t \leq t_0$. If t_0 is finite, it follows from (5.5.4) that

$$0 = z(t_0) = e^{-at_0} + \int_0^{t_0} e^{-a(t_0-s)} f(s)\,ds \geq e^{-at_0} > 0,$$

giving a contradiction. Therefore $z(t) > 0$ for all $t \geq 0$. Employing the positivity of k, $f(t) \geq 0$ for all $t \geq 0$, and hence (5.5.4) implies that $z(t) \geq e^{-at}$ for all $t \geq 0$. Consequently

$$f(t) = \int_0^t k(t-s)z(s)\,ds \geq \int_0^t k(t-s)e^{-as}\,ds.$$

Thus $f(t) \geq g(t)$ for all $t \geq 0$, where $g(t) = e^{-at} \int_0^t e^{as} k(s) \, ds$ is independent of z. Hence using (5.5.4) again,

$$z(t) \geq e^{-at} \int_0^t e^{as} f(s) \, ds \geq e^{-at} \int_0^t e^{as} g(s) \, ds, \quad t \geq 0,$$

and consequently, using the positivity of $k(t)$,

$$\frac{z(t)}{k(t)} \geq \frac{\int_0^t e^{as} g(s) \, ds}{e^{at} k(t)}, \quad t \geq 0. \tag{5.5.6}$$

By L'Hôpital's rule, (5.5.2) and (5.5.3),

$$\lim_{t \to \infty} \frac{g(t)}{k(t)} = \lim_{t \to \infty} \frac{\int_0^t e^{as} k(s) \, ds}{e^{at} k(t)} = \lim_{t \to \infty} \frac{1}{\left(\frac{k'(t)}{k(t)} + a \right)} = \frac{1}{a}.$$

Using L'Hôpital's rule again,

$$\lim_{t \to \infty} \frac{\int_0^t e^{as} g(s) \, ds}{e^{at} k(t)} = \lim_{t \to \infty} \frac{\frac{g(t)}{k(t)}}{\left(\frac{k'(t)}{k(t)} + a \right)} = \frac{1}{a^2}.$$

This and (5.5.6) establish that (5.5.5) holds. Due to (5.5.3) and (5.5.5),

$$z(t) e^{\varepsilon t} = \frac{z(t)}{k(t)} k(t) e^{\varepsilon t} \to \infty$$

as $t \to \infty$ if $\varepsilon > 0$, completing the proof.

We conclude with some remarks.

Remark 5.5.1. (5.5.3) implies that, for each $T > 0$,

$$\frac{k(t-s)}{k(t)} \to 1 \quad \text{as } t \to \infty, \text{ uniformly for } 0 \leq s \leq T. \tag{5.5.7}$$

If this condition is assumed instead of (5.5.3) and $a \geq \int_0^\infty k(s) \, ds$, the lower bound in (5.5.5) can be improved to

$$\liminf_{t \to \infty} \frac{z(t)}{k(t)} \geq \frac{1}{a \left(a - \int_0^\infty k(s) \, ds \right)},$$

where the right hand side is interpreted as infinity if $a = \int_0^\infty k(s) \, ds$ [see Appleby and Reynolds (2004)]. It turns out that (5.5.7) also implies (5.5.3).

Remark 5.5.2. Theorem 5.5.1 asserts that the solution z does not decay to zero faster than the kernel k. Positive, integrable, continuous functions satisfying (5.5.7) and

$$\frac{\int_0^t k(t-s)k(s)\,ds}{k(t)} \to 2\int_0^\infty k(s)\,ds \quad \text{as} \quad t \to \infty, \tag{5.5.8}$$

are called *subexponential* in Appleby and Reynolds (2002). It is shown in Appleby and Reynolds (2002, 2003) that if the kernel k is subexponential and $a > \int_0^\infty k(s)\,ds$, then z and k decay at exactly the same rate: indeed

$$\lim_{t\to\infty} \frac{z(t)}{k(t)} = \frac{1}{\left(a - \int_0^\infty k(s)\,ds\right)^2}\,.$$

Remark 5.5.3. At first glance the conditions (5.5.7) and (5.5.8) seem very restrictive. However if k is a positive, continuous and integrable function which obeys $k(\lambda t)k(t)^{-1} \to \lambda^\alpha$ as $t \to \infty$ for all $\lambda > 0$ for some $\alpha < -1$, then k is subexponential. An example is $k(t) = (1+t^2)^{-1}$. Another example outside this class is $k(t) = \exp(-(t+1)^\beta)$ with $0 < \beta < 1$.

Chapter 6

Stability and Boundedness

6.1 Stability Theory for Ordinary Differential Equations

Consider a system of ordinary differential equations

$$\mathbf{x}'(t) = \mathbf{G}(t, \mathbf{x}(t)), \quad \mathbf{G}(t, \mathbf{0}) = \mathbf{0}, \tag{6.1.1}$$

in which $\mathbf{G} : [0, \infty) \times D \to R^n$ is continuous and D is an open set in R^n with $\mathbf{0}$ in D.

We review the basic definitions for stability.

Definition 6.1.1. *The solution* $\mathbf{x}(t) = \mathbf{0}$ *of* (6.1.1) *is*

(a) stable *if, for each* $\varepsilon > 0$ *and* $t_0 \geq 0$, *there is a* $\delta > 0$ *such that*

$$|\mathbf{x}_0| < \delta \quad and \quad t \geq t_0$$

imply $|\mathbf{x}(t, t_0, \mathbf{x}_0)| < \varepsilon$,

(b) uniformly stable *if it is stable and* δ *is independent of* $t_0 \geq 0$,

(c) asymptotically stable *if it is stable and if, for each* $t_0 \geq 0$, *there is an* $\eta > 0$ *such that* $|\mathbf{x}_0| < \eta$ *implies* $|\mathbf{x}(t, t_0, \mathbf{x}_0)| \to 0$ *as* $t \to \infty$ (*If, in addition, all solutions tend to zero, then* $\mathbf{x} = \mathbf{0}$ *is* asymptotically stable in the large *or is* globally asymptotically stable.),

(d) uniformly asymptotically stable *if it is uniformly stable and if there is an* $\eta > 0$ *such that, for each* $\gamma > 0$, *there is a* $T > 0$ *such that*

$$|\mathbf{x}_0| < \eta, \quad t_0 \geq 0, \quad and \quad t \geq t_0 + T$$

imply $|\mathbf{x}(t, t_0, \mathbf{x}_0)| < \gamma$. (*If* η *may be arbitrarily large, then* $\mathbf{x} = \mathbf{0}$ *is* uniformly asymptotically stable in the large.)

171

Under suitable smoothness conditions on $\mathbf{G}$, all of the stability properties except (c) have been characterized by Liapunov functions.

Definition 6.1.2.   *A continuous function* $W : [0, \infty) \to [0, \infty)$ *with* $W(0) = 0$, $W(s) > 0$ *if* $s > 0$, *and* W *strictly increasing is called a* wedge. *(In this book wedges are always denoted by* W *or* W_i, *where* i *is an integer.)*

Definition 6.1.3.   *A function* $U : [0, \infty) \times D \to [0, \infty)$ *is called*

(a) positive definite *if* $U(t, \mathbf{0}) = 0$ *and if there is a wedge* W_1 *with* $U(t, \mathbf{x}) \geq W_1(|\mathbf{x}|)$,

(b) decrescent *if there is a wedge* W_2 *with* $U(t, \mathbf{x}) \leq W_2(|\mathbf{x}|)$,

(c) negative definite *if* $-U(t, \mathbf{x})$ *is positive definite.*

(d) radially unbounded *if* $D = R^n$ *and there is a wedge* $W_3(|\mathbf{x}|) \leq U(t, \mathbf{x})$ *and* $W_3(r) \to \infty$ *as* $r \to \infty$, *and*

(e) mildly unbounded *if* $D = R^n$ *and if, for each* $T > 0$, $U(t, \mathbf{x}) \to \infty$ *as* $|\mathbf{x}| \to \infty$ *uniformly for* $0 \leq t \leq T$.

Definition 6.1.4.   *A continuous function* $V : [0, \infty) \times D \to [0, \infty)$ *that is locally Lipschitz in* x *and satisfies*

$$V'_{(6.1.1)}(t, \mathbf{x}) = \limsup_{h \to 0^+} \left[V\left(t + h, \mathbf{x} + h\mathbf{G}(t, \mathbf{x})\right) - V(t, \mathbf{x}) \right]/h \leq 0 \quad (6.1.2)$$

on $[0, \infty) \times D$ *is called a* Liapunov function *for (6.1.1).*

If V has continuous first partial derivatives, then (6.1.2) becomes

$$V'(t, \mathbf{x}) = \operatorname{grad} V(t, \mathbf{x}) \cdot \mathbf{G}(t, \mathbf{x}) + \partial V / \partial t \leq 0.$$

Theorem 6.1.1.   *Suppose there is a Liapunov function* V *for (6.1.1).*

(a) *If* V *is positive definite, then* $\mathbf{x} = 0$ *is stable.*

(b) *If* V *is positive definite and decrescent, then* $x = 0$ *is uniformly stable.*

(c) *If* V *is positive definite and decrescent and* $V'_{(6.1.1)}(t, \mathbf{x})$ *is negative definite, then* $\mathbf{x} = \mathbf{0}$ *is uniformly asymptotically stable. Moreover, if* $D = R^n$ *and if* V *is radially unbounded, then* $\mathbf{x} = \mathbf{0}$ *is uniformly asymptotically stable in the large.*

(d) *If* $D = R^n$ *and if* V *is radially unbounded, then all solutions of (6.1.1) are bounded.*

(e) *If* $D = R^n$ *and if* V *is mildly unbounded, then each solution can be continued for all future time.*

Proof. (a)   We have $V'_{(6.1.1)}(t, \mathbf{x}) \leq 0$, V continuous, $V(t, \mathbf{0}) = 0$, and $W_1(|\mathbf{x}|) \leq V(t, \mathbf{x})$. Let $\varepsilon > 0$ and $t_0 \geq 0$ be given. We must find $\delta > 0$ such that

$$|\mathbf{x}_0| < \delta \quad \text{and} \quad t \geq t_0$$

imply $|\mathbf{x}(t, t_0, \mathbf{x}_0)| < \varepsilon$. (Throughout these proofs we assume ε so small that $|\mathbf{x}| < \varepsilon$ implies $\mathbf{x} \in D$.) As $V(t_0, \mathbf{x})$ is continuous and $V(t_0, \mathbf{0}) = 0$, there is a $\delta > 0$ such that $|\mathbf{x}| < \delta$ implies $V(t_0, \mathbf{x}) < W_1(\varepsilon)$. Thus, if $t \geq t_0$, then $V' \leq 0$ implies that, for $|\mathbf{x}_0| < \delta$ and $\mathbf{x}(t) = \mathbf{x}(t, t_0, \mathbf{x}_0)$, we have

$$\begin{aligned} W_1(|\mathbf{x}(t)|) &\leq V(t, \mathbf{x}(t)) \\ &\leq V(t_0, \mathbf{x}_0) < W_1(\varepsilon) \,, \end{aligned}$$

or $|\mathbf{x}(t)| < \varepsilon$ as required.

(b)   To prove uniform stability, for a given $\varepsilon > 0$, we select $\delta > 0$ such that $W_2(\delta) < W_1(\varepsilon)$, where $W_1(|\mathbf{x}|) \leq V(t, \mathbf{x}) \leq W_2(|\mathbf{x}|)$. Now, if $t_0 \geq 0$ and $|\mathbf{x}_0| < \delta$, then, for $\mathbf{x}(t) = \mathbf{x}(t, t_0, \mathbf{x}_0)$ and $t \geq t_0$, we have

$$\begin{aligned} W_1(|\mathbf{x}(t)|) &\leq V(t, \mathbf{x}(t) \leq V(t_0, \mathbf{x}_0) \\ &\leq W_2(|\mathbf{x}_0| < W_2(\delta) < W_1(\varepsilon) \,, \end{aligned}$$

or $|\mathbf{x}(t)| < \varepsilon$ as required.

(c)   The conditions for uniform stability are satisfied. Thus, for $\varepsilon = 1$ we find the δ of uniform stability and call it η in the definition of U.A.S. Now, let $\gamma > 0$ be given. We must find $T > 0$ such that

$$|\mathbf{x}_0| < \eta \,, \quad t_0 \geq 0 \,, \quad \text{and} \quad t \geq t_0 + T$$

imply $|\mathbf{x}(t, t_0, \mathbf{x}_0)| < \gamma$. Set $\mathbf{x}(t) = \mathbf{x}(t, t_0, \mathbf{x}_0)$. Pick $\mu > 0$ with $W_2(\mu) < W_1(\gamma)$, so that if there is a $t_1 \geq t_0$ with $|\mathbf{x}(t_1)| < \mu$, then, for $t \geq t_1$, we have

$$\begin{aligned} W_1(|\mathbf{x}(t)|) &\leq V(t, \mathbf{x}(t)) \leq V(t_1, \mathbf{x}(t_1)) \\ &\leq W_2(|\mathbf{x}(t_1)|) < W_2(\mu) < W_1(\gamma) \,, \end{aligned}$$

or $|\mathbf{x}(t)| < \gamma$. Now $V'(t, \mathbf{x}) \leq -W_3(|\mathbf{x}|)$, so that as long as $|\mathbf{x}(t)| \geq \mu$, then $V'(t, \mathbf{x}(t)) \leq -W_3(\mu)$; thus

$$\begin{aligned} V(t, \mathbf{x}(t)) &\leq V(t_0, \mathbf{x}_0) - \int_{t_0}^{t} W_3(|\mathbf{x}(s)|) \, ds \\ &\leq W_2(|\mathbf{x}_0|) - W_3(\mu)(t - t_0) \\ &\leq W_2(\eta) - W_3(\mu)(t - t_0) \,, \end{aligned}$$

which vanishes at

$$t = t_0 + W_2(\eta)/W_3(\mu) \stackrel{\text{def}}{=} t_0 + T \,.$$

Hence, if $t > t_0 + T$, then $|\mathbf{x}(t)| > \mu$ fails, and we have $|\mathbf{x}(t)| < \gamma$ for all $t \geq t_0 + T$. This proves U.A.S. The proof for U.A.S. in the large is accomplished in the same way.

(d) Because V is radially unbounded, we have $V(t, \mathbf{x}) \geq W_1(|\mathbf{x}|) \to \infty$ as $|\mathbf{x}| \to \infty$. Thus, given $t_0 \geq 0$ and $\mathbf{x}_0$, there is an $r > 0$ with $W_1(r) > V(t_0, \mathbf{x}_0)$. Hence, if $t \geq t_0$ and $\mathbf{x}(t) = \mathbf{x}(t, t_0, \mathbf{x}_0)$, then

$$W_1(|\mathbf{x}(t)|) \leq V(t, \mathbf{x}(t)) \leq V(t, \mathbf{x}_0) < W_1(r) \,,$$

or $|\mathbf{x}(t)| < r$.

(e) To prove continuation of solutions it will suffice to show that if $\mathbf{x}(t)$ is a solution on any interval $[t_0, T)$, then there is an M with $|\mathbf{x}(t)| < M$ on $[t_0, T)$. Now $V(t, \mathbf{x}) \to \infty$ as $|\mathbf{x}| \to \infty$ uniformly for $0 \leq t \leq T$. Thus, there is an $M > 0$ with $V(t, \mathbf{x}) > V(t_0, \mathbf{x}_0)$ if $0 \leq t \leq T$ and $|\mathbf{x}| > M$. Hence, for $0 \leq t < T$ we have $V(t, \mathbf{x}(t)) \leq V(t_0, \mathbf{x}_0)$, so that $|\mathbf{x}(t)| < M$.

The proof of Theorem 6.1.1 is complete.

Theorem 6.1.1 is an attempt to bring the book into focus as we look back at earlier chapters and forward to Chapters 6 and 8. The continuation question treated in Theorem 6.1.1(e) was considered in detail in Section 3.3 and will be seen throughout the remainder of the book. Next, in Chapters 2 and 5 we have seen examples of (a) and (b) extended to Volterra equations. For ordinary differential equations the concept of a Liapunov function being decrescent is simple, natural, and we can readily find examples. However, it is still not known what type of decrescent condition might be necessary and sufficient for asymptotic stability in this scheme. Suppose we have a Liapunov function which is positive definite, decrescent, and whose derivative is not positive. The decrescent condition does two things. First, it allows us to give an argument yielding uniform stability, as we have seen several times in Chapters 2 and 5. Involved in this is the fact that it allows us to show that a solution which gets close to zero will then stay close to zero; that property does not hold for functional differential equations.

For Volterra equations and general functional differential equations the decrescent concept becomes elusive and, to this day, a uniformly satisfactory form is unknown. The concept in (2.5.11) of $V' \leq -\mu[|\mathbf{x}| + |\mathbf{x}'|]$ was a major step in avoiding the decrescent question, as was an upcoming Marachkov condition which regulates the speed of a solution. Sections 8.3 and 8.7 are mainly devoted to additional steps in that direction.

We have chosen our wedges to simplify proofs. But that choice makes examples more difficult. One can define a wedge as a continuous function $W : D \to [0, \infty)$ with $W(\mathbf{0}) = 0$ and $W(\mathbf{x}) > 0$ if $\mathbf{x} \neq 0$. That choice makes examples easier, but proofs more difficult. The following device is helpful in constructing a wedge $W_1(|\mathbf{x}|)$ from a function $W(\mathbf{x})$.

Suppose $W : D \to [0, \infty)$, $D = \{ \mathbf{x} \in R^n : |\mathbf{x}| \leq 1 \}$, $W(\mathbf{0}) = 0$, and $W(\mathbf{x}) > 0$ if $\mathbf{x} \neq \mathbf{0}$. We suppose there is a function $V(t, \mathbf{x}) \geq W(\mathbf{x})$ and we wish to construct a wedge $W_1(|\mathbf{x}|) \leq V(t, \mathbf{x})$. First, define $\alpha(r) = \min_{r \leq |\mathbf{x}| \leq 1} W(\mathbf{x})$, so that $\alpha : [0, 1] \to [0, \infty)$ and α is nondecreasing. Next, define $W_1(r) = \int_0^r \alpha(s)\, ds$ and note that $W_1(0) = 0$, $W_1'(r) = \alpha(r) > 0$ if $r > 0$, and $W_1(r) \leq r\alpha(r) \leq \alpha(r)$. Thus, if $|\mathbf{x}_1| \leq 1$, then

$$V(t, \mathbf{x}_1) \geq W(\mathbf{x}_1) \geq \min_{|\mathbf{x}_1| \leq |\mathbf{x}| \leq 1} W(\mathbf{x}) = \alpha(|\mathbf{x}_1|) \geq W_1(|\mathbf{x}_1|) \,.$$

The next result is the fundamental boundedness result for Liapunov's direct method. It is our view that the extension of this result to Liapunov functionals for Volterra equations is one of the most important unsolved problems of the theory at this time.

Theorem 6.1.2. *Let $D = R^n$ and let $H = \{ \mathbf{x} \in R^n : |\mathbf{x}| \geq M, M > 0 \}$. Suppose that $V : [0, \infty) \times H \to [0, \infty)$ is continuous, locally Lipschitz in $\mathbf{x}$, radially unbounded, and $V'_{(6.1.1)}(t, \mathbf{x}) \leq 0$ if $|\mathbf{x}| \geq M$. If there is a constant $P > 0$ with $V(t, \mathbf{x}) \leq P$ for $|\mathbf{x}| = M$, then all solutions of (6.1.1) are bounded.*

Proof. As in the proof of Theorem 6.1.1(d), if a solution $\mathbf{x}(t)$ satisfies $|\mathbf{x}(t)| \geq M$ for all t, then it is bounded. Suppose $\mathbf{x}(t)$ is a solution with $|\mathbf{x}(t_1)| = M$ and $|\mathbf{x}(t)| \geq M$ on an interval $[t_1, T]$. Then

$$W_1(|\mathbf{x}(t)|) \leq V(t, \mathbf{x}(t)) \leq V(t_1, \mathbf{x}(t_1)) \leq P \,,$$

so that $|\mathbf{x}(t)| \leq W_1^{-1}(P)$ on $[t_1, T]$. As we may repeat this argument on any such interval $[t_1, T]$, it follows that $W_1^{-1}(P)$ is a future bound for any solution entering H^C. This completes the proof.

The notable part of Theorem 6.1.1 is that it contains no result on asymptotic stability that is not uniform. That problem is not solved to our satisfaction, although much has been written on it.

The following example is fundamental for understanding the difficulties in driving solutions to zero.

Example 6.1.1.   Let $g : [0, \infty) \to (0, 1]$ be a differentiable function with $g(n) = 1$ and $\int_0^\infty g(s)\,ds < \infty$. We wish to construct a function $V(t, x) = a(t)x^2$ with $a(t) > 0$ and with the derivative of V along any solution of

$$x' = \left[g'(t)/g(t)\right]x \qquad\qquad (6.1.3)$$

satisfying

$$V'(t, x) = -x^2 .$$

We shall, thereby, see that $V \geq 0$ and V' negative definite do not imply that solutions tend to zero, because $x(t) = g(t)$ is a solution of (6.1.3).

To this end, we compute

$$V'_{(6.1.3)}(t, x) = a'(t)x^2 + 2a(t)\left[g'(t)/g(t)\right]x^2$$

and set $V' = -x^2$. That yields

$$a'(t) + 2a(t)\left[g'(t)/g(t)\right] = -1$$

or

$$a'(t) = -2a(t)\left[g'(t)/g(t)\right] - 1$$

with the solution

$$a(t) = \left[a(0)g^2(0) - \int_0^t g^2(s)\,ds\right]\bigg/ g^2(t) .$$

Because $0 < g(t) \leq 1$ and g is in $L^1[0, \infty)$, we may pick $a(0)$ so large that $a(t) > 1$ on $[0, \infty)$. Notice that $V(t, x) \geq 0$ and positive definite; however, V is not decrescent.

The first real progress on the problem of asymptotic stability was made by Marachkov [see Antosiewicz (1958, Theorem 7, p. 149)].

Theorem 6.1.3.   Marachkov   *If* $\mathbf{G}(t, \mathbf{x})$ *is bounded for* $|\mathbf{x}|$ *bounded and if there is a positive definite Liapunov function for* (6.1.1) *with negative definite derivative, then the zero solutions of* (6.1.1) *is asymptotically stable.*

Proof. There is a function $V : [0, \infty) \times D \to [0, \infty)$ with $W_1(|\mathbf{x}|) \leq V(t, \mathbf{x})$ and $V'_{(6.1.1)}(t, \mathbf{x}) \leq -W_2(|\mathbf{x}|)$ for wedges W_1 and W_2. Also, there is a constant P with $\mathbf{G}(t,\mathbf{x})| \leq P$ if $|\mathbf{x}| \leq m$, where m is chosen so that $|\mathbf{x}| \leq m$ implies $\mathbf{x}$ is in D.

Because V is positive definite and $V' \leq 0$, then $\mathbf{x} = 0$ is stable. To show asymptotic stability, let $t_0 \geq 0$ be given and let $W_1(m) = \alpha > 0$. Because $V(t_0,\mathbf{x})$ is continuous and $V(t_0,\mathbf{0}) = 0$, there is an $\eta > 0$ such that $|\mathbf{x}_0| < \eta$ implies $V(t_0,\mathbf{x}_0) < \alpha$. Now for $\mathbf{x}(t) = \mathbf{x}(t,t_0,\mathbf{x}_0)$, we have $V'(t,\mathbf{x}(t)) \leq 0$, so

$$W_1(|\mathbf{x}(t)|) \leq V(t,\mathbf{x}(t)) \leq V(t_0,\mathbf{x}_0) < W_1(m) \,,$$

implying $|\mathbf{x}(t)| < m$ if $t \geq t_0$. Notice that $V'(t,\mathbf{x}(t)) \leq -W_2(|\mathbf{x}(t)|)$, so that

$$0 \leq V(t,\mathbf{x}(t)) \leq V(t_0,\mathbf{x}_0) - \int_{t_0}^{t} W_2(|\mathbf{x}(s)|)\,ds \,,$$

from which we conclude that there is a sequence $\{t_n\} \to \infty$ with $\mathbf{x}(t_n) \to 0$.

Now if $\mathbf{x}(t) \nrightarrow 0$, there is an $\varepsilon > 0$ and a sequence $\{s_n\}$ with $|\mathbf{x}(s_n)| \geq \varepsilon$ and $s_n \to \infty$. But because $\mathbf{x}(t_n) \to 0$ and $\mathbf{x}(t)$ is continuous, there is a pair of sequences $\{U_n\}$ and $\{J_n\}$ with $U_n < J_n < U_{n+1}$, $|\mathbf{x}(U_n)| = \varepsilon/2$, $|\mathbf{x}(J_n)| = \varepsilon$, and $\varepsilon/2 \leq |\mathbf{x}(t)| \leq \varepsilon$ if $U_n \leq t \leq J_n$. Integrating (6.1.1) from U_n to J_n we have

$$\mathbf{x}(J_n) = \mathbf{x}(U_n) + \int_{U_n}^{J_n} G(s,\mathbf{x}(s))\,ds \,,$$

so that

$$\varepsilon/2 \leq |\mathbf{x}(J_n) - \mathbf{x}(U_n)| \leq P(J_n - U_n)$$

or

$$J_n - U_n \geq \varepsilon/2P \,.$$

Also, if $t > J_n$, then

$$0 \leq V(t,\mathbf{x}(t))$$
$$\leq V(t_0,\mathbf{x}_0) - \int_{t_0}^{t} W_2(|\mathbf{x}(s)|)\,ds$$
$$\leq V(t_0,\mathbf{x}_0) - \sum_{i=1}^{n} \int_{U_i}^{J_i} W_2(|\mathbf{x}(s)|)\,ds$$
$$\leq V(t_0,\mathbf{x}_0) - \sum_{i=1}^{n} \int_{U_i}^{J_i} W_2(\varepsilon/2)\,ds$$
$$\leq V(t_0,\mathbf{x}_0) - nW_2(\varepsilon/2)\varepsilon/2P \to -\infty$$

as $n \to \infty$, a contradiction.

Definition 6.1.5. *The argument given in the final paragraph of the proof of Theorem 6.1.3 is called the* annulus argument.

That argument is central to 40 years of research on conditions needed to conclude asymptotic stability in ordinary and functional differential equations. Marachkov's work was done in 1940 and was extended in 1952 by Barbashin and Krasovskii [see Barbashin (1968, p. 1099)] in a very significant manner that allows V' to be zero on certain sets. Somewhat similar extensions were given independently by La Salle, Levin and Nohel, and Yoshizawa. The following is essentially Yoshizawa's formulation of one of those results, all of which, incidentally, will play a central role in construction of Liapunov functionals for Volterra equations.

For our purposes a scalar function $f : R^n \to [0, \infty)$ is *positive definite with respect to a set A* if $f(\mathbf{x}) = 0$ for $\mathbf{x} \in A$ and for each $\varepsilon > 0$ and each compact set Q in R^n there exists $\delta = \delta(Q, \varepsilon)$ such that $f(\mathbf{x}) \geq \delta$ for $\mathbf{x} \in Q \cap U(A, \varepsilon)^c$, where $U(A, \varepsilon)$ is the ε-neighborhood of A.

Theorem 6.1.4. Yoshizawa (1963)   *Let $D = R^n$ and let $\mathbf{G}(t, \mathbf{x})$ be bounded for $\mathbf{x}$ bounded. Also suppose that all solutions of (6.1.1) are bounded. If there is a continuous function $V : [0, \infty) \times R^n \to [0, \infty)$ that is locally Lipschitz in $\mathbf{x}$, if there is a continuous function $W : R^n \to [0, \infty)$ that is positive definite with respect to a closed set Ω, and if $V'_{(6.1.1)}(t, \mathbf{x}) \leq -W(\mathbf{x})$, then every solution of (6.1.1) approaches Ω as $t \to \infty$.*

Proof. Consider a solution $\mathbf{x}(t)$ on $[t_0, \infty)$ that, being bounded, remains in some compact set Q for $t \geq t_0$. If $\mathbf{x}(t) \not\to \Omega$, then there is an $\varepsilon > 0$ and a sequence $\{t_n\} \to \infty$ with $\mathbf{x}(t_n) \in U(\Omega, \varepsilon)^c \cap Q$. Because $\mathbf{G}(t, \mathbf{x})$ is bounded for $\mathbf{x}$ in Q, there is a K with $|\mathbf{G}(t, \mathbf{x}(t))| \leq K$. Thus, there is a $T > 0$ with $\mathbf{x}(t) \in U(\Omega, \varepsilon/2)^c \cap Q$ for $t_n \leq t \leq t_n + T$. By taking a subsequence, if necessary, we may suppose these intervals disjoint. Now, for this $\varepsilon/2$ there is a $\delta > 0$ with

$$V'(t, \mathbf{x}) \leq -\delta \quad \text{on} \quad [t_n, t_n + T].$$

Thus for $t \geq t_n + T$ we have

$$0 \leq V(t, \mathbf{x}(t)) \leq V(t_0, \mathbf{x}(t_0)) - \int_{t_0}^t W(\mathbf{x}(s))\, ds$$

$$\leq V(t_0, \mathbf{x}(t_0)) - \sum_{i=1}^n \int_{t_i}^{t_i + T} W(\mathbf{x}(s))\, ds$$

$$\leq V(t_0, \mathbf{x}(t_0)) - nT\delta \to -\infty,$$

a contradiction.

Usually, boundedness is proved by showing V to be radially unbounded. Also, we understand from the proof that the requirement that all solutions be bounded can be dropped and the conclusion changed to read that all bounded solutions approach Ω. Moreover, some authors let $V : [0, \infty) \times R^n \to (-\infty, \infty)$ and ask that V be bounded from below for $\mathbf{x}$ bounded, concluding again that bounded solutions approach Ω [see Haddock (1974)].

Example 6.1.2. Consider the scalar system (i.e., x and y are scalars)

$$x' = y,$$
$$y' = -(x^2 + 1)y - x \tag{6.1.4}$$

with Liapunov function $V(x, y) = x^2 + y^2$, so that

$$V'_{(6.1.4)}(x, y) = -2(x^2 + 1)y^2 \stackrel{\text{def}}{=} -W(x, y)$$

and Ω is the x axis. Because V is radially unbounded and (6.1.4) is bounded for (x, y) bounded, all solutions are bounded and approach the x axis.

When (6.1.1) is independent of t, say,

$$\mathbf{x}' = \mathbf{G}(\mathbf{x}) \tag{6.1.5}$$

and solutions are unique, then much more can be said.

A point y is an *ω-limit point* of a solutions $\mathbf{x}(t)$ of (6.1.5) if there is a sequence $\{t_n\} \to \infty$ with $\mathbf{x}(t_n) \to \mathbf{y}$. The set of ω-limit points of a solution of (6.1.5) is called the *ω-limit set.* By uniqueness, if $\mathbf{y}$ is in the ω-limit set of $\mathbf{x}(t)$, then the orbit through $\mathbf{y}$, say,

$$\left\{ \mathbf{z} \in R^n : \mathbf{z} = \mathbf{x}(t, 0, \mathbf{y}), \ t \geq 0 \right\}$$

is also in the ω-limit set. (Actually, this follows from continual dependence on initial conditions, which, in turn, follows from uniqueness.)

A set A is *positively invariant* if $\mathbf{y} \in A$ implies $\mathbf{x}(t, 0, \mathbf{y}) \in A$ for $t \geq 0$.

Theorem 6.1.5. *Let the conditions of Theorem 6.1.4 hold for (6.1.5) and let $V = V(\mathbf{x})$. Also, let M be the largest invariant set in Ω. Then every solution of (6.1.5) approaches M as $t \to \infty$.*

Proof. If $\mathbf{x}(t)$ is a solution of (6.1.5), then it approaches Ω. Suppose there is a point $\mathbf{y}$ in the ω-limit set of $\mathbf{x}(t)$ not in M. Certainly, $\mathbf{y} \in \Omega$, and as $\mathbf{y} \notin M$, there is a $t_1 > 0$ with $\mathbf{x}(t_1, 0, \mathbf{y}) \notin \Omega$. Also, there is a sequence $\{t_n\} \to \infty$ with $\mathbf{x}(t_n) \to \mathbf{x}(t_1, 0, \mathbf{y})$, a contradiction to $\mathbf{x}(t) \to \Omega$ as $t \to \infty$. This completes the proof.

The result can be refined further by noticing that $V(\mathbf{x}(t)) \to c$ so the set M is restricted still more by satisfying $V(\mathbf{x}) = c$ for some $\mathbf{c} \geq 0$.

The ideas in the last two theorems were extended by Hale to autonomous functional differential equations using Liapunov functionals and by Haddock and Terjéki using a Razumikhin technique. These will be discussed in Chapter 8. They were also extended to certain classes of partial differential equations by Dafermos.

Example 6.1.3. Consider Example 6.1.2 once more with Ω being the x axis. Notice that if a solution starts in Ω with $x_1 \neq 0$, then $y' = -x_1 \neq 0$, so the solution leaves Ω. Hence, $M = \{(0,0)\}$.

Theorems 6.1.4 and 6.1.5 frequently enable us to conclude asymptotic stability (locally or in the large) using a "poor" Liapunov function. But when (6.1.1) is perturbed, we need a superior Liapunov function so we can analyze the behavior of solutions. For example, suppose $D = R^n$ and there is a continuous function $V : [0, \infty) \times R^n \to [0, \infty)$ with

(a) $\left| V(t, \mathbf{x}_1) - V(t, \mathbf{x}_2) \right| \leq K |\mathbf{x}_1 - \mathbf{x}_2|$ on $[0, \infty) \times R^n$ with K constant,

(b) $V'_{(6.1.1)}(t, \mathbf{x}) \leq -cV(t, \mathbf{x})$, $c > 0$, and

(c) $V(t, \mathbf{x}) \geq W_1(|\mathbf{x}|) \to \infty$ as $|\mathbf{x}| \to \infty$.

Then for a perturbed form of (6.1.1), say,

$$\mathbf{x}' = \mathbf{G}(t, \mathbf{x}) + \mathbf{F}(t, \mathbf{x}) \tag{6.1.6}$$

with $\mathbf{G}, \mathbf{F} : [0, \infty) \times R^n \to R^n$ being continuous, we have

$$\begin{aligned} V'_{(6.1.6)}(t, \mathbf{x}) &\leq V'_{(6.1.1)}(t, \mathbf{x}) + K|\mathbf{F}(t, \mathbf{x})| \\ &\leq -cV(t, \mathbf{x}) + K|\mathbf{F}(t, \mathbf{x})| , \end{aligned}$$

so that

$$\begin{aligned} V(t, \mathbf{x}(t)) \leq {} & V(t_0, \mathbf{x}(t_0)) e^{-c(t - t_0)} \\ & + \int_{t_0}^{t} e^{-c(t - s)} K|\mathbf{F}(s, \mathbf{x}(s))| \, ds , \end{aligned} \tag{6.1.7}$$

which is a "poor man's" variation of parameter formula.

Although it may be extremely hard to find a V satisfying (a), (b), and (c), once we establish that (6.1.1) is uniformly asymptotically stable in the large (possibly by using Theorems 6.1.4 and 6.1.5), then the next result assures us that this superior V does exist. And, although it does not specify

c exactly, it does claim that $c > 0$ so that (6.1.7) will yield boundedness when $\mathbf{F}$ is bounded; additional results are obtained when $|\mathbf{F}(t, \mathbf{x})| \leq \lambda(t)$, where λ is a continuous function tending to zero or in $L^p[0, \infty)$.

The following result may be found in Yoshizawa (1966, p. 100). We offer it here without proof.

Theorem 6.1.6. *Consider*

$$\mathbf{x}' = \mathbf{G}(t, \mathbf{x}), \quad \mathbf{G}(t, \mathbf{0}) = \mathbf{0} \tag{6.1.1}$$

with $\mathbf{G} : [0, \infty) \times R^n \to R^n$ *continuous and locally Lipschitz in* $\mathbf{x}$. *If* $\mathbf{x} = \mathbf{0}$ *is uniformly asymptotically stable in the large, then there is a* $V : [0, \infty) \times R^n \to [0, \infty)$ *with*

$$W_1(|\mathbf{x}|) \leq V(t, \mathbf{x}) \leq W_2(|\mathbf{x}|), \tag{a}$$

where the W_i *are wedges and* $W_1(r) \to \infty$ *as* $r \to \infty$, *such that*

$$V'_{(6.1.1)}(t, \mathbf{x}) \leq -cV(t, \mathbf{x}), \quad c > 0. \tag{b}$$

If, in addition, for any compact set $K \subset R^n$ *there is a constant* $L(K) > 0$ *such that* $\left|\mathbf{G}(t, \mathbf{x}_1) - \mathbf{G}(t, \mathbf{x}_2)\right| \leq L(K)|\mathbf{x}_1 - \mathbf{x}_2|$ *whenever* $(t, \mathbf{x}_i) \in [0, \infty) \times K$, *then*

$$\left|V(t, \mathbf{x}_1) - V(t, \mathbf{x}_2)\right| \leq h(K)|\mathbf{x}_1 - \mathbf{x}_2|, \tag{c}$$

where $h(K)$ *is a constant depending on* K.

There are two important alternatives to asking that $|\mathbf{G}(t, \mathbf{x})|$ be bounded for x bounded when we seek to establish asymptotic stability for a V with a negative-definite derivative. These two alternatives are very effective for Volterra equations.

Suppose there is a Liapunov function $V : R^n \to [0, \infty)$ for (6.1.1). Notice that

$$\begin{aligned} V'_{(6.1.1)}(\mathbf{x}) &= \operatorname{grad} V(\mathbf{x}) \cdot \mathbf{G}(t, \mathbf{x}) \\ &= \left|\operatorname{grad} V(\mathbf{x})\right| \left|\mathbf{G}(t, \mathbf{x})\right| \cos\theta, \end{aligned} \tag{6.1.8}$$

where θ is the angle between the vectors $\operatorname{grad} V(\mathbf{x})$ and $\mathbf{G}(t, \mathbf{x})$ and $| \cdot |$ denote Euclidean length. If V is shrewdly chosen, it may be possible to deduce that

$$V'_{(6.1.1)}(\mathbf{x}) \leq -\delta|\mathbf{G}(t, \mathbf{x})| \quad \delta > 0. \tag{6.1.9}$$

In that case, we have

$$0 \leq V(\mathbf{x}(t)) \leq V(\mathbf{x}(t_0)) - \delta \int_{t_0}^{t} |\mathbf{x}'(s)| \, ds, \tag{6.1.10}$$

which implies that any solution has finite arc length. The annulus argument (see Definition 6.1.5) would become trivial in that case. Note, however, that

(6.1.10) alone does not imply that solutions tend to zero. For example, $x' = 0$ and $V(x) = x^2$ satisfy (6.1.9) and (6.1.10), but all solutions are constant.

We shall find that in Theorems 6.1.3 and 6.1.4 we may replace $|\mathbf{G}(t, \mathbf{x})|$ bounded for $|\mathbf{x}|$ bounded by

$$V'_{(6.1.1)}(t, \mathbf{x}) \leq -\delta |\mathbf{G}(t, \mathbf{x})|, \quad \delta > 0.$$

Example 6.1.4. Let

$$\mathbf{x}' = A\mathbf{x}, \tag{6.1.11}$$

where A is an $n \times n$ constant, real matrix whose characteristic roots all have negative, real parts. Let

$$V(x) = [\mathbf{x}^T B \mathbf{x}]^{1/2},$$

where $B^T = B$ and

$$A^T B + BA = -I. \tag{6.1.12}$$

Note that the zero solution is unique and that V has continuous first partial derivatives for $\mathbf{x} \neq 0$. Thus

$$V'_{(6.1.11)}(\mathbf{x}) = \left(\mathbf{x}^T A^T B \mathbf{x} + \mathbf{x}^T B A \mathbf{x}\right) / \left\{2[\mathbf{x}^T B \mathbf{x}]^{1/2}\right\}$$
$$= -\mathbf{x}^T \mathbf{x} / \left\{2[\mathbf{x}^T B \mathbf{x}]^{1/2}\right\},$$

and there is a $k > 0$ with

$$|\mathbf{x}| / \left\{2[\mathbf{x}^T B \mathbf{x}]^{1/2}\right\} \leq k,$$

so that

$$V'_{(6.1.11)}(\mathbf{x}) \leq -k|\mathbf{x}| \leq -\delta |A\mathbf{x}|, \quad \delta > 0. \tag{6.1.13}$$

These ideas have been developed extensively by Erhart (1973), Haddock (1977a,b), Hatvani (1978) and Burton (1977). They were employed in the proof of Theorem 2.5.1.

In many results, such as Theorem 6.1.3, we can weaken the condition $V'(t, x) \leq -W(|x|)$. Here is a typical way.

Definition 6.1.6. *A scalar function $p : [0, \infty) \to [0, \infty)$ is said to be integrally positive if for each $\delta > 0$ and each sequence $\{t_n\} \to \infty$ monotonically,*

$$\liminf_{n \to \infty} \int_{t_n}^{t_n + \delta} p(t)\, dt > 0.$$

Thus, we are allowed to have $p(t) = \sin^2 t$, but $p(t) = |\sin t| + \sin t$ would not qualify as being integrally positive.

Exercise 6.1.1. In Theorem 6.1.3, drop the condition that the derivative of V is negative definite. Instead, ask that $V'(t,x) \leq -p(t)W(|x|)$ where p is integrally positive. Show that the zero solution is still asymptotically stable.

6.2   Construction of Liapunov Functions

Beyond any doubt, construction of Liapunov functions is an art. But like any other art, there are guidelines and there are masters to emulate. Whereas the previous section concentrated on formal theorems concerning consequences of Liapunov functions, this section contains a detailed account of the construction of somewhat special Liapunov functions. Such constructions are fundamental in the construction of Liapunov functionals for Volterra equations.

A. $\mathbf{x}' = A\mathbf{x}$

As we have seen, given

$$\mathbf{x}' = A\mathbf{x} \tag{6.2.1}$$

we try $V(\mathbf{x}) = \mathbf{x}^T B\mathbf{x}$ with $B = B^T$ and

$$A^T B + BA = -I \tag{6.2.2}$$

when no characteristic root of A has a zero real part. [In Chapter 5 we explored the possibility of solving (6.2.2).]

Also, if all characteristic roots of A have a zero real part and if the elementary divisors are simple, then the equation

$$A^T B + BA = 0 \tag{6.2.3}$$

may be solved for $B = B^T$. The reader may wish to try this for

$$A = \begin{pmatrix} 0 & 1 \\ -1 & 0 \end{pmatrix}.$$

Equation (6.2.3) has important consequences for perturbed forms of (6.2.1).

B. $x' = y,\ y' = f(x, y)y - g(x)$

Long before Liapunov, the mathematician Lagrange noted that equilibrium was stable when the total energy of the system was at a minimum. That idea, applied to the scalar equation

$$x'' + f(x, x')x' + g(x) = 0 \tag{6.2.4}$$

with $f(x, x') > 0$ and $xg(x) > 0$ for $x \neq 0$, produced the Liapunov function

$$V(x, y) = \frac{1}{2}\, y^2 + \int_0^x g(s)\, ds \tag{6.2.5}$$

for the system of the form

$$\begin{aligned}
x' &= y, \\
y' &= -f(x, y)y - g(x).
\end{aligned} \tag{6.2.6}$$

The result is

$$V'_{(6.2.6)}(x, y) = -f(x, y)y^2 \leq 0. \tag{6.2.7}$$

Equation (6.2.4) may be thought of as a fairly general statement of Newton's second law of motion for an object with one degree of freedom. Equations (6.2.5)–(6.2.7) generated scores, if not hundreds, of research articles between 1940 and the present. Bibliographies may be found in the work of Graef (1972) and Burton-Townsend (1971).

C. $x' = y,\ y' = -c(t)f(x)$

It was recognized very early in the development of Liapunov's direct method that a first integral might serve as a Liapunov function. For example, consider

$$x'' + x^3 = 0 \tag{6.2.8}$$

and the equivalent system

$$\begin{aligned}
x' &= y, \\
y' &= -x^3.
\end{aligned} \tag{6.2.9}$$

We have

$$dy/dx = -x^3/y,$$

so that

$$y \, dy + x^3 \, dx = 0 \,,$$

yielding

$$2y^2 + x^4 = \text{constant} \,.$$

If we take

$$V(x, y) = 2y^2 + x^4 \,,$$

then

$$V'_{(6.2.9)}(x, y) = 0 \,.$$

Because V is positive definite and $V' \leq 0$, the zero solution is stable.
   It is but a small jump, then, to try the Liapunov function

$$V(t, x, y) = 2y^2 + c(t)x^4$$

for the system

$$\begin{aligned} x' &= y \,, \\ y' &= -c(t)x^3 \end{aligned} \tag{6.2.10}$$

when $c(t) \geq c_0 > 0$ and $c'(t) \leq 0$.
   From there it is natural to deduce that a Liapunov function for

$$\begin{aligned} x' &= y \,, \\ y' &= -c(t)f(x) \,, \end{aligned} \tag{6.2.11}$$

with $xf(x) > 0$, if $x \neq 0$ and differentiable $c(t) > 0$, may be obtained as follows. Write $c(t) = a(t)b(t)$ with $a(t)$ nondecreasing and $b(t)$ nonincreasing. Then

$$V(t, x, y) = \left[y^2/2b(t)\right] + a(t) \int_0^x f(s) \, ds \tag{6.2.12}$$

is a Liapunov function, from which investigators have derived reams of information. For a bibliography consult Burton-Grimmer (1972).

D. $V' \leq f(t, V)$

Corduneanu was the first to notice that "poor" Liapunov functions might be made into "good" ones by use of differential inequalities. Consider (6.2.10) again where $c(t) > 0$ but $c'(t) \leq 0$ fails. We let

$$V(t, x, y) = 2y^2 + c(t)x^4$$

and find

$$V'_{(6.2.10)}(t, x, y) = c'(t)x^4 \leq \left[c'_+(t)/c(t)\right] V(t, x, y) \,,$$

where $c'_+(t) = \max[c'(t), 0]$. Thus, an integration yields

$$V(t, x, y) = V(t_0, x_0, y_0) \exp \int_{t_0}^{t} \left[c'_+(s)/c(s)\right] ds \,,$$

implying V bounded if the integral is bounded. For extensive use of differential inequalities see Lakshmikantham and Leela (1969).

E. $x = G(t, x) + F(t, x)$

If $V(t, \mathbf{x})$ is globally Lipschitz in $\mathbf{x}$ for a constant L with the derivative of V along

$$\mathbf{x}' = \mathbf{G}(t, \mathbf{x}) \tag{6.2.13}$$

satisfying $V'_{(6.2.13)}(t, \mathbf{x}) \leq 0$, then one may perturb (6.2.13) and write

$$\mathbf{x}' = \mathbf{G}(t, \mathbf{x}) + \mathbf{F}(t, \mathbf{x}) \,, \tag{6.2.14}$$

where $|\mathbf{F}(t, \mathbf{x})| \leq \lambda(t)$ and $\int_0^\infty \lambda(t)\, dt < \infty$. Then use

$$W(t, \mathbf{x}) = [V(t, \mathbf{x}) + 1] \exp \left[- L \int_0^t \lambda(s)\, ds \right].$$

We have $W'_{(6.2.14)}(t, \mathbf{x}) \leq 0$, so that if V is radially unbounded, so is W and solutions of (6.2.14) are bounded.

A continuation result is obtained in the same way, as is seen in Section 3.3, Theorem 3.3.3. For if V is a mildly unbounded function satisfying $V'_{(6.2.13)}(t, \mathbf{x}) \leq 0$ and if $|\mathbf{F}(t, \mathbf{x})| \leq \lambda(t)$ with λ continuous, then $W'_{(6.2.14)}(t, \mathbf{x}) \leq 0$ and W is mildly unbounded. Hence, solutions of (6.2.14) are continuable. This is an important principle. If $\mathbf{G}$ is smooth enough and if all solutions of (6.2.13) are continuable, the converse theorems show the existence of a mildly unbounded V with $V'_{(6.2.13)}(t, \mathbf{x}) \leq 0$. Hence, we deduce that, for V Lipschitz, continuability of (6.2.13) plus $|\mathbf{F}(t, \mathbf{x})| \leq \lambda(t)$ imply continuability of (6.2.14). For converse theorems on continuability see Kato and Strauss (1967).

F.   First Integral Solutions

We see that investigators began with a much simplified equation, obtained a first integral, and used that first integral as a spring board to attack their actual problem. That progress can be clearly seen by reviewing scores of papers that proceed from

$$x'' + g(x) = 0, \quad xg(x) > 0,$$

through the series

$$x'' + f(x)x' + g(x) = 0, \qquad f(x) \geq 0,$$
$$x'' + h(x, x')x' + g(x) = 0, \qquad h(x, x') \geq 0,$$
$$x'' + h(x, x')x' + g(x) = e(t), \quad e(t + T) = e(t),$$

and

$$x'' + k(t, x, x')x' + a(t)g(x) = e(t, x, x'),$$

e bounded, $k \geq 0$, and $a(t) > 0$. For bibliographies see Graef (1972), Sansone-Conti (1964), and Reissig *et al.* (1963).

G. $x' = P(x, y),\ y' = Q(x, y)$

Nor is one restricted to a first integral of a given system. From the point of view of subsequent perturbations the very best Liapunov functions are obtained as follows. Consider a pair of first-order scalar equations

$$x' = P(x, y),$$
$$y' = Q(x, y),$$

$$(6.2.15)$$

so that

$$dy/dx = Q(x, y)/P(x, y).$$

Then the orthogonal trajectories are obtained from

$$dy/dx = -P(x, y)/Q(x, y)$$

or

$$P(x, y)\, dx + Q(x, y)\, dy = 0.$$

If we can find an integration factor $\mu(x, y)$ so that

$$\mu(x, y)P(x, y)\, dx + \mu(x, y)Q(x, y)\, dy = 0$$

is exact, then there is a function $V(x, y)$ with $\partial V/\partial x = \mu P$ and $\partial V/\partial y = \mu Q$, so that

$$V'_{(6.2.15)}(x, y) = \mu(x, y)\left[P^2(x, y) + Q^2(x, y)\right]. \qquad (6.2.16)$$

If V and μ are each of one sign and $V\mu \leq 0$, then $\pm V$ is a Liapunov function for (6.2.15). Moreover, if we review Eqs. (6.1.8)–(6.1.10), we have

$$V'_{(6.2.15)}(x, y) = \left|\operatorname{grad} V(x, y)\right|\left|(P(x, y), Q(x, y))\right|\cos\theta, \qquad (6.2.17)$$

and because V is obtained from the orthogonal trajectories, we have

$$\cos\theta = \pm 1. \qquad (6.2.18)$$

For this reason, (6.2.15) can be perturbed with comparatively large functions without disturbing stability properties of the zero solution.

H. $\mathbf{x}' = A\mathbf{x} + \mathbf{b}f(\sigma),\ \sigma' = \mathbf{c}^T\mathbf{x} - rf(\sigma)$

In view of (6.2.2) and (6.2.12), one can quickly see how to proceed with the $(n + 1)$-dimensional control problem

$$\begin{aligned}
\mathbf{x}' &= A\mathbf{x} + \mathbf{b}f(\sigma), \\
\sigma' &= \mathbf{c}^T\mathbf{x} - rf(\sigma),
\end{aligned} \qquad (6.2.19)$$

in which A is an $n \times n$ matrix of constants whose characteristic roots all have negative real parts, $\mathbf{b}$ and $\mathbf{c}$ are constant vectors, r is a positive constant, σ and f are scalars, and $\sigma f(\sigma) > 0$ if $\sigma \neq 0$. This is called the problem of Lurie and it concerns automatic control devices. The book by Lefschetz (1965) is devoted entirely to it and considers several interesting Liapunov functions. Lurie used the Liapunov function

$$V(\mathbf{x}, \sigma) = \mathbf{x}^T B\mathbf{x} + \int_0^\sigma f(s)\, ds, \qquad (6.2.20)$$

in which $B = B^T$ and $A^T B + BA = -D$, where $D = D^T$ is positive definite. Then we have

$$V'_{(6.2.19)}(\mathbf{x}, \sigma) = -\mathbf{x}^T D\mathbf{x} + f(\sigma)[2\mathbf{b}^T B + \mathbf{c}^T]\mathbf{x} - rf^2(\sigma).$$

And Lefschetz (1965) showed that this is negative definite if and only if

$$r > (B\mathbf{b} + \mathbf{c}/2)^T D^{-1}(B\mathbf{b} + \mathbf{c}/2). \qquad (6.2.21)$$

It is interesting to see how (6.2.19) and (6.2.20) are modified to take into account the time delay, which is always present, in the feedback system. Such modifications were done by Somolinos (1977).

I. $\mathbf{x}' = A(t)\mathbf{x}$

It is natural to attempt to investigate

$$\mathbf{x}' = A(t)\mathbf{x} \tag{6.2.22}$$

in the same way that (6.2.1) was treated. Suppose that A is an $n \times n$ matrix of functions continuous for $0 \leq t < \infty$. A common procedure may be described as follows.

If all characteristic roots of $A(t)$ have negative real parts for every value of $t \geq 0$, then for each t the equation

$$A^T(t)B(t) + B(t)A(t) = -I \tag{6.2.23}$$

may be uniquely solved for a positive definite matrix $B(t) = B^T(t)$. For brevity, let us suppose $B(t)$ is differentiable on $[0, \infty)$. We then seek a differentiable function $b : [0, \infty) \to [0, \infty)$ such that

$$V(t, \mathbf{x}) = b(t)\mathbf{x}^T B(t)\mathbf{x} \tag{6.2.24}$$

will be a Liapunov function for (6.2.22). Thus,

$$\begin{aligned}
V'_{(6.2.22)}(t, \mathbf{x}) &= b'(t)\mathbf{x}^T B(t)\mathbf{x} + b(t)\mathbf{x}^T [A^T B + BA + B']\mathbf{x} \\
&= \mathbf{x}^T \big[b(t)(A^T B + BA + B') + b'(t)B \big]\mathbf{x} \\
&\overset{\text{def}}{=} b(t)\mathbf{x}^T H(t)\mathbf{x},
\end{aligned}$$

where

$$H(t) = -I + B'(t) + [b'(t)/b(t)]B(t).$$

If we take $\bar{\alpha}(t)$ to be the largest root of the equation

$$\det \big[-I + B'(t) + \alpha(t)B(t) \big] = 0,$$

and

$$\beta(t) = b(0) \exp \left[-\int_0^t \bar{\alpha}(s)\, ds \right],$$

then the condition

$$[b'(t)/b(t)] < [\beta'(t)/\beta(t)] = -\bar{\alpha}(t)$$

is necessary and sufficient for $H(t)$ to have only negative characteristic roots. In that case, stability and asymptotic stability may be determined from V' and V. For more details see Lebedev (1957), Hahn (1963, pp. 29–32), and Krasovskii (1963, pp. 56–62).

J. $\mathbf{x}' = \mathbf{F}(\mathbf{x})$

The most common method of attack on a nonlinear system

$$\mathbf{x}' = \mathbf{F}(\mathbf{x}), \quad \mathbf{F}(\mathbf{0}) = \mathbf{0}, \tag{6.2.25}$$

is by way of the linear approximation. If $\mathbf{F}$ is differentiable at $\mathbf{x} = \mathbf{0}$, then it may be approximated by a linear function there. One may write (6.2.25) as

$$\mathbf{x}' = A\mathbf{x} + \mathbf{G}(\mathbf{x}), \tag{6.2.26}$$

in which A is the Jacobian matrix of $\mathbf{F}$ at $\mathbf{x} = \mathbf{0}$ and $\lim_{\mathbf{x} \to 0} |G(\mathbf{x})|/|\mathbf{x}| = 0$. For example, if $f(x) = f(x_1, \ldots, x_n)$ is a differentiable scalar function at $\mathbf{x} = \mathbf{0}$, then

$$f(\mathbf{x}) = f(\mathbf{0}) + (\partial f / \partial x_1)x_1 + (\partial f / \partial x_2)x_2$$
$$+ \cdots + (\partial f / \partial x_n)x_n + \text{ higher-order terms,}$$

where the partials are evaluated at $\mathbf{x} = \mathbf{0}$. One expands each component of $\mathbf{F}$ in this way and selects the matrix A from the coefficients of the x_i.

It is more efficient to consider

$$\mathbf{x}' = A\mathbf{x} + \mathbf{H}(t, \mathbf{x}), \tag{6.2.27}$$

where A is a constant $n \times n$ matrix, $H : [0, \infty) \times D \to R^n$ is continuous, D is an open set in R^n with 0 in D, and

$$\lim_{|\mathbf{x}| \to 0} |\mathbf{H}(t, \mathbf{x})|/|\mathbf{x}| = 0 \quad \text{uniformly for } 0 \le t < \infty. \tag{6.2.28}$$

Theorem. Liapunov   *If (6.2.27) and (6.2.28) hold and if all characteristic roots of A have negative real parts, then the zero solution of (6.2.27) is uniformly asymptotically stable.*

Proof. By our assumption on A we can solve $A^T B + BA = -I$ for a unique positive definite matrix $B = B^T$. We form

$$V(\mathbf{x}) = \mathbf{x}^T B \mathbf{x}$$

and obtain

$$\begin{aligned} V'_{(6.2.27)}(\mathbf{x}) &= (\mathbf{x}^T A^T + H^T)B\mathbf{x} + \mathbf{x}^T B(A\mathbf{x} + H) \\ &= -\mathbf{x}^T \mathbf{x} + 2H^T B\mathbf{x} \\ &\le -|\mathbf{x}|^2 + 2|H| \, |B| \, |\mathbf{x}|, \end{aligned}$$

so that for $\mathbf{x} \neq 0$ we have

$$V'(\mathbf{x})/|\mathbf{x}|^2 \leq -1 + 2|B|\,|\mathbf{H}(t,\mathbf{x})|/|\mathbf{x}| < -1/2\,,$$

if $|\mathbf{x}|$ is small enough, in consequence of (6.2.28). The conditions of Theorem 6.1.1(c) are satisfied and $\mathbf{x} = 0$ is U.A.S.

K. $A(x) = \int_0^1 J(sx)\,ds$

Much may be lost by evaluating the Jacobian of $\mathbf{F}$ in (6.2.25) only at $\mathbf{x} = \mathbf{0}$. If we write the Jacobian of $\mathbf{F}$ as $J(\mathbf{x}) = (\partial F_i / \partial x_k)$, evaluated at $\mathbf{x}$, then for

$$A(\mathbf{x}) = \int_0^1 J(s\mathbf{x})\,ds$$

we have

$$\mathbf{F}(\mathbf{x}) = A(\mathbf{x})\mathbf{x}\,.$$

Investigators have discovered many simple Liapunov functions from $A(\mathbf{x})$ yielding global stability. A summary may be found in Hartman (1964, pp. 537–555). Excellent collections of Liapunov functions for specific equations are found in the work of Reissig *et al.* (1963) and Barbashin (1968).

6.3   A First Integral Liapunov Functional

We consider a system of Volterra equations

$$\mathbf{x}' = A(t)\mathbf{x} + \int_0^t C(t,s)\mathbf{x}(s)\,ds\,, \tag{6.3.1}$$

with A and C being $n\times n$ matrices continuous on $[0,\infty)$ and $0 \leq s \leq t < \infty$, respectively.

To arrive at a Liapunov functional for (6.3.1), integrate it from 0 to t and interchange the order of integration to obtain

$$\mathbf{x}(t) = \mathbf{x}(0) + \int_0^t A(s)\mathbf{x}(s)\,ds + \int_0^t \int_s^t C(u,s)\,du\,\mathbf{x}(s)\,ds\,.$$

We then have

$$\mathbf{h}(t,\mathbf{x}(\cdot)) = \mathbf{x}(t) + \int_0^t \left[-A(s) - \int_s^t C(u,s)\,du \right]\mathbf{x}(s)\,ds\,, \tag{6.3.2}$$

which is identically equal to $\mathbf{x}(0)$. Hence, the derivative of $\mathbf{h}$ along a solution of (6.3.1) is zero. It is reasonable to think of $\mathbf{h}$ as a first integral functional for (6.3.1). Compare this with Eqs. (6.2.8)–(6.2.12) for constructing Liapunov functions.

Now **h** may serve as a suitable Liapunov functional for (6.3.1) as it stands. Moreover, the changes necessary to convert **h** to an outstanding Liapunov functional are quite minimal.

Suppose that (6.3.1) is scalar,

$$A(t) \leq 0, \; C(t,s) \geq 0, \quad \text{and} \quad -A(s) - \int_s^t C(u,s)\, du \geq 0$$

for $0 \leq s \leq t < \infty$. Consider solutions of (6.3.1) on the entire interval $[0, \infty)$ (as opposed to solutions on some $[t_0, \infty)$ with $t_0 > 0$). Because $-x(t)$ is a solution whenever $x(t)$ is a solution, we need only consider solutions $x(t)$ with $x(0) > 0$. Notice that when $x(0) > 0$ and $C(t,s) \geq 0$, the solutions all remain positive. Hence, *along these solutions* the scalar equation

$$h(t, x(\cdot)) = x(t) + \int_0^t \left[-A(s) - \int_s^t C(u,s)\, du \right] x(s)\, ds$$

is a positive definite functional. In fact, we may write it as

$$H(t, x(\cdot)) = |x(t)| + \int_0^t \left[|A(s)| - \int_s^t |C(u,s)|\, du \right] |x(s)|\, ds, \quad (6.3.3)$$

and the derivative of H along these solutions of (6.3.1) is zero. Under the conditions of this paragraph, we see that solutions of (6.3.1) are bounded. However, much more can be said. Notice that if

$$|A(s)| - \int_s^t |C(u,s)|\, du \geq \alpha > 0,$$

then boundedness of H implies that $x(t)$ must be $L^1[0, \infty)$.

Definition 6.3.1. *A scalar functional $H(t, \mathbf{x}(\cdot))$ expands relative to zero if there is a $t_1 \geq 0$ and $\alpha > 0$ such that if $|\mathbf{x}(t)| \geq \alpha$ on $[t_2, \infty)$ with $t_2 \geq t_1$, then $H(t, \mathbf{x}(\cdot)) \to \infty$ as $t \to \infty$.*

We formally state and prove these observations.

Theorem 6.3.1. *Let (6.3.1) be a scalar equation with $A(s) \leq 0$ and $|A(s)| - \int_s^t |C(u,s)|\, du \geq 0$ for $0 \leq s \leq t < \infty$. Then the zero solution of (6.3.1) is stable. If, in addition, there is a $t_2 \geq 0$ and an $\alpha > 0$ with $|A(s)| - \int_s^t |C(u,s)|\, du \geq \alpha$ for $t_2 \leq s \leq t < \infty$ and both $\int_0^t |C(t,s)|\, ds$ and $A(t)$ are bounded, then $x = 0$ is asymptotically stable.*

Proof. Let $x(t) = x(t, t_0, \phi)$ be any solution of (6.3.1). We compute

$$H'_{(6.3.1)}(t, x(\cdot)) \leq A(t)|x| + \int_0^t |C(t, s)| \, |x(s)| \, ds$$

$$+ |A(t)| \, |x| - \int_0^t |C(t, s)| \, |x(s)| \, ds$$

$$\equiv 0 \, .$$

(Notice that once we formed H, then it was no longer necessary to ask $C(t, s) \geq 0$, nor that only solutions on $[0, \infty)$ be considered.) The stability is now clear; for if we are given $\varepsilon > 0$ and $t_0 \geq 0$, we let $\phi : [0, t_0] \to R$ be continuous and satisfy $|\phi(s)| < \delta$ on $[0, t_0]$, where δ is to be determined. Then for $x(t) = x(t, t_0, \phi)$ and $t > t_0$ we have

$$|x(t)| \leq H(t, x(\cdot))$$
$$\leq H(t_0, \phi(\cdot))$$
$$\leq |\phi(t_0)| + \int_0^{t_0} \left[|A(s)| - \int_s^{t_0} |C(u, s)| \, du \right] |\phi(s)| \, ds$$
$$\leq \delta \left\{ 1 + \int_0^{t_0} \left[|A(s)| - \int_s^{t_0} |C(u, s)| \, du \right] ds \right\}$$
$$< \varepsilon$$

from which $\delta = \delta(\varepsilon, t_0)$ is readily obtained.

If t_2 and α exist, then

$$H(t_0, \phi(\cdot)) \geq H(t, x(\cdot))$$

$$\geq |x(t)| + \int_{t_2}^t \alpha |x(s)| \, ds$$

for $t \geq t_2$, so that x is in $L^1[0, \infty)$. As $\int_0^t |C(t, s)| \, ds$, $A(t)$, and $x(t)$ are bounded, it follows that $x'(t)$ is bounded. Hence, $x(t) \to 0$. This completes the proof.

We recall from Section 6.1 that there are two alternatives to asking $x'(t)$ bounded. Whereas the requirement that $\int_0^t |C(t, s)| \, ds$ be bounded is consistent with the other assumptions, the requirement that $A(t)$ be bounded is not only severe but it conflicts with the intuition that the more negative $A(t)$ is, the more stable (6.3.1) should be.

Let us return to the vector equation (6.3.2). If we wish to pass from (6.3.2) to a scalar functional analogous to (6.3.3), we have several options for the norms and each option will yield different results.

Let us suppose there is a constant positive definite matrix $D = D^T$ and a continuous scalar function $\mu : [0, \infty) \to [0, \infty)$ with

$$\mathbf{x}^T [A^T D + DA] \mathbf{x} \le -\mu(t) \mathbf{x}^T \mathbf{x}. \tag{6.3.4}$$

The norm we will take on the solution $\mathbf{x}(t)$ will be $[\mathbf{x}^T D \mathbf{x}]^{1/2}$ and bounds will be needed. There are positive constants s, k, and K with

$$|\mathbf{x}| \ge 2k [\mathbf{x}^T D \mathbf{x}]^{1/2}, \tag{6.3.5}$$

$$|D\mathbf{x}| \le K [\mathbf{x}^T D \mathbf{x}]^{1/2}, \tag{6.3.6}$$

and

$$s|\mathbf{x}| \le [\mathbf{x}^T D \mathbf{x}]^{1/2}. \tag{6.3.7}$$

With this norm, if we replace (6.3.3) by

$$P(t, \mathbf{x}(\cdot)) = [\mathbf{x}^T D \mathbf{x}]^{1/2} + \int_0^t \left[|A(s)| - \int_s^t |C(u, s)| \, du \right] |\mathbf{x}(s)| \, ds$$

and differentiate along solutions of (6.3.1), then we readily see that we need to refine P and write

$$P(t, \mathbf{x}(\cdot)) = [\mathbf{x}^T D \mathbf{x}]^{1/2} + \int_0^t \left[k\mu(s) - K \int_s^t |C(u, s)| \, du \right] |\mathbf{x}(s)| \, ds, \tag{6.3.8}$$

with

$$k\mu(s) - \int_s^t K|C(u, s)| \, du \ge 0 \quad \text{for} \quad 0 \le s \le t < \infty. \tag{6.3.9}$$

Theorem 6.3.2. *Let (6.3.4)–(6.3.9) hold.*

(a) *The zero solution of (6.3.1) is stable.*

(b) *If $P(t, \mathbf{x}(\cdot))$ expands relative to zero and if $\mathbf{x}'$ is bounded for $\mathbf{x}$ bounded, then $\mathbf{x} = \mathbf{0}$ is asymptotically stable.*

(c) *If there is an $M > 0$ with*

$$\int_0^t \left[k\mu(s) - K \int_s^t |C(u, s)| \, du \right] ds \le M \tag{6.3.10}$$

for $0 \le t < \infty$, then $\mathbf{x} = \mathbf{0}$ is uniformly stable.

Proof. For $\mathbf{x} \neq \mathbf{0}$ we compute

$$
\begin{aligned}
P'_{(6.3.1)}(t, \mathbf{x}(\cdot)) = \Bigg\{ & \left[\mathbf{x}^T A^T + \int_0^t \mathbf{x}^T(s) C^T(t,s)\, ds \right] D\mathbf{x} \\
& + \mathbf{x}^T D \left[A\mathbf{x} + \int_0^t C(t,s)\mathbf{x}(s)\, ds \right] \Bigg\} \Big/ \{ 2[\mathbf{x}^T D\mathbf{x}]^{1/2} \} \\
& + k\mu(t)|\mathbf{x}| - \int_0^t K|C(t,s)|\,|\mathbf{x}(s)|\, ds \\
\leq & -\{ \mu(t)\mathbf{x}^T \mathbf{x}/2[\mathbf{x}^T D\mathbf{x}]^{1/2} \} \\
& + \int_0^t |C(t,s)|\,|\mathbf{x}(s)|\, ds \, \{ |D\mathbf{x}|/[\mathbf{x}^T D\mathbf{x}]^{1/2} \} \\
& + k\mu(t)|\mathbf{x}| - \int_0^t K|C(t,s)|\,|\mathbf{x}(s)|\, ds \leq 0\,.
\end{aligned}
$$

Stability is clear and asymptotic stability follows [See the proof of Theorem 6.3.1].

For uniform stability, let $\varepsilon > 0$ be given and $\delta > 0$ still to be determined. If $t_0 \geq 0$, if $\phi : [0, t_0] \to R^n$ with $|\phi(s)| < \delta$ on $[0, t_0]$, and if $\mathbf{x}(t) = \mathbf{x}(t, t_0, \phi)$, then for $t \geq t_0$ we have

$$
\begin{aligned}
s|\mathbf{x}(t)| \leq P(t, \mathbf{x}(\cdot)) & \leq P(t_0, \phi(\cdot)) \\
& \leq [\phi(t_0)^T D\phi(t_0)]^{1/2} + M\delta \\
& \leq [\delta/2k] + M\delta = [(1/2k) + M]\delta < \varepsilon
\end{aligned}
$$

if $\delta < \varepsilon/[(1/2k) + M]$. This completes the proof.

Three things attract our attention concerning Theorem 6.3.2. First, in view of our work with

$$
\mathbf{x}' = A(t)\mathbf{x} \tag{6.2.22}
$$

and the selection of $B(t)$ by

$$
A^T(t)B(t) + B(t)A(t) = -I\,, \tag{6.2.23}
$$

followed by defining

$$
V(t, \mathbf{x}) = b(t)\mathbf{x}^T B(t)\mathbf{x}\,, \tag{6.2.24}
$$

it seems clear that one may replace $[x^T Dx]^{1/2}$ in (6.3.8) by $[b(t)x^T B(t)x]^{1/2}$ and generalize Theorem 6.3.2. We leave the details as an exercise.

Next, to secure (6.3.10), we could replace $\mu(s)$ by

$$(K/k) \int_s^\infty |C(u,s)| \, du \,,$$

so that the integrand in (6.3.10) becomes

$$K \left[\int_s^\infty |C(u,s)| \, du - \int_s^t |C(u,s)| \, du \right] = K \int_t^\infty |C(u,s)| \, du$$

and (6.3.8) becomes

$$W(t, \mathbf{x}(\cdot)) = [\mathbf{x}^T D \mathbf{x}]^{1/2} + K \int_0^t \int_t^\infty |C(u,s)| \, du \, |\mathbf{x}(s)| \, ds \,. \qquad (6.3.11)$$

Hypotheses (b) and (c) of Theorem 6.3.2 are mutually exclusive. However, by replacing $\mu(s)$ as we did, the derivative of W may become negative. Thus, we may obtain asymptotic stability and uniform stability at the same time.

Finally, if we write W as

$$V(t, \mathbf{x}(\cdot)) = [\mathbf{x}^T D \mathbf{x}]^{1/2} + \bar{K} \int_0^t \int_t^\infty |C(u,s)| \, du \, |\mathbf{x}(s)| \, ds \qquad (6.3.12)$$

with $\bar{K} > K$, then we may be able to drop the requirement that $\mathbf{x}'$ be bounded for $\mathbf{x}$ bounded and perform the simplified annulus argument as noted following Eq. (6.1.10). However, with $A(t)$ variable there may still be problems with the annulus argument. Those problems will evaporate when we consider the one-sided Lipschitz conditions introduced in (6.1.16) and (6.1.17) and to be developed in Definition 6.4.1.

The foregoing explanation shows in detail how we arrive at the Liapunov functional used to prove Theorem 2.5.1. The reader is urged to review Theorem 2.5.1 and its proof carefully at this time. Moreover, the functional

$$\int_0^t \int_t^\infty |C(u,s)| \, du \, |\mathbf{x}(s)| \, ds$$

of (6.3.11) turns out to be a fundamental part of each Liapunov functional, with, at most, minimal changes needed.

The method outlined for constructing a Liapunov functional for the linear system can be extended without difficulty to nonlinear equations.

Consider the system

$$\mathbf{x}' = \mathbf{g}(t, \mathbf{x}) + \int_0^t \mathbf{p}(t, s, \mathbf{x}(s)) \, ds \,, \tag{6.3.13}$$

in which $\mathbf{g}$ and $\mathbf{p}$ are continuous when $\mathbf{g} : [0, \infty) \times U \to R^n$, $\mathbf{p} : [0, \infty) \times [0, \infty) \times U \to R^n$, and $U = \{\mathbf{x} \in R^n : |x| < \varepsilon \,, \ \varepsilon > 0\}$. We integrate (6.3.13) from 0 to t and interchange the order of integration to obtain

$$\begin{aligned}
\mathbf{x}(t) &= \mathbf{x}(0) + \int_0^t \mathbf{g}(s, \mathbf{x}(s)) \, ds + \int_0^t \int_0^u \mathbf{p}(u, s, \mathbf{x}(s)) \, ds \, du \\
&= \mathbf{x}(0) + \int_0^t \mathbf{g}(s, \mathbf{x}(s)) \, ds + \int_0^t \int_s^t \mathbf{p}(u, s, \mathbf{x}(s)) \, du \, ds \\
&= \mathbf{x}(0) + \int_0^t \left[\mathbf{g}(s, \mathbf{x}(s)) + \int_s^t \mathbf{p}(u, s, \mathbf{x}(s)) \, du \right] ds \,,
\end{aligned}$$

so that

$$\begin{aligned}
\mathbf{r}(t, \mathbf{x}(\cdot)) &= \mathbf{x}(t) + \int_0^t \left[- \mathbf{g}(s, \mathbf{x}(s)) - \int_s^t \mathbf{p}(u, s, \mathbf{x}(s)) \, du \right] ds \\
&= \mathbf{x}(0)
\end{aligned}$$

and hence $\mathbf{r}'_{(6.3.13)}(t, \mathbf{x}(\cdot)) \equiv 0$. The same sequence following Eq. (6.3.2) may be repeated. Briefly, in the scalar case we write

$$R(t, x(\cdot)) = |x| + \int_0^t \left[|g(s, x(s))| - \int_s^t |p(u, s, x(s))| \, du \right] ds \,. \tag{6.3.14}$$

If $xg(t, x) \le 0$, then we obtain

$$\begin{aligned}
R'_{(6.3.13)}(t, \mathbf{x}(\cdot)) &\le -|g(t, x)| + \int_0^t |p(t, s, x(s))| \, ds \\
&\quad + |g(t, x)| - \int_0^t |p(t, s, x(s))| \, ds \equiv 0 \,.
\end{aligned}$$

If

$$|g(s, x(s))| \ge \int_s^t |p(u, s, x(s))| \, ds$$

for $0 \le s \le t < \infty$ and x an arbitrary continuous function, $x : [0, \infty) \to U$, then the zero solution of (6.3.13) is stable. If, in addition, the functional expands relative to zero, then asymptotic stability may be obtained. Finally, under proper convergence assumptions we write

$$V(t, x(\cdot)) = |x| + \int_0^t \int_t^\infty |p(u, s, x(s))| \, du \, ds \,.$$

The situation becomes much more interesting in the vector case. We then interpret the norm of $\mathbf{x}$ in (6.3.14) as a norm of solutions of

$$\mathbf{y}' = \mathbf{g}(t, \mathbf{y}) \,. \tag{6.3.15}$$

That is, we seek a Liapunov function $W(t, \mathbf{y})$ for (6.3.15) with

$$W'_{(6.3.15)}(t, \mathbf{y}) \leq -Z(t, \mathbf{y}) \leq 0 \,.$$

If $W_1(|\mathbf{y}|) \leq W(t, \mathbf{y}) \leq W_2(|\mathbf{y}|)$ for wedges W_i, then W acts as a norm for $\mathbf{y}$. We then write (6.3.14) as

$$V(t, \mathbf{x}(\cdot)) = W(t, \mathbf{x}) + \int_0^t \left[|\mathbf{g}(s, \mathbf{x}(s))| - \int_s^t |\mathbf{p}(u, s, \mathbf{x}(s))| \, du \right] ds \,. \tag{6.3.16}$$

Specific results for this V are developed in the next section.

6.4 Nonlinearities and an Annulus Argument

Consider the system

$$\mathbf{x}' = \mathbf{g}(t, \mathbf{x}) + \int_0^t \mathbf{p}(t, s, \mathbf{x}(s)) \, ds + \mathbf{F}(t) \,, \tag{6.4.1}$$

in which $\mathbf{g} : [0, \infty) \times U \to R^n$, $\mathbf{p} : [0, \infty) \times [0, \infty) \times U \to R^n$, $\mathbf{F} : [0, \infty) \to R^n$, and where $U = \{ x \in R^n : |\mathbf{x}| < \varepsilon, \ \varepsilon > 0 \}$. We suppose that $\mathbf{F}$, $\mathbf{g}$, and $\mathbf{p}$ are continuous and $\mathbf{g}(t, \mathbf{0}) = \mathbf{0}$. Moreover, we suppose there is a continuous function $W : [0, \infty) \times U \to [0, \infty)$ and a constant $L > 0$ with

$$|W(t, \mathbf{x}_1) - W(t, \mathbf{x}_2)| \leq L|\mathbf{x}_1 - \mathbf{x}_2|$$

on $[0, \infty) \times U$, along with a wedge W_1 and a continuous function $Z : [0, \infty) \times U \to [0, \infty)$, such that $W(t, \mathbf{0}) = 0$, $W_1(|\mathbf{x}|) \leq W(t, \mathbf{x})$, and the derivative of W along solutions of

$$\mathbf{y}' = \mathbf{g}(t, \mathbf{y}) \tag{6.4.2}$$

satisfies

$$W'(t, \mathbf{y}) \leq -Z(t, \mathbf{y}) \,.$$

Theorem 6.4.1. *Let the conditions in the last paragraph hold and let* $\int_0^\infty |\mathbf{F}(t)|\, dt < \infty$. *Suppose there are constants* c_1 *and* c_2 *with* $0 < c_1 < 1$ *and* $L < c_2$ *such that*

$$\int_0^{t-s} |\mathbf{p}(u+s, s, \mathbf{x}(s))|\, du \leq c_1 Z(s, \mathbf{x}(s))/c_2$$

if $0 \leq s \leq t < \infty$ *and* $\mathbf{x} : [0, \infty) \to U$ *is continuous. Then for each* $t_0 \geq 0$ *and each* $\bar{\varepsilon} > 0$ *there exists* $\eta > 0$ *such that if* $\int_0^\infty |\mathbf{F}(t)|\, dt < \eta$ *and* $|\boldsymbol{\phi}(t)| < \eta$ *on* $[0, t_0]$, *then any solution* $\mathbf{x}(t) = \mathbf{x}(t, t_0, \boldsymbol{\phi})$ *of* (6.4.1) *on* $[t_0, \infty)$ *satisfies* $|\mathbf{x}(t)| < \bar{\varepsilon}$ *for* $t \geq t_0$. *If, in addition,* $Z(t, \mathbf{x}) \geq b|\mathbf{g}(t, \mathbf{x})|$ *for some* $b > 0$ *and if for each* $\mathbf{x}_0 \in U - \{\mathbf{0}\}$ *there exists* $\delta > 0$ *and a continuous function* $h : [0, \infty) \to [0, \infty)$ *with* $Z(t, \mathbf{x}) \geq h(t)$ *for* $|\mathbf{x} - \mathbf{x}_0| < \delta$ *and* $\int_0^\infty h(t)\, dt = \infty$, *then* $\mathbf{x}(t) \to 0$ *as* $t \to \infty$. *(Note that* h *depends on* $\mathbf{x}_0$.*)*

Proof. Define a functional

$$V(t, \mathbf{x}(\cdot)) = \exp\left[-(1+L)\int_0^t |\mathbf{F}(s)|\, ds\right]\left\{1 + W(t, \mathbf{x}(t))\right.$$
$$\left. + \int_0^t \left[c_1 Z(s, \mathbf{x}(s)) - c_2 \int_0^{t-s} |\mathbf{p}(u+s, s, \mathbf{x}(s))|\, du\right] ds\right\},$$

so that by the Lipschitz condition on W we have

$$V'_{(6.4.1)}(t, \mathbf{x}(\cdot)) \leq \left\{-Z(t, \mathbf{x}) + L|\mathbf{F}(t)| + c_1 Z(t, \mathbf{x})\right.$$
$$+ (L - c_2)\int_0^t |\mathbf{p}(t, s, \mathbf{x}(s))|\, ds$$
$$\left. - (1+L)|\mathbf{F}(t)|\right\} \exp\left[-(1+L)\int_0^t |\mathbf{F}(s)|\, ds\right]$$
$$\leq \left\{-(1-c_1)Z(t, \mathbf{x}) - (c_2 - L)\int_0^t |\mathbf{p}(t, s, \mathbf{x}(s))|\, ds\right.$$
$$\left. - |\mathbf{F}(t)|\right\} \exp\left[-(1+L)\int_0^\infty |\mathbf{F}(t)|\, dt\right]$$

or

$$V'_{(6.4.1)}(t, \mathbf{x}(\cdot)) \leq -\mu\left[Z(t, \mathbf{x}) + \int_0^t |\mathbf{p}(t, s, \mathbf{x}(s))|\, ds + |\mathbf{F}(t)|\right] \quad (6.4.3)$$

for some $\mu > 0$.

Because W is positive definite and $W'_{(6.4.2)}(t, \mathbf{y}) \leq -Z(t, \mathbf{y}) \leq 0$, it follows that $Z(t, \mathbf{0}) = \mathbf{0}$. Thus because

$$\int_0^{t-s} \big| \mathbf{p}(u + s, s, \mathbf{x}(s)) \big|\, du \leq c_1 Z(s, \mathbf{x}(s))/c_2\,,$$

we have $\mathbf{p}(t, s, \mathbf{0}) = \mathbf{0}$. Hence, if $t_0 \geq 0$ and if $|\boldsymbol{\phi}(s)|$ is sufficiently small on $[0, t_0]$, then

$$V(t_0, \boldsymbol{\phi}(\cdot)) \leq (1 + r) \exp\left[(-1 + L)\int_0^{t_0} |\mathbf{F}(s)|\, ds\right],$$

where r is an arbitrarily small preassigned number. Thus, if $t \geq t_0$, then

$$\{1 + W(t, \mathbf{x}(t))\} \exp\left[-(1 + L)\int_0^t |\mathbf{F}(s)|\, ds\right]$$
$$\leq V(t, \mathbf{x}(\cdot)) \leq V(t_0, \boldsymbol{\phi}(\cdot))$$
$$\leq \{1 + r\}\exp\left[-(1 + L)\int_0^{t_0} |\mathbf{F}(s)|\, ds\right],$$

so that

$$1 + W(t, \mathbf{x}(t)) \leq \{1 + r\}\exp\left[(1 + L)\int_{t_0}^t |\mathbf{F}(s)|\, ds\right]$$
$$\leq \{1 + r\}\exp[(1 + L)\eta]\,.$$

Then

$$W_1(|\mathbf{x}(t)|) \leq W(t, \mathbf{x}(t))$$
$$\leq \{1 + r\}\{\exp[(1 + L)\eta]\} - 1 \to 0$$

as $r \to 0$ and $\eta \to 0$. Thus, for r and η small, $|\mathbf{x}(t)| < \bar{\varepsilon}$ if $t \geq t_0$.

We now show that the additional assumptions imply that $\mathbf{x}(t) \to \mathbf{0}$. For in that case we have

$$V'_{(6.4.1)}(t, \mathbf{x}(\cdot)) \leq -\bar{\mu}\big[|\mathbf{x}'(t)| + Z(t, \mathbf{x})\big]$$

for some $\bar{\mu} > 0$.

Let $|\mathbf{x}(t)| < \varepsilon$ on $[t_0, \infty)$ and suppose $\mathbf{x}(t) \not\to \mathbf{0}$. Then there exists a $\delta > 0$ and a sequence $\{t_n\}$ with $|\mathbf{x}(t_n)| \geq \delta$ and $t_n \to \infty$.

First, if $\mathbf{x}(t)$ has a limit $\mathbf{y}$, then for large t we have $Z(t, \mathbf{x}) \geq h(t)$, so that $V'(t, \mathbf{x}(\cdot)) \leq -\bar{\mu}h(t)$. An integration sends V to $-\infty$. If $\mathbf{x}(t)$ does not have a limit, then there is a sequence $\{T_n\}$ with $|\mathbf{x}(t_n) - \mathbf{x}(T_n)| \geq q$

for some $q > 0$ and $T_n \to \infty$. We may suppose $t_n < T_n < t_{n+1}$. Thus, if $t > T_n$, then

$$V(t, \mathbf{x}(\cdot)) \leq V(t_0, \boldsymbol{\phi}(\cdot)) - \bar{\mu} \sum_{i=1}^{n} \left| \int_{t_i}^{T_i} \mathbf{x}'(s)\, ds \right|$$

$$+ V(t_0, \boldsymbol{\phi}(\cdot)) - \bar{\mu} n q \to -\infty$$

as $n \to \infty$, completing the proof.

Exercise 6.4.1. Define

$$V(t, \mathbf{x}(\cdot)) = \left\{ \exp\left[-(1+L) \int_0^t |\mathbf{F}(s)|\, ds \right] \right\}$$

$$\times \left\{ 1 + W(t, \mathbf{x}) + c \int_0^t \int_t^{\infty} |\mathbf{p}(u, s, \mathbf{x}(s))|\, du\, ds \right\},$$

differentiate V along a solution of (6.4.1), and then reformulate Theorem 6.4.1 to suit this Liapunov functional. Prove your new result.

Exercise 6.4.2. In Eq. (6.4.1) let $\mathbf{F}(t) \equiv \mathbf{0}$, let $W_1(|\mathbf{x}|) \leq W(t, \mathbf{x}) \leq W_2(|\mathbf{x}|)$, let $W'_{(6.4.2)}(t, \mathbf{y}) \leq -W_3(|\mathbf{y}|)$, and define

$$V(t, \mathbf{x}(\cdot)) = W(t, \mathbf{x}) + c \int_0^t \int_t^{\infty} |\mathbf{p}(u, s, \mathbf{x}(s))|\, du\, ds.$$

Using this V, formulate a theorem for (6.4.1) yielding

(a) stability,

(b) uniform stability,

(c) asymptotic stability, and

(d) uniform asymptotic stability.

[Refer to Theorem 2.5.1.]

Exercise 6.4.3. Consider the scalar equation

$$x' = -2x + x^3 + \int_0^t [1 + (t-s)^2]^{-1} x^2(s)\, ds.$$

Construct the V of Exercise 6.4.1 and decide if the zero solution is stable.

Theorem 6.4.1 is one of the weakest forms for an equation like (6.4.1). In asking

$$W'_{(6.4.2)}(t, \mathbf{y}) \leq -Z(t, \mathbf{y})$$

with W positive definite and $Z \geq 0$, we ask only that the zero solution of (6.4.2) be stable. As we demand more and more of (6.4.2), converse theorems will bestow more and more properties on W. In Theorem 6.4.1 we ultimately demanded that $Z(t, \mathbf{x})$ be related to $|\mathbf{g}(t, \mathbf{x})|$ to obtain

$$V' \leq -\bar{\mu}[|\mathbf{x}'|] \,,$$

and thus, to obtain an annulus argument.

We noted in Section 6.3 that such a relation is sometimes hard to obtain and we now present yet a third technique for accomplishing the annulus argument.

Consider the system

$$\mathbf{x}' = \mathbf{g}(t, \mathbf{x}) + \int_0^t \mathbf{p}(t, s, \mathbf{x}(s)) \, ds \qquad (6.4.4)$$

with $\mathbf{g}$ and $\mathbf{p}$ continuous, $\mathbf{g} : [0, \infty) \times U \to R^n$, $\mathbf{p} : [0, \infty) \times [0, \infty) \times U \to R^n$, and $U = \{\mathbf{x} \in R^n : |\mathbf{x}| < \varepsilon, \ \varepsilon > 0\}$.

Let $P(t, \mathbf{x}(\cdot))$ be a continuous functional when $0 \leq t < \infty$ and $\mathbf{x} : [0, \infty) \to U$ is continuous.

Definition 6.4.1. *The scalar functional $P(t, \mathbf{x}(\cdot))$ satisfies a one-sided Lipschitz condition with constant $L > 0$ if, whenever $\mathbf{x} : [0, \infty) \to U$ is continuous and $0 \leq t_1 < t_2$, either*

$$P(t_2, \mathbf{x}(\cdot)) - P(t_1, \mathbf{x}(\cdot)) \leq L(t_2 - t_1) \qquad (6.4.5)$$

or

$$P(t_2, \mathbf{x}(\cdot)) - P(t_1, \mathbf{x}(\cdot)) \geq L(t_1 - t_2) \,. \qquad (6.4.6)$$

Example 6.4.1. Let U be the interval $(-1, 1)$, $C(t, s)$ be a scalar function continuous for $0 \leq s \leq t < \infty$, and define

$$P(t, x(\cdot)) = \int_0^t \int_t^\infty |C(u, s)| \, du \, |x(s)| \, ds \,.$$

Then

$$P'(t, x(\cdot)) = \int_t^\infty |C(u, t)| \, du \, |x| - \int_0^t |C(t, s)| \, |x(s)| \, ds \,,$$

so that if $\int_t^\infty |C(u, t)| \, du$ is bounded, then (6.4.5) is satisfied, whereas if $\int_0^t |C(t, s)| \, ds$ is bounded, then (6.4.6) is satisfied.

Definition 6.4.2.   *The functional $P(t, \mathbf{x}(\cdot))$ is positive semidefinite if $P(t, \mathbf{x}(\cdot)) \geq 0$ for $0 \leq t < \infty$ and $\mathbf{x} : [0, \infty) \to U$ and if for each $t_0 \geq 0$ and each $\eta > 0$ there is a $\delta > 0$ such that if $\phi : [0, t_0] \to U$ and $|\phi(t)| < \delta$, then $P(t_0, \phi(\cdot)) < \eta$.*

Definition 6.4.3.   *The functional $P(t, \mathbf{x}(\cdot))$ is decrescent on $[0, \infty) \times U$ if there is a wedge W_4 such that for each $\varepsilon > 0$ and each $t_0 \geq 0$, if $\phi : [0, t_0] \to U$ is continuous and satisfied $|\phi(s)| < \varepsilon$ on $[0, t_0]$, then $P(t_0, \phi(\cdot)) < W_4(\varepsilon)$.*

Theorem 6.4.2.   *Suppose there is a functional*

$$V(t, \mathbf{x}(\cdot)) = W(t, \mathbf{x}) + P(t, \mathbf{x}(\cdot)) \tag{6.4.7}$$

with P satisfying Definitions 6.4.1 and 6.4.2, $W : [0, \infty) \times U \to [0, \infty)$ continuous and locally Lipschitz in $\mathbf{x}$,

$$W_1(|\mathbf{x}|) \leq W(t, \mathbf{x}) \leq W_2(|\mathbf{x}|),$$

and

$$V'_{(6.4.4)}(t, \mathbf{x}(\cdot)) \leq -W_3(|\mathbf{x}|) \tag{6.4.8}$$

for wedges W_1, W_2, and W_3. Then the zero solution of (6.4.4) is asymptotically stable. If we replace (6.4.8) by

$$V'_{(6.4.4)}(t, \mathbf{x}(\cdot)) \leq 0$$

and if P is decrescent, then $\mathbf{x} = \mathbf{0}$ is uniformly stable.

Proof. To show that $\mathbf{x} = \mathbf{0}$ is stable, let $\varepsilon > 0$ and $t_0 \geq 0$ be given. We must find $\delta > 0$ such that $\phi : [0, t_0] \to U$ with $|\phi(s)| < \delta$ on $[0, t_0]$ and $t \geq t_0$ imply $|\mathbf{x}(t, t_0, \phi)| < \varepsilon$.

Use Definition 6.4.2 to find $\delta > 0$ such that $|\phi(t)| < \delta$ on $[0, t_0]$ implies $W_2(\delta) + P(t_0, \phi(\cdot)) < W_1(\varepsilon)$. Then, if $|\phi(t)| < \delta$, if $\mathbf{x}(t) = \mathbf{x}(t, t_0, \phi)$, and if $t \geq t_0$, we have $V' \leq 0$, so that

$$W_1(|\mathbf{x}(t)|) \leq V(t, \mathbf{x}(\cdot)) \leq V(t_0, \phi(\cdot)) - \int_{t_0}^{t} W_3(|\mathbf{x}(s)|)\, ds$$
$$\leq W_2(|\phi(t_0)|) + P(t_0, \phi(\cdot)) < W_1(\varepsilon)$$

and we conclude that $|\mathbf{x}(t)| < \varepsilon$. Thus, $\mathbf{x} = \mathbf{0}$ is stable.

Let (6.4.8) hold and let $\mathbf{x}(t) = \mathbf{x}(t, t_0, \phi)$ where $|\phi(t)| < \delta$ on $[0, t_0]$. We suppose $\mathbf{x}(t) \nrightarrow 0$ as $t \to \infty$. Then there is a $\mu > 0$ and a sequence $\{t_n\} \to \infty$ with $|\mathbf{x}(t_n)| \geq \mu$. To be definite, we suppose (6.4.5) holds.

Now determine $\alpha > 0$ so that $W_1(\mu) > 2W_2(\alpha)$. Because $V'(t, \mathbf{x}(\cdot)) \leq -W_3(|\mathbf{x}|)$ there is a sequence $\{T_n\} \to \infty$ with $|\mathbf{x}(T_n)| \leq \alpha$. In fact, we may suppose $|\mathbf{x}(T_n)| = \alpha$, $|\mathbf{x}(t_n)| = \mu$, and $\alpha \leq |\mathbf{x}(t)| \leq \mu$ if $t_n \leq t \leq T_n$, by renaming t_n and T_n if necessary. Now

$$P(T_n, \mathbf{x}(\cdot)) - P(t_n, \mathbf{x}(\cdot)) \leq L(T_n - t_n). \tag{6.4.9}$$

Also, $V'(t, \mathbf{x}(\cdot)) \leq 0$ implies that $V(t, \mathbf{x}(\cdot)) \to c$, a positive constant, as $\mathbf{x}(t) \nrightarrow 0$. Thus, $\big| V(t_n, \mathbf{x}(\cdot)) - V(T_n, \mathbf{x}(\cdot)) \big|$ may be made arbitrarily small by taking n large. But

$$
\begin{aligned}
V(t_n, & \mathbf{x}(\cdot)) - V(T_n, \mathbf{x}(\cdot)) \\
&= W(t_n, \mathbf{x}(t_n)) - W(T_n, \mathbf{x}(T_n)) \\
&\quad + P(t_n, \mathbf{x}(\cdot)) - P(T_n, \mathbf{x}(\cdot)) \\
&\geq W_1(\mu) - W_2(\alpha) - \big[P(T_n, \mathbf{x}(\cdot)) - P(t_n, \mathbf{x}(\cdot)) \big] \\
&\geq 2W_2(\alpha) - W_2(\alpha) - L(T_n - t_n)
\end{aligned}
$$

or

$$V(t_n, \mathbf{x}(\cdot)) - V(T_n, \mathbf{x}(\cdot)) \geq W_2(\alpha) - L(T_n - t_n).$$

As the left side tends to zero, for each $\eta > 0$, there exists N such that $n \geq N$ implies

$$
\begin{aligned}
\eta &\geq V(t_n, \mathbf{x}(\cdot)) - V(T_n, \mathbf{x}(\cdot)) \\
&\geq W_2(\alpha) - L(T_n - t_n)
\end{aligned}
$$

or

$$\eta + L(T_n - t_n) \geq W_2(\alpha),$$

so that

$$L(T_n - t_n) \geq W_2(\alpha) - \eta > W_2(\alpha)/2$$

if $\eta < W_2(\alpha)/2$. Hence, for $n > N$, we have

$$T_n - t_n \geq W_2(\alpha)/2L \overset{\text{def}}{=} T.$$

Because $V'(t, \mathbf{x}(\cdot)) \leq -W_3(|\mathbf{x}(t)|)$, if $T_n < t$, then

$$0 \leq V(t, \mathbf{x}(\cdot)) \leq V(t_N, \mathbf{x}(\cdot)) - \int_{t_N}^t W_3(|\mathbf{x}(s)|)\, ds$$

$$\leq V(t_N, \mathbf{x}(\cdot)) - \sum_{i=N}^{i=n} \int_{t_i}^{T_i} W_3(|\mathbf{x}(s)|)\, ds$$

$$\leq V(t_N, \mathbf{x}(\cdot)) - \sum_{i=N}^{i=n} \int_{t_i}^{T_i} W_3(\alpha)\, ds$$

$$= V(t_N, \mathbf{x}(\cdot)) - (n - N)TW_3(\alpha) \to -\infty$$

as $n \to \infty$. This proves asymptotic stability.

We now prove the uniform stability. Let $\varepsilon > 0$ be given. We must find $\delta > 0$ such that

$$t_0 \geq 0\,, \quad \boldsymbol{\phi} : [0, t_0] \to U \text{ with } |\boldsymbol{\phi}(t)| < \delta \text{ on } [0, t_0]\,, \quad \text{and} \quad t \geq t_0$$

imply $|\mathbf{x}(t, t_0, \boldsymbol{\phi})| < \varepsilon$.

Because P is decrescent, select W_4 so that for $\varepsilon > 0$ we may find $\delta > 0$ such that

$$W_2(\delta) + W_4(\delta) < W_1(\varepsilon)\,.$$

Thus, if $t_0 \geq 0$ and $\boldsymbol{\phi} : [0, t_0] \to U$ with $|\boldsymbol{\phi}(t)| < \delta$ on $[0, t_0]$, then $V' \leq 0$, so

$$W_1(|\mathbf{x}(t)|) \leq V(t, \mathbf{x}(\cdot)) \leq V(t_0, \boldsymbol{\phi}(\cdot))$$
$$= W(t_0, \boldsymbol{\phi}(t_0)) + P(t_0, \boldsymbol{\phi}(\cdot))$$
$$\leq W_2(\delta) + W_4(\delta) < W_1(\varepsilon)$$

implying $|\mathbf{x}(t)| < \varepsilon$ for $t \geq t_0$. This completes the proof.

Exercise 6.4.4. In Theorem 6.4.2 replace (6.4.7) by

$$W_1(|\mathbf{x}|) \leq V(t, \mathbf{x}(\cdot)) \leq W_1(|\mathbf{x}|) + P(t, \mathbf{x}(\cdot)) \tag{6.4.7$'$}$$

and prove the result. See Burton (1979a) for details.

The conclusion of this exercise is that V need not be too well behaved, as long as it is bounded above and below by well-behaved functions. Note that it is the same W_1 above and below V.

Consider again

$$\mathbf{y}' = \mathbf{g}(t, \mathbf{y}) \tag{6.4.2}$$

and suppose there is a Liapunov function W with

$$W'_{(6.4.2)}(t, \mathbf{y}) \leq -cW(t, \mathbf{y}). \tag{6.4.10}$$

Theorem 6.4.3.   *Consider Eq. (6.4.4) and suppose there is a function $W : [0, \infty) \times U \to [0, \infty)$ with $W_1(|\mathbf{x}|) \leq W(t, \mathbf{x}) \leq W_2(|\mathbf{x}|)$, $|W(t, \mathbf{x}_1) - W(t, \mathbf{x}_2)| \leq L|\mathbf{x}_1 - \mathbf{x}_2|$ on $[0, \infty) \times U$ for some $L > 0$, and suppose W satisfies (6.4.10) for some $c > 0$. Suppose also that $\int_t^\infty |\mathbf{p}(u, s, \mathbf{x}(s))| \, du$ is defined for $0 \leq t < \infty$ whenever $\mathbf{x} : [0, \infty) \to U$ is continuous.*

(a)   *If there is a wedge W_3 with*

$$-cW(t, \mathbf{x}) + L \int_t^\infty |\mathbf{p}(u, t, \mathbf{x}(t))| \, du \leq -W_3(|\mathbf{x}(t)|)$$

and if

$$\int_0^t \int_t^\infty |\mathbf{p}(u, s, \mathbf{x}(s))| \, du \, ds$$

satisfies Definition 6.4.2, then $\mathbf{x} = \mathbf{0}$ is asymptotically stable.

(b)   *If*

$$-cW(t, \mathbf{x}) + L \int_t^\infty |\mathbf{p}(u, t, \mathbf{x}(t))| \, du \leq 0$$

and if

$$\int_0^t \int_t^\infty |\mathbf{p}(u, s, \mathbf{x}(s))| \, du \, ds$$

satisfies Definition 6.4.3, then $\mathbf{x} = \mathbf{0}$ is uniformly stable.

Proof. Define

$$V(t, \mathbf{x}(\cdot)) = W(t, \mathbf{x}) + L \int_0^t \int_t^\infty |\mathbf{p}(u, s, \mathbf{x}(s))| \, du \, ds,$$

so that

$$V'_{(6.4.4)}(t, \mathbf{x}(\cdot)) \leq -cW(t, \mathbf{x}) + L \int_0^t |\mathbf{p}(t, s, \mathbf{x}(s))| \, ds$$

$$+ L \int_t^\infty |\mathbf{p}(u, t, \mathbf{x}(t))| \, du - L \int_0^t |\mathbf{p}(t, s, \mathbf{x}(s))| \, ds$$

$$\leq -cW(t, \mathbf{x}) + L \int_t^\infty |\mathbf{p}(u, t, \mathbf{x}(t))| \, du.$$

If (a) holds, then $V' \leq -W_3(|\mathbf{x}(t)|)$, but if (b) holds, then $V' \leq 0$. The conditions of Theorem 6.4.2 are readily satisfied.

In Chapter 5 we employed functionals for

$$\mathbf{x}' = A\mathbf{x} + \int_0^t C(t,s)\mathbf{x}(s)\,ds$$

of the form

$$V(t,\mathbf{x}(\cdot)) = \mathbf{x}^T B\mathbf{x} + \int_0^t \int_t^\infty |C(u,s)|\,du\,|\mathbf{x}(s)|^2\,ds\,.$$

The quadratic term $\mathbf{x}^T B\mathbf{x}$ resulted in the requirement that

$$\int_t^\infty |C(u,t)|\,du + \int_0^t |C(t,s)|\,ds$$

be small. That is, both coordinates of C are integrated.

When our functional uses $[\mathbf{x}^T B\mathbf{x}]^{1/2}$, then only $\int_t^\infty |C(u,t)|\,du$ is required to be small, at the expense of requiring that B be positive definite.

In Chapter 2 we used a Razumikhin-type argument to place the smallness burden on $\int_0^t |C(t,s)|\,ds$ with B positive definite. We now present a modified Razumikhin argument that places the smallness requirement on the second coordinate of C.

Consider again

$$\mathbf{x}' = \mathbf{g}(t,\mathbf{x}) + \int_0^t \mathbf{p}(t,s,\mathbf{x}(s))\,ds \qquad (6.4.4)$$

with $\mathbf{g}$ and $\mathbf{p}$ continuous, $\mathbf{g} : [0,\infty) \times U \to R^n$, $\mathbf{p} : [0,\infty) \times [0,\infty) \times U \to R^n$, and $U = \{x \in R^n : |\mathbf{x}| < \varepsilon,\ \varepsilon > 0\}$. Let $W : [0,\infty) \times U \to [0,\infty)$ with $|W(t,\mathbf{x}_1) - W(t,\mathbf{x}_2)| \leq L|\mathbf{x}_1 - \mathbf{x}_2|$ on $[0,\infty) \times U$ for $L > 0$, $W_1(|\mathbf{x}|) \leq W(t,\mathbf{x})$, $W(t,\mathbf{0}) = 0$, $W'_{(6.4.2)}(t,\mathbf{y}) \leq -Z(t,\mathbf{y})$, where $Z : [0,\infty) \times U \to [0,\infty)$ is continuous.

Theorem 6.4.4.   *Let the conditions of the preceding paragraph hold. Suppose there is a continuous function $q : [0,\infty) \times U \to [0,\infty)$ with $|\mathbf{p}(t,s,\mathbf{x}(s))|/W(s,\mathbf{x}(s)) \leq q(t,s)$ if $\mathbf{x}(\cdot)$ is any continuous function in U and if $0 \leq s \leq t < \infty$. Also suppose that $Z(t,\mathbf{x}) \geq cW(t,\mathbf{x})$ for some $c > 0$ and that there are constants c_1 and c_2 with $0 < c_1 < c$ and $c_2 > L$, so that $\int_0^t q(t,s)\,ds \leq c_1/c_2$ if $0 \leq t < \infty$. Then for each $t_0 \geq 0$ and each $\varepsilon > 0$ there exists $\delta > 0$ such that if $|\phi(t)| < \delta$ on $[0,t_0]$ and $\mathbf{x}(t,t_0,\phi)$ is a solution of (6.4.4), then $|\mathbf{x}(t,t_0,\phi)| < \varepsilon$ for $t \geq t_0$.*

Proof. Define

$$V(t, \mathbf{x}(\cdot)) = W(t, \mathbf{x}) + \int_0^t \left[c_1 W(u, \mathbf{x}(u)) - c_2 \int_0^u |\mathbf{p}(u, s, \mathbf{x}(s))| \, ds \right] du \,,$$

so that along a solution $\mathbf{x}(t)$ of (6.4.4) we have

$$V'(t, \mathbf{x}(\cdot)) \leq -cW(t, \mathbf{x}) + L \int_0^t |\mathbf{p}(t, s, \mathbf{x}(s))| \, ds$$

$$+ c_1 W(t, \mathbf{x}) - c_2 \int_0^t |\mathbf{p}(t, s, \mathbf{x}(s))| \, ds < 0$$

if $\mathbf{x} \neq \mathbf{0}$. Because V is not necessarily positive, boundedness of $\mathbf{x}(t)$ may not yet be concluded.

Suppose there is a solution $\mathbf{x}(t)$ in U on $[t_0, T]$ with the property that $W(s, \mathbf{x}(s)) < W(T, \mathbf{x}(T))$ if $0 \leq s < T$. Then $W'(t, \mathbf{x}(t)) \geq 0$ at $t = T$ so that $V'(T, \mathbf{x}(\cdot)) < 0$ implies that, for $t = T$, we have

$$(d/dt) \int_0^t \left[c_1 W(u, \mathbf{x}(u)) - c_2 \int_0^u |\mathbf{p}(u, s, \mathbf{x}(s))| \, ds \right] du$$

$$= c_1 W(T, \mathbf{x}(T)) - c_2 \int_0^T |\mathbf{p}(T, s, \mathbf{x}(s))| \, ds < 0 \,.$$

This will be a contradiction because we have

$$\int_0^T |\mathbf{p}(T, s, \mathbf{x}(s))| \, ds = \int_0^T \left[W(s, \mathbf{x}(s)) \, |\mathbf{p}(T, s, \mathbf{x}(s))| \, / \, W(s, x(s)) \right] ds$$

$$\leq W(T, \mathbf{x}(T)) \int_0^T q(T, s) \, ds$$

$$\leq W(T, \mathbf{x}(T)) c_1 / c_2 \,.$$

Because W is continuous, $W(t, \mathbf{0}) = 0$, and $W(t, \mathbf{x}) > W_1(|\mathbf{x}|)$, the result now follows.

Exercise 6.4.5. Construct another alternative type of Razumikhin result. Consider the system

$$\mathbf{x}' = A\mathbf{x} + \int_0^t C(t, s) \mathbf{x}(s) \, ds$$

with A stable and find $B = B^T$ satisfying $A^T B + BA = -I$ and $|\alpha| |\mathbf{x}| \leq [\mathbf{x}^T B \mathbf{x}]^{1/2} \leq \beta |\mathbf{x}|$ for α and β positive. Take $c_2 > \beta K/\alpha$ and $k/c_2 \geq$

$\int_0^u |C(u,s)|\,ds$ if $0 \le s \le u < \infty$ with k and K defined in Theorem 2.5.1. Define

$$V(t, \mathbf{x}(\cdot)) = [\mathbf{x}^T B \mathbf{x}]^{1/2} + \int_0^t \left[k - c_2 \int_0^u |C(u,s)|\,ds \right] |\mathbf{x}(u)|\,du\,,$$

so that

$$V'(t, \mathbf{x}(\cdot)) \le K \int_0^t |C(t,s)|\,|\mathbf{x}(s)|\,ds - c_2 \int_0^t |C(t,s)|\,ds\,|\mathbf{x}|$$

$$\le (K/\alpha) \int_0^t |C(t,s)|\,[\mathbf{x}^T(s) B \mathbf{x}(s)]^{1/2}\,ds$$

$$- (c_2/\beta)[\mathbf{x}^T(t) B \mathbf{x}(t)]^{1/2} \int_0^t |C(t,s)|\,ds\,.$$

Argue that if $\mathbf{x}^T(s) B \mathbf{x}(s) < \mathbf{x}^T(T) B \mathbf{x}(T)$ for $0 \le s < T$, then $[\mathbf{x}^T(t) B \mathbf{x}(t)]' \ge 0$ at T and $V'(t, \mathbf{x}(\cdot)) < 0$ at T, so that

$$(d/dt) \int_0^t \left[k - c_2 \int_0^u |C(u,s)|\,ds \right] |\mathbf{x}(u)|\,du < 0$$

at T, a contradiction. Argue now that this yields stability. What must be added for asymptotic stability?

In his monograph Yoshizawa (1966, pp. 118–153) showed in great detail how the existence of a Lipschitz-Liapunov function with a negative definite derivative implied stability under many types of perturbations. The same types of results hold for Volterra equations, and owing to the greater complexity of functions, there is even more variety.

We showed in Theorem 2.5.1 that under certain conditions the derivative of

$$V(t, \mathbf{x}(\cdot)) = [\mathbf{x}^T B \mathbf{x}]^{1/2} + \bar{K} \int_0^t \int_t^\infty |C(u,s)|\,du\,|\mathbf{x}(s)|\,ds$$

along solutions of

$$\mathbf{x}' = A\mathbf{x} + \int_0^t C(t,s)\mathbf{x}(s)\,ds$$

satisfies

$$V'(t, \mathbf{x}(\cdot) \le -\mu[|\mathbf{x}| + |\mathbf{x}'|]\,, \quad \mu > 0\,.$$

Because V is Lipschitz in $\mathbf{x}(t)$, V' will also be negative along solutions of

$$\mathbf{x}' = A\mathbf{x} + \int_0^t C(t,s)\mathbf{x}(s)\,ds + \int_0^t D(t,s)\mathbf{r}(\mathbf{x}(s))\,ds$$
$$+ \mathbf{q}(t,\mathbf{x}) + \mathbf{H}(t,\mathbf{x}(\cdot)) + \mathbf{p}(t,\mathbf{x}) \tag{6.4.11}$$

under the following assumptions:

(i) $|\mathbf{H}(t,\mathbf{x}(\cdot))| \leq \beta|\mathbf{x}(t)|\left|\int_0^t E(t,s)\mathbf{m}(\mathbf{x}(s))\,ds\right|$ where $\beta > 0$, with E and $\mathbf{m}$ defined below.

(ii) D and E are continuous, $n \times n$ matrices on $[0,\infty) \times [0,\infty)$ with $|D(t,s)| \leq \alpha|C(t,s)|$ and $|E(t,s)| \leq \alpha|C(t,s)|$ for some $\alpha > 0$ and $0 \leq s \leq t < \infty$.

(iii) $\mathbf{q} : [0,\infty) \times U \to R^n$ is continuous, $U = \{\mathbf{x} \in R^n : |\mathbf{x}| < \varepsilon,\ \varepsilon > 0\}$, and $|\mathbf{q}(t,\mathbf{x})|\,/\,|\mathbf{x}| \to 0$ as $|\mathbf{x}| \to 0$ uniformly for $0 \leq t < \infty$.

(iv) $\mathbf{r}$ and $\mathbf{m} : U \to R^n$ are continuous, $|\mathbf{r}(\mathbf{x})|/|\mathbf{x}| \to 0$ as $|\mathbf{x}| \to 0$, and $|\mathbf{m}(\mathbf{x})| \leq \omega|\mathbf{x}|$ for some $\omega > 0$.

(v) $\mathbf{p} : [0,\infty) \times U \to R^n$ is continuous and $|\mathbf{p}(t,\mathbf{x})| \leq \lambda(t)|\mathbf{x}|$ where $\lambda : [0,\infty) \to [0,\infty)$ is continuous and $\int_0^\infty \lambda(t)\,dt < \infty$.

(vi) A is an $n \times n$ constant matrix whose characteristic roots all have negative real parts and there is a matrix $B = B^T$ satisfying $A^T B + BA = -I$, $|\mathbf{x}| \geq 2k[\mathbf{x}^T B\mathbf{x}]^{1/2}$, $|B\mathbf{x}| \leq K[\mathbf{x}^T B\mathbf{x}]^{1/2}$, and $\gamma|\mathbf{x}| \leq [\mathbf{x}^T B\mathbf{x}]^{1/2}$ for γ, k, and K positive.

(vii) $C(t,s)$ is continuous for $0 \leq s \leq t < \infty$ and $\int_t^\infty |C(u,s)|\,du$ is continuous for $0 \leq s \leq t < \infty$.

(viii) There exists $\bar{K} > K$ and $\bar{k} > 0$ with $\bar{k} \leq k - \bar{K}\int_t^\infty |C(u,t)|\,du$.

Theorem 6.4.5. *Consider (6.4.11) and suppose that (i)–(viii) hold. Then for each $\varepsilon > 0$ and each $t_0 \geq 0$ there exists $\delta > 0$ such that if $|\phi(t)| < \delta$ on $[0,t_0]$, then $|\mathbf{x}(t,t_0,\phi)| < \varepsilon$ on $[t_0,\infty)$ and $|\mathbf{x}(t,t_0,\phi)| \to 0$ as $t \to \infty$.*

Proof. Define

$$V(t,\mathbf{x}(\cdot)) = \left\{[\mathbf{x}^T B\mathbf{x}]^{1/2} + \bar{K}\int_0^t \int_t^\infty |C(u,s)|\,du\,|\mathbf{x}(s)|\,ds\right\}$$
$$\times \exp\left[-L\int_0^t \lambda(s)\,ds\right],$$

where L satisfies $[\mathbf{x}^T B \mathbf{x}]^{1/2} L \geq K|\mathbf{x}|$. Then

$$V'_{(6.4.11)}(t, \mathbf{x}(\cdot))$$

$$\leq \left\{ -L\lambda(t)[\mathbf{x}^T B \mathbf{x}]^{1/2} - \left[k - \bar{K} \int_t^\infty |C(u,t)|\, du \right] |\mathbf{x}| \right.$$

$$- (\bar{K} - K) \int_0^t |C(t,s)|\, |\mathbf{x}(s)|\, ds + K \int_0^t \left| D(t,s) \mathbf{r}(\mathbf{x}(s)) \, ds \right.$$

$$+ K|\mathbf{q}(t,\mathbf{x})| + K|\mathbf{H}(t,\mathbf{x}(\cdot))|$$

$$\left. + K|\mathbf{p}(t,\mathbf{x})| \right\} \exp \left[-L \int_0^t \lambda(s)\, ds \right]$$

$$\leq \left\{ -\bar{k}|\mathbf{x}| - (\bar{K} - K) \int_0^t |C(t,s)|\, |\mathbf{x}(s)|\, ds \right.$$

$$+ K\alpha \int_0^t |C(t,s)|\, |\mathbf{r}(\mathbf{x}(s))|\, ds + K|\mathbf{q}(t,\mathbf{x})|$$

$$\left. + K\beta\alpha\omega|\mathbf{x}| \int_0^t |C(t,s)|\, |\mathbf{x}(s)|\, ds \right\} \exp \left[-L \int_0^t \lambda(s)\, ds \right].$$

By the conditions on $\mathbf{q}$ and $\mathbf{r}$, this is nonpositive for $|\mathbf{x}(\cdot)|$ small. Thus

$$[\mathbf{x}^T(t) B \mathbf{x}(t)]^{1/2} \exp \left[-L \int_0^t \lambda(s)\, ds \right] \leq V(t, \mathbf{x}(\cdot)) \leq V(t_0, \boldsymbol{\phi}(\cdot)),$$

and because $V(t_0, \boldsymbol{\phi}(\cdot)) \to 0$ as $|\boldsymbol{\phi}| \to 0$, we see that the zero solution is stable. The proof that solutions tend to zero proceeds as in Theorem 2.5.1.

Exercise 6.4.6. Show that Theorem 6.4.5 is still true if we add the term $\mathbf{b}(t, \mathbf{x})$ to (6.4.11), where $\mathbf{b} : [0, \infty) \times U \to R^n$ is continuous, $|\mathbf{b}(t, \mathbf{x})| \leq a(t)$ with $a : [0, \infty) \to [0, \infty)$ continuous, and $\int_0^\infty a(t)\, dt < a_0$ where a_0 is small. We then speak of the stability of the zero function, because $\mathbf{x} = \mathbf{0}$ is not a solution.

Most of the results of this section are found in Burton (1979a,c, and 1980a).

6.5   A Functional in the Unstable Case

The construction in Sections 6.3 and 6.4 are based on the assumption that

$$y' = g(t, y)$$

is stable and that

$$x' = g(t, x) + \int_0^t p(t, s, \mathbf{x}(s)) \, ds \tag{6.5.1}$$

will inherit that stability for small p. If we pursue the ideas developed in Chapter 5, we obtain stability from a combination of the properties of g and p.

Our discussion here centers on the scalar case. Thus, we consider (6.5.1) with g and p continuous, $g : [0, \infty) \times U \to R$, $p : [0, \infty) \times [0, \infty) \times U \to R$, and $U = \{x \in R : |x| < \varepsilon, \ \varepsilon > 0\}$. We suppose there is a function $P(t, s, x)$ with $\partial P(t, s, x)/\partial t = p(t, s, x)$, so that (6.5.1) may be written as

$$x' = Q(t, x) + (d/dt) \int_0^t P(t, s, x(s)) \, ds \,, \tag{6.5.2}$$

where

$$Q(t, x) = g(t, x) - P(t, t, x) \,. \tag{6.5.3}$$

Let $L : [0, \infty) \times U \to [0, \infty)$ be continuous and define

$$V(t, x(\cdot)) = \left[x - \int_0^t P(t, s, x(s)) \, ds \right]^2$$
$$+ \int_0^t \int_t^\infty |P(u, s, x(s))| \, du \, L(s, x(s)) \, ds \tag{6.5.4}$$

under the assumption that $\int_t^\infty |P(u, s, x(s))| \, du$ is continuous for $0 \le s \le t < \infty$ and all continuous $x(t)$ in U. Then

$$V'_{(6.5.2)}(t, x(\cdot)) = 2 \left[x - \int_0^t P(t, s, x(s)) \, ds \right] Q(t, x)$$
$$+ \int_t^\infty |P(u, t, x(t))| \, du \, L(t, x)$$
$$- \int_0^t |P(t, s, x(s))| \, L(s, x(s)) \, ds \,,$$

so that

$$V'_{(6.5.2)}(t, x(\cdot)) \le 2x Q(t, x) + 2|Q(t, x)| \int_0^t |P(t, s, x(s))| \, ds$$
$$+ \int_t^\infty |P(u, t, x(t))| \, du \, L(t, x)$$
$$- \int_0^t |P(t, s, x(s))| \, L(s, x(s)) \, ds \tag{6.5.5}$$

from which stability results are readily drawn, as in Chapter 5. The basic assumption must be that

$$xQ(t,x) \leq 0.$$
(6.5.6)

We then distinguish three cases:

$$\lim_{|x|\to 0} |Q(t,x)|/|x| = 0 \quad \text{uniformly for } 0 \leq t < \infty,$$
(6.5.7)

$$|Q(t,x)|/|x| \geq A > 0 \quad \text{for} \quad 0 < t < \infty \quad \text{and} \quad \text{all } x \in U,$$
(6.5.8)

and

$$\lim_{|x|\to 0} |x|/|Q(t,x)| = 0 \quad \text{uniformly for } 0 < t < \infty.$$
(6.5.9)

Theorem 6.5.1. *Consider the scalar equation* (6.5.2) *with* (6.5.3) *and* (6.5.6) *holding. Suppose there is a constant $M > 0$ such that*

$$\int_0^t |P(t,s,x(s))/Q(s,x(s))| \, ds < M$$

if $0 \leq t < \infty$ and x is any continuous function in U. Also suppose that there is an $\alpha < 2$ and a wedge W_1 such that $W_1(|x|)/|x| \to 0$ as $|x| \to 0$, $W_1(|x|) \geq |Q(t,x)|$, and

$$\int_t^\infty |P(u,t,x(t))| \, du \geq \alpha |x(t)|$$

for any continuous $x(t)$ in U and $0 \leq t < \infty$. then the zero solution is stable.

Proof. Define $L(t,x) = |Q(t,x)|$ so that (6.5.5) yields

$$V'_{(6.5.2)}(t,x(\cdot)) \leq 2xQ(t,x) + \int_0^t \left| P(t,s,x(s))/Q(s,x(s)) \right| \left(Q^2(s,x(s)) \right.$$

$$+ Q^2(t,x(t))) \, ds + \int_t^\infty |P(u,t,x(t))| \, du \, |Q(t,x)|$$

$$- \int_0^t |P(t,s,x(s))| \, |Q(s,x(s))| \, ds$$

$$\leq 2xQ(t,x) + Q^2(t,x(t))M$$

$$+ |Q(t,x)| \int_t^\infty |P(u,t,x(t))| \, du$$

$$\leq 2xQ(t,x) + Q^2(t,x)M + \alpha|x| \, |Q(t,x)|$$

$$\leq 0$$

if $|x|$ is small enough. Now

$$\left| \int_0^t P(t,s,x(s))\,ds \right| \le \int_0^t \left| P(t,s,x(s))/Q(s,x(s)) \right| \left| Q(s,x(s)) \right| ds$$

$$\le \int_0^t \left| P(t,s,x(s))/Q(s,x(s)) \right| W_1(|x(s)|)\,ds\,.$$

Suppose that a given solution $x(t) = x(t,t_0,\phi)$ satisfies $|\phi(t)| < \delta$ on $[0,t_0]$ for some $\delta > 0$, but there is a $t_1 > t_0$ with $|x(t_1)| > |x(s,t_0,\phi)|$ if $0 \le s < t_1$. Then

$$\left| x(t_1) - \int_0^{t_1} P(t_1,s,x(s))\,ds \right| \ge |x(t_1)| - \int_0^{t_1} |P(t_1,s,x(s))|\,ds$$

$$\ge |x(t_1)|$$

$$- \int_0^{t_1} \left| \frac{P(t_1,s,x(s))}{Q(s,x(s))} \right| W_1(|x(s)|)\,ds$$

$$\ge |x(t_1)| \left[1 - (W_1(|x(t_1)|)/|x(t_1)|) \right.$$

$$\left. \times \int_0^{t_1} \left| \frac{P(t_1,s,x(s))}{Q(s,x(s))} \right| ds \right]$$

$$\ge |x(t_1)| \left[1 - (W_1(|x(t_1)|)/|x(t_1)|)M \right]$$

$$\ge |x(t_1)|/2$$

if $|x(t_1)| < \rho$, for some $\rho > 0$, because $W_1(|x|)/|x| \to 0$ as $|x| \to 0$.

Also, if $0 < \varepsilon < \rho$ and $t_0 \ge 0$, there is a $\delta > 0$, $\delta < \varepsilon$, such that $\phi : [0,t_0] \to R$ and $|\phi(t)| < \delta$ on $[0,t_0]$ imply

$$V(t_0,\phi(\cdot)) = \left[\phi(t_0) - \int_0^{t_0} P(t_0,s,\phi(s))\,ds \right]^2$$

$$+ \int_0^{t_0} \int_{t_0}^{\infty} |P(u,s,\phi(s))|\,du\,|Q(s,\phi(s))|\,ds < \varepsilon^2/4\,.$$

Suppose that $t_1 > t_0$ has the property that $|x(t,t_0,\phi)| < |x(t_1,t_0,\phi)| = \varepsilon$ if $0 \le t < t_1$. Then for $x(t) = x(t,t_0,\phi)$ we have

$$x^2(t_1)/4 \le V(t_1,x(\cdot))$$

$$\le V(t_1,\phi(\cdot)) < \varepsilon^2/4\,,$$

or $|x(t_1)| < \varepsilon$, a contradiction to $|x(t_1)| = \varepsilon$. This completes the proof.

Theorem 6.5.2. *Let (6.5.2) be a scalar equation with (6.5.3), (6.5.6), and (6.5.8) holding. Suppose that for x in U and $0 \leq t < \infty$ there are positive constants R, N, and $M < 1$ with*

$$\int_0^t \left| P(t,s,x(s))/x(s) \right| ds \leq M \,,$$

$$|Q(t,s)| \leq N|x| \,,$$

$$\int_0^t |P(u,t,x(t))| \, du \leq R|x(t)| \,,$$

and

$$-2A + MN^2 + R \leq 0 \,.$$

Then the zero solution of (6.5.2) is stable.

Proof. Let $L(t,x) = |x|$ and have

$$
\begin{aligned}
V'_{(6.5.2)}(t,x(\cdot)) &\leq 2xQ(t,x) + 2|Q(t,x)| \int_0^t |P(t,s,x(s))| \, ds \\
&\quad + \int_t^\infty |P(u,t,x(t))| \, du \, |x| - \int_0^t |P(t,s,x(s))| \, |x(s)| \, ds \\
&\leq 2xQ(t,x) + \int_0^t \left| P(t,s,x(s))/x(s) \right| \left(Q^2(t,x) + x^2(s) \right) ds \\
&\quad + |x| \int_t^\infty |P(u,t,x(t))| \, du - \int_0^t |P(t,s,x(s))| \, |x(s)| \, ds \\
&\leq 2xQ(t,x) + MQ^2(t,x) + |x| \int_t^\infty |P(u,t,x(t))| \, du \\
&\leq -2Ax^2 + MN^2|x|^2 + Rx^2 \leq 0, \, .
\end{aligned}
$$

Next, we note that

$$\left| \int_0^t P(t,s,x(s)) \, ds \right| \leq \int_0^t \left| P(t,s,x(s))/x(s) \right| |x(s)| \, ds$$

$$\leq M \max_{0 \leq s \leq t} |x(s)| \,.$$

Because $M < 1$, the stability follows as in the proof of Theorem 6.5.1.

Exercise 6.5.1. Formulate $L(t,x)$ and the appropriate stability result when (6.5.9) holds.

We will return to many of these stability questions for functional differential equations in Chapter 8. Avoiding the Marachkov condition will occupy almost all of Section 8.3 and we will see several interesting ways of doing so. In Section 8.7 we will avoid the Marachkov condition by using one Liapunov function and one Liapunov functional.

Chapter 7

The Resolvent

7.1 General Theory

We briefly mentioned the resolvent in Section 2.3. It is used to obtain a variation of parameters formula. We noted, in the convolution case, that it was quite effective for dealing with perturbations because it employs the solutions of the unforced equations about which we frequently know a great deal.

The nonconvolution case presents many new difficulties. But recently we have been able to make some good progress through use of fixed point theory and Becker's form of the resolvent. In preparation for later work we will give a brief sketch of the resolvent for both integral and integro-differential equations. Some of the results are only stated and the proofs are left to the references.

Given the integral equation

$$\mathbf{x}(t) = \mathbf{f}(t) + \int_0^t C(t,s)\mathbf{x}(s)\,ds \qquad (7.1.1)$$

with $\mathbf{f} : [0,a] \to R^n$ being continuous and C continuous for $0 \le s \le t \le a$, we define the formal *resolvent* equation as

$$R(t,s) = -C(t,s) + \int_s^t R(t,u)C(u,s)\,du. \qquad (7.1.2)$$

Assuming that a solution $R(t,s)$ exists as a continuous function for $0 \le s \le t \le a$, we note that $\mathbf{x}(t)$ may be found with the aid of $R(t,s)$ to be

$$\mathbf{x}(t) = \mathbf{f}(t) - \int_0^t R(t,u)\mathbf{f}(u)\,du, \qquad (7.1.3)$$

a variation of parameters formula.

To verify (7.1.3), left multiply (7.1.1) by $R(t,s)$ and integrate from 0 to t:

$$\int_0^t R(t,u)\mathbf{x}(u)\,du - \int_0^t R(t,u)\mathbf{f}(u)\,du$$

$$= \int_0^t R(t,u)\int_0^u C(u,s)\mathbf{x}(s)\,ds\,du$$

$$= \int_0^t \int_s^t R(t,u)C(u,s)\,du\,\mathbf{x}(s)\,ds$$

$$= \int_0^t \big[R(t,s) + C(t,s)\big]\mathbf{x}(s)\,ds$$

by (7.1.2). Thus

$$-\int_0^t R(t,u)\mathbf{f}(u)\,du = \int_0^t C(t,s)\mathbf{x}(s)\,ds\,,$$

which, together with (7.1.1) yields

$$\mathbf{x}(t) = \mathbf{f}(t) - \int_0^t R(t,u)\mathbf{f}(u)\,du\,,$$

as required.

Miller (1971a, p. 200) shows that

$$\int_s^t R(t,u)C(u,s)\,du = \int_s^t C(t,u)R(u,s)\,du\,,$$

so that (7.1.2) may be written as

$$R(t,s) = -C(t,s) + \int_s^t C(t,u)R(u,s)\,du\,, \tag{7.1.4}$$

and if we replace t by $t+s$ in (7.1.4), we have

$$R(t+s,s) = -C(t+s,s) + \int_s^{t+s} C(t+s,u)R(u,s)\,du$$

or

$$R(t+s,s) = -C(t+s,s) + \int_0^t C(t+s,u+s)R(u+s,s)\,du\,. \tag{7.1.5}$$

In this form s is simply a parameter and we may write (7.1.5) as

$$L(t) = -D(t) + \int_0^t C(t+s,u+s)L(u)\,du\,, \tag{7.1.6}$$

and the proof of existence and uniqueness may be applied directly to it.

Equation (7.1.2) is conceptually much more complicated than (7.1.1). Thus, one is often inclined to believe that more progress can be made by attacking (7.1.1) directly without going through a variation of parameters argument. We have already indicated that, in the case of an integro-differential equation, one may use differential inequalities and Liapunov functionals to bypass the resolvent.

Nevertheless, much has been discovered about (7.1.2), both theoretically and technically. The interested reader is referred to Miller [(1968), (1971a, Chapter IV)], Nohel (1973), Becker (1979), and Corduneanu (1971).

In particular, when (7.1.1) is perturbed with a nonlinear term, then (7.1.4) can be used to rewrite the equation into a much more manageable form. Recall that the ordinary differential equation

$$\mathbf{x}' = A\mathbf{x} + \mathbf{f}(t, \mathbf{x})$$

may be expressed as

$$\mathbf{x}(t) = e^{At}\mathbf{x}_0 + \int_0^t e^{A(t-s)}\mathbf{f}(s, \mathbf{x}(s)) \, ds \,.$$

Similarly, the solution of

$$\mathbf{x}(t) = \int_0^t C(t, s)\mathbf{x}(s) \, ds + \mathbf{h}(t, \mathbf{x}(\cdot)) \,,$$

where h is an appropriate functional, may be expressed with the aid of (7.1.3) as

$$\mathbf{x}(t) = \mathbf{h}(t, \mathbf{x}(\cdot)) - \int_0^t R(t, u)\mathbf{h}(u, \mathbf{x}(\cdot)) \, du \,.$$

For special functionals this may be simplified, as may be seen in Miller (1971a, Chapter IV).

Whereas we have seen that integro-differential equations may be expressed as integral equations, there are certain advantages to considering resolvents of integro-differential equations directly. We consider

$$\mathbf{x}'(t) = \mathbf{f}(t) + A(t)\mathbf{x}(t) + \int_0^t B(t, s)\mathbf{x}(s) \, ds \,, \quad \mathbf{x}(0) = \mathbf{x}_0 \,, \qquad (7.1.7)$$

in which $\mathbf{f} : [0, a] \to R^n$ is continuous, A an $n \times n$ matrix continuous on $[0, a]$, and B an $n \times n$ matrix continuous for $0 \le s \le t \le a$.

Then we seek a solution $R(t, s)$ of the formal resolvent (or adjoint equation)

$$R_s(t, s) = -R(t, s)A(s) - \int_s^t R(t, u)B(u, s)\, du\,, \quad R(t, t) = I\,, \quad (7.1.8)$$

on the interval $0 \le s \le t$. (Here $R_s = \partial R/\partial s$.) A proof of the existence of R may be found in Grossman and Miller (1970).

Given $R(t, s)$, the solution of the initial-value problem (7.1.7) is given by

$$\mathbf{x}(t) = R(t, 0)\mathbf{x}_0 + \int_0^t R(t, s)\mathbf{f}(s)\, ds\,, \qquad\qquad (7.1.9)$$

a variation of parameters formula.

Assuming the existence of $R(t, s)$, (7.1.9) may be verified as follows. Let $\mathbf{x}(t)$ be the solution of (7.1.7) and integrate by parts. We have

$$\int_0^t R(t, s)\mathbf{x}'(s)\, ds = R(t, s)\mathbf{x}(s)\big|_{s=0}^{s=t} - \int_0^t R_s(t, s)\mathbf{x}(s)\, ds$$

or

$$\int_0^t \big[R(t, s)\mathbf{x}'(s) + R_s(t, s)\mathbf{x}(s)\big]\, ds = R(t, t)\mathbf{x}(t) - R(t, 0)\mathbf{x}_0$$

$$= \mathbf{x}(t) - R(t, 0)\mathbf{x}_0$$

as $R(t, t) = I$. Now, because $\mathbf{x}(t)$ satisfies (7.1.7) we write this as

$$\mathbf{x}(t) = R(t, 0)\mathbf{x}_0 + \int_0^t \left\{ R(t, s)\left[\mathbf{f}(s) + A(s)\mathbf{x}(s) + \int_0^s B(s, u)\mathbf{x}(u)\, du\right] \right.$$

$$\left. + R_s(t, s)\mathbf{x}(s) \right\} ds\,.$$

Changing the order of integration we have

$$\int_0^t \int_0^s R(t, s)B(s, u)\mathbf{x}(u)\, du\, ds = \int_0^t \int_u^t R(t, s)B(s, u)\mathbf{x}(u)\, ds\, du$$

$$= \int_0^t \int_s^t R(t, u)B(u, s)\mathbf{x}(s)\, du\, ds\,.$$

Then

$$\mathbf{x}(t) - R(t, 0)\mathbf{x}_0 - \int_0^t R(t, s)\mathbf{f}(s)\, ds$$

$$= \int_0^t \left[R(t, s)A(s) + R_s(t, s) + \int_s^t R(t, u)B(u, s)\, du\right] \mathbf{x}(s)\, ds\,.$$

The integral on the right is zero according to (7.1.8), so (7.1.9) is verified.

If (7.1.7) is perturbed by a nonlinear functional, then (7.1.9) may simplify the equation. Proceeding formally again, if $R(t, s)$ satisfies (7.1.8), then the solution of

$$\mathbf{x}'(t) = \mathbf{f}(t) + A(t)\mathbf{x}(t) + \int_0^t B(t, s)\mathbf{x}(s)\, ds + \mathbf{h}(t, \mathbf{x}(\cdot)), \quad \mathbf{x}(0) = \mathbf{x}_0, \quad (7.1.10)$$

for an appropriate functional $\mathbf{h}$, may be expressed by (7.1.9) as

$$\mathbf{x}(t) = R(t, 0)\mathbf{x}_0 + \int_0^t R(t, s)[\mathbf{h}(s, \mathbf{x}(\cdot)) + \mathbf{f}(s)]\, ds. \quad (7.1.11)$$

Such results are considered in detail by Grossman and Miller (1970).

But Becker (1979) took a different view. We noted that (7.1.2) is conceptually much more complicated than (7.1.1) and the same case can be made that (7.1.8) is more complicated than (7.1.7). Becker's idea was that $R(t, s)$ could be obtained from (7.1.7) with $\mathbf{f} = 0$ and the lower limit changed. This would allow many of the same techniques used to derive information from (7.1.7) to obtain information about R. Indeed, that is exactly the case for convolution constant coefficient equations and it proves to be correct here. We follow his presentation, but more detail and applications can be found in his work.

Return to (7.1.7) with a view to showing that its solution can be expressed in terms of solutions of

$$\mathbf{y}'(t) = A(t)\mathbf{y}(t) + \int_s^t B(t, u)\mathbf{y}(u)\, du \quad (7.1.12)$$

where $0 \leq s \leq t < \infty$. We will show that for each $s \geq 0$ there is a unique $n \times n$ matrix $Z(t, s)$ such that $Z(s, s) = I$ which satisfies (7.1.12). A particular solution of (7.1.7) will then be expressed in terms of this matrix and f. In fact, it will be true that $Z(t, s) = R(t, s)$.

We will use a contraction mapping argument here. A proof of the contractive mapping principle was given in Section 3.1.

Proposition 7.1.1. *The solution* $\mathbf{x}(t)$ *of*

$$\mathbf{x}'(t) = A(t)\mathbf{x}(t) + \int_s^t B(t, u)\mathbf{x}(u)\, du, \quad \mathbf{x}(s) = \mathbf{x}_0, \quad (7.1.13)$$

is unique and exists on $[s, \infty)$.

Proof. Write (7.1.13) as

$$\mathbf{x}(t) = \mathbf{x}_0 + \int_s^t \left[A(v)\mathbf{x}(v) + \int_s^v B(v,u)\mathbf{x}(u)du \right] dv$$

$$= \mathbf{x}_0 + \int_s^t A(u)\mathbf{x}(u)\, du + \int_s^t \int_u^t B(v,u)dv\, \mathbf{x}(u)\, du$$

$$= \mathbf{x}_0 + \int_s^t \left[A(u) + \int_u^t B(v,u)\, dv \right] \mathbf{x}(u)\, du$$

where we have changed the order of integration. For a given $T > s$ and an $n \times n$ matrix $C(t,u)$ defined and continuous for $s \le u \le t \le T$, we define a matrix norm by $|C|$ to be $\sup_{s \le u \le t \le T, |\mathbf{x}| \le 1} |C(t,u)\mathbf{x}|$. Find a number r with

$$\left| A(u) + \int_u^t B(v,u)\, dv \right| \le r - 1, \quad s \le u \le t \le T.$$

Let $(M, | \cdot |_r)$ be the complete metric space of continuous functions $\phi : [s,T] \to R^n$ with $\phi(s) = \mathbf{x}_0$ and with the metric induced by the norm

$$|\phi|_r := \sup_{s \le t \le T} |\phi(t)| e^{-rt}.$$

Define $\mathbf{P} : M \to M$ by $\phi \in M$ implies that

$$(\mathbf{P}\phi)(t) = \mathbf{x}_0 + \int_s^t \left[A(u) + \int_u^t B(v,u)\, dv \right] \phi(u)\, du.$$

It is clear that $\mathbf{P}\phi$ is continuous and we will show that $\mathbf{P}$ is a contraction. To see that $\mathbf{P}$ is a contraction, let $\phi, \eta \in M$. Then

$$\left| (\mathbf{P}\phi)(t) - (\mathbf{P}\eta)(t) \right| e^{-rt} \le \int_s^t (r-1) e^{-rt+ru} |\phi(u) - \eta(u)| e^{-ru}\, du$$

$$\le |\phi - \eta|_r \int_s^t (r-1) e^{-r(t-u)}\, du$$

$$\le \frac{r-1}{r} |\phi - \eta|_r.$$

This means that $\mathbf{P}$ has a unique fixed point and it is the required solution. This completes the proof.

We will call the matrix $Z(t,s)$ which satisfies (7.1.12) with $Z(s,s) = I$ the *principal matrix solution*.

Note that $Z(t,s)$ is the solution of the initial value problem.

$$\frac{\partial}{\partial t} Z(t,s) = A(t)Z(t,s) + \int_s^t B(t,u)Z(u,s)\, du, \quad Z(s,s) = I. \quad (7.1.13)$$

Theorem 7.1.1. *The solution of (7.1.7) such that* $\mathbf{x}(0) = \mathbf{x}_0$ *is given by the variation of parameters formula*

$$\mathbf{x}(t) = Z(t,0)\mathbf{x}_0 + \int_0^t Z(t,s)\mathbf{f}(s)\,ds\,. \tag{7.1.14}$$

Proof. Define $y : [0,T] \to R^n$ by $\mathbf{y}(t) = \int_0^t Z(t,s)\mathbf{f}(s)\,ds$. Differentiating and using (7.1.13), we have

$$\mathbf{y}'(t) = Z(t,t)\mathbf{f}(t) + \int_0^t \frac{\partial}{\partial t} Z(t,s)\mathbf{f}(s)\,ds$$

$$= I\mathbf{f}(t) + \int_0^t \left(A(t)Z(t,s) + \int_s^t B(t,u)Z(u,s)\,du \right)\mathbf{f}(s)\,ds$$

$$= \mathbf{f}(t) + A(t)\int_0^t Z(t,s)\mathbf{f}(s)\,ds + \int_0^t \int_s^t B(t,u)Z(u,s)\mathbf{f}(s)\,du\,ds$$

$$= \mathbf{f}(t) + A(t)\mathbf{y}(t) + \int_0^t \int_0^u B(t,u)Z(u,s)\mathbf{f}(s)\,ds\,du$$

$$= \mathbf{f}(t) + A(t)\mathbf{y}(t) + \int_0^t B(t,u)\int_0^u Z(u,s)\mathbf{f}(s)\,ds\,du$$

$$= A(t)\mathbf{y}(t) + \int_0^t B(t,u)\mathbf{y}(u)\,du + \mathbf{f}(t)\,.$$

Thus, $\mathbf{y}(t)$ is a solution of (7.1.7) for $0 \le t \le T$. Since T is arbitrary it follows that it is a solution for all $t > s$. Moreover, since $Z(t,0)\mathbf{x}_0$ satisfies the homogeneous equation, $Z(t,0)\mathbf{x}_0 + \mathbf{y}(t)$ is the desired solution of the nonhomogeneous equation. This completes the proof.

If the lower limit of integration in (7.1.7) is replaced by τ, then the solution with $\mathbf{x}(\tau) = \mathbf{x}_0$ is given by

$$\mathbf{x}(t) = Z(t,\tau)\mathbf{x}_0 + \int_\tau^t Z(t,s)\mathbf{f}(s)\,ds\,.$$

7.2   A Floquet Theory

In Section 2.6 we had a brief look at Floquet theory for ordinary differential equations. Suppose that $Z(t)$ is the principal matrix solution of

$$\mathbf{x}' = A(t)\mathbf{x} \tag{7.2.1}$$

so that $Z'(t) = A(t)Z(t)$ and $Z(0) = I$. Then the variation of parameters formula for

$$\mathbf{x}' = A(t)x + \mathbf{f}(t) \tag{7.2.2}$$

is given by

$$\mathbf{x}(t, 0, \mathbf{x}_0) = Z(t)\mathbf{x}_0 + \int_0^t Z(t)Z^{-1}(s)\mathbf{f}(s)\,ds\,. \tag{7.2.3}$$

Suppose that $\mathbf{f}(t)$ is bounded and we want to show that solutions of (7.2.2) are bounded. Even if we know that $Z(t) \to 0$ and that $Z \in L^1[0, \infty)$, $Z^{-1}(s)$ has terms of Z divided by the determinant of Z so that $Z^{-1}(s)$ can be very large. It requires Draconian conditions on $A(t)$ to ensure boundedness of solutions.

But if $A(t + T) = A(t)$ for all t and some $T > 0$, it is possible to find a constant matrix R and a periodic matrix P with

$$Z(t) = P(t)e^{Rt}\,, \quad P(0) = I\,. \tag{7.2.4}$$

Now, the variation of parameters formula becomes

$$\mathbf{x}(t) = P(t)e^{Rt}x_0 + \int_0^t P(t)e^{R(t-s)}P^{-1}(s)\mathbf{f}(s)\,ds\,. \tag{7.2.5}$$

The critical term e^{Rt} is preserved. The matrix P is periodic and nonsingular so those terms are bounded. Thus, for bounded f we are asking that

$$\int_0^\infty |e^{Rt}|\,dt < \infty\,. \tag{7.2.6}$$

In the last section we studied the resolvent and saw that in the variation of parameters formula we would need to integrate $Z(t, s)$ with respect to s, rather than with respect to t. That did not happen in the convolution case with A constant

$$\mathbf{y}' = A\mathbf{y} + \int_0^t B(t - s)\mathbf{y}(s)\,ds + \mathbf{f}(t)\,. \tag{7.2.7}$$

For in that case we had the resolvent equation as

$$Z' = AZ + \int_0^t B(t - s)Z(s)\,ds \tag{7.2.8}$$

and the variation of parameters formula as

$$\mathbf{y}(t, 0, \mathbf{y}_0) = Z(t)\mathbf{y}_0 + \int_0^t Z(t - s)\mathbf{f}(s)\, ds\,, \qquad (7.2.9)$$

so that for bounded f, in order to get a bounded solution we needed

$$\int_0^\infty |Z(u)|\, du < \infty\,, \qquad (7.2.10)$$

a condition equivalent to

$$\int_0^\infty |z(t, 0, e_j)|\, dt < \infty\,, \quad j = 1, \ldots, n\,.$$

The goal of this section is to show that the variation of parameters formula for periodic Volterra equations allows integration of the resolvent with respect to t instead of with respect to s except on the interval $0 \leq s \leq T$. This work may be found in Becker-Burton-Krisztin (1988).

Consider the system of Volterra equations

$$\frac{d\mathbf{y}(t)}{dt} = A(t)\mathbf{y}(t) + \int_0^t B(t, s)\mathbf{y}(s)\, ds + \mathbf{f}(t) \qquad (7.2.11)$$

in which A and B are $n \times n$ matrices, $\mathbf{y}$ and $\mathbf{f}$ are vectors, $A(t+T) = A(t)$, and $B(t+T, s+T) = B(t, s)$ for some $T > 0$. It is also assumed that A and $\mathbf{f}$ are continuous on $(-\infty, \infty)$ while B is continuous for $-\infty < s \leq t < \infty$.

The main problem on which we focus is that of showing that (7.2.11) has bounded solutions for $\mathbf{f}$ bounded.

We have used Z in so many contexts above, that here we will use R when denoting the resolvent. In the last section we showed that a solution of (7.2.11) can be written as

$$\mathbf{y}(t, 0, \mathbf{y}_0) = R(t, 0)\mathbf{y}_0 + \int_0^t R(t, s)\mathbf{f}(s)\, ds\,, \qquad (7.2.12)$$

where $R(t, s)$ is an $n \times n$ matrix which is the unique solution of Becker's resolvent

$$\frac{\partial R(t, s)}{\partial t} = A(t)R(t, s) + \int_s^t B(t, u)R(u, s)\, du\,, \quad R(s, s) = I\,. \quad (7.2.13)$$

In (7.2.12) it is a point of major concern that we are integrating with respect to s. Our goal is to follow the ideas in Floquet theory and change that into integration with respect to t as we saw in (7.2.6) and (7.2.10). In particular, we want to characterize the condition

$$\sup_{t \geq 0} \int_0^t |R(t,s)|\, ds < \infty. \tag{7.2.14}$$

As a corollary we show that (7.2.14) holds if there is an $E > 0$ such that

$$\int_s^\infty |R(t,s)|\, dt \leq E \quad \text{for all} \quad s \in [0,T]. \tag{7.2.15}$$

In preparation for the main result we first prove a special form of Sobolev's inequality.

If $g : [a,b] \to R^n$ has a continuous derivative, then

$$\int_a^t \big(|g(u)| + (b-a)|g'(u)|\big)\, du \geq (b-a) \max_{a \leq u \leq b} |g(u)|. \tag{7.2.16}$$

A simple proof proceeds as follows. Let $u_0, u_1 \in [a,b]$ and $m, M \in R$ be defined by

$$m = \min_{a \leq u \leq b} |g(u)| = |g(u_0)|, \quad M = \max_{a \leq u \leq b} |g(u)| = |g(u_1)|.$$

Then

$$\int_a^b \big(|g(u)| + (b-a)|g'(u)|\big)\, du$$
$$\geq (b-a)m + (b-a)\left| \int_{u_0}^{u_1} g'(u)\, du \right|$$
$$\geq (b-a)m + (b-a)|g(u_1) - g(u_0)|$$
$$\geq (b-a)m + (b-a)\big(|g(u_1)| - |g(u_0)|\big) = (b-a)M.$$

One may verify that (7.2.16) holds for $n \times n$ matrices using the induced matrix norm.

Theorem 7.2.1. *Let A and B be continuous, $R(t,s)$ satisfy (7.2.13),*

$$B(t+T, s+T) = B(t,s) \quad \text{and} \quad A(t+T) = A(t). \tag{7.2.17}$$

(i) *If there is $J > 0$ such that*

$$\int_s^t |B(u,s)| \, du \leq J \quad \text{for} \quad 0 \leq s \leq t < \infty \tag{7.2.18}$$

and

$$\int_0^T \int_s^\infty |R(t,s)| \, dt \, ds < \infty, \tag{7.2.19}$$

then

$$\sup_{t \geq 0} \int_0^t |R(t,s)| \, ds = M \quad \text{for some} \quad M > 0. \tag{7.2.20}$$

(ii) *If there is a $K > 0$ such that*

$$\int_0^t |B(t,s)| \, ds \leq K \quad \text{for} \quad t \geq 0, \tag{7.2.21}$$

then (7.2.20) implies (7.2.19).

Proof. If we integrate (7.2.13) from s to t, $s \leq t$, we obtain

$$\int_s^t |\partial R(u,s)/\partial u| \, du$$

$$\leq \|A\| \int_s^t |R(u,s)| \, du + \int_s^t \int_s^u |B(u,v)| \, |R(v,s)| \, dv \, du$$

$$\leq (\|A\| + J) \int_s^t |R(u,s)| \, du,$$

where $\|A\| = \max_{0 \leq t \leq T} |A(t)|$, as may be seen by interchanging the order of integration.

We will shortly be needing the fact that, under the stated conditions,

$$R(t+T, s+T) = R(t,s),$$

as may be seen in Burton (1985; p. 105).

Let $t > T$ be fixed. There exists an integer $k \geq 1$ and $\eta \in [0, T)$ with $t = kT + \eta$. Then,

$$\int_0^t |R(t, s)| \, ds = \int_0^{kT} |R(kT + \eta, s)| \, ds + \int_{kT}^{kT+\eta} |R(kT + \eta, s)| \, ds$$

$$= \int_0^{kT} |R(kT + \eta, s)| \, ds + \int_0^{\eta} |R(\eta, u)| \, du$$

(using $R(t + T, s + T) = R(t, s)$

and a variable change)

$$= \int_0^T \sum_{i=1}^k |R(iT + \eta, s)| \, ds + \int_0^{\eta} |R(\eta, u)| \, du$$

(by induction)

$$\leq \alpha + \int_0^T \sum_{i=1}^k \max_{iT \leq u \leq (i+1)T} |R(u, s)| \, ds \,,$$

where $\alpha = \sup_{0 \leq u \leq T} \int_0^u |R(u, s)| \, ds$. Applying (7.2.16), we have

$$\int_0^t |R(t, s)| \, ds$$

$$\leq \alpha + \int_0^T \sum_{i=1}^k \int_{iT}^{(i+1)T} \left[\frac{1}{T} |R(u, s)| + \left| \frac{\partial R(u, s)}{\partial u} \right| \right] \, du \, ds$$

$$\leq \alpha + \int_0^T \int_s^{\infty} \left[\frac{1}{T} |R(u, s)| + \left| \frac{\partial R(u, s)}{\partial u} \right| \right] \, du \, ds$$

$$\leq \alpha + \left(\|A\| + J + \frac{1}{T} \right) \int_0^T \int_s^{\infty} |R(u, s)| \, du \, ds \,.$$

Since t is arbitrary, (7.2.19) implies (7.2.20).

Now assume that (7.2.20) and (7.2.21) hold. In order to prove (7.2.19), by Fubini's theorem and the continuity of $R(t, s)$, it suffices to show that

$$\int_T^{\infty} \int_0^T |R(t, s)| \, ds \, dt < \infty \,.$$

Let

$$r(t) = \int_0^T |R(t, s)| \, ds \quad \text{for} \quad t \geq T \,.$$

Then for $t_2 \geq t_1 \geq T$ we have

$$|r(t_2) - r(t_1)| = \left| \int_0^T \left(|R(t_2, s)| - |R(t_1, s)| \right) ds \right|$$

$$\leq \int_0^T |R(t_2, s) - R(t_1, s)| \, ds$$

$$= \int_0^T \left| \int_{t_1}^{t_2} (\partial R(t, s)/\partial t) \, dt \right| ds$$

$$\leq \int_0^T \int_{t_1}^{t_2} |\partial R(t, s)/\partial t| \, dt \, ds \, .$$

Changing the order of integration yields

$$|r(t_2) - r(t_1)| \leq \int_{t_1}^{t_2} \int_0^T |\partial R(t, s)/\partial t| \, ds \, dt$$

$$= \int_{t_1}^{t_2} \int_0^T \left| A(t) R(t, s) + \int_s^t B(t, v) R(v, s) \, dv \right| ds \, dt$$

$$\leq \int_{t_1}^{t_2} \|A\| \int_0^T |R(t, s)| \, ds \, dt + \int_{t_1}^{t_2} \int_0^t \int_s^t |B(t, v)| \, |R(v, s)| \, dv \, ds \, dt$$

$$\leq \int_{t_1}^{t_2} \|A\| \int_0^t |R(t, s)| \, ds \, dt + \int_{t_1}^{t_2} \int_0^t \int_0^v |R(v, s)| \, ds \, |B(t, v)| \, dv \, dt$$

$$\leq \int_{t_1}^{t_2} (\|A\| M + KM) \, dt \leq (\|A\| M + KM)(t_2 - t_1) \, .$$

This shows that $r(t)$ is Lipschitz continuous with Lipschitz constant $L = (\|A\| + K)M$.

Now (7.2.20) and $R(t + T, s + T) = R(t, s)$ imply that

$$\sum_{i=1}^{\infty} r(iT + \eta) \leq M$$

for all $\eta \in [0, T)$.

Let $k > 0$ be an integer. It follows from the Lipschitz condition on r that

$$r(iT + \eta + u) \leq r(iT + \eta) + L(T/k)$$

for $i = 1, 2, \ldots, \eta \in [0, T)$, $u \in [0, T/k]$. Thus,

$$
\int_T^{kT} r(t)dt = \sum_{i=1}^{k-1}\sum_{j=0}^{k-1} \int_{iT+j(T/k)}^{iT+(j+1)(T/k)} r(t)\,dt
$$

$$
\leq \sum_{i=1}^{k-1}\sum_{j=0}^{k-1} (T/k)\{r(iT + j[T/k]) + L(T/k)\}
$$

$$
\leq \left\{ (T/k)\sum_{j=0}^{k-1}\sum_{i=1}^{k-1} r(iT + j[T/k]) \right\} + (k-1)k(T/k)L(T/k)
$$

$$
\leq TM + LT^2 < \infty.
$$

Since k is arbitrary, $r \in L^1[T, \infty)$ and the proof is complete.

Corollary. *Let A and B be continuous, $R(t, s)$ satisfy (7.2.13), and let (7.2.17) and (7.2.18) hold. If there is $E > 0$ such that*

$$
\int_s^\infty |R(t, s)|\, dt \leq E \quad \text{for all} \quad s \in [0, T], \tag{7.2.22}
$$

then

$$
\sup_{t \geq 0} \int_0^t |R(t, s)|\, ds = M \quad \text{for some} \quad M > 0. \tag{7.2.20}
$$

is satisfied.

Proof. If we integrate (7.2.22) from 0 to T, the value is bounded by ET.

Theorem 7.2.2. *Suppose that (7.2.17) and (7.2.18) hold with*

$$
\int_0^\infty |R(t, 0)|\, dt < \infty.
$$

Then $R(t, 0) \to 0$ as $t \to \infty$.

Proof. We showed in the proof of Theorem 7.2.1 that

$$\int_0^\infty |\partial R(u,0)/\partial u|\, du \le (\|A\| + J) \int_0^\infty |R(u,0)|\, du\,.$$

A similar result holds for each jth column of $R(t,0)$, say $z(t,0,e_j)$. If the theorem is false, there is a j, an $\epsilon > 0$, and a sequence $\{t_n\} \to \infty$ with $|z(t_n,0,e_j)| \ge \epsilon$. Also,

$$z(t,0,e_j) = e_j + \int_0^t z'(u,0,e_j)\, du$$

so that $t_n \le t \le t_n + 1$ implies that

$$\left| z(t,0,e_j) - z(t_n,0,e_j) \right| \le \int_{t_n}^t |z'(u,0,e_j)|\, du < \epsilon/2$$

for large n. Hence $|z(t,0,e_j)| \ge \epsilon/2$ for $t_n \le t \le t_n + 1$, contradicting $z(t,0,e_j) \in L^1$. This completes the proof.

Example 7.2.1. Consider the scalar equation

$$z' = -A(t)z + \int_s^t B(t,u)z(u)\, du$$

in which (7.2.17) and (7.2.18) hold with A and B continuous. If there is an $\alpha > 0$ and $s^* \in [0,T]$ with

$$|B(t - s + s^*, s^*)| \ge |B(t,s)| \quad \text{for} \quad 0 \le s \le t < \infty$$

and with

$$-A(t) + \int_0^\infty |B(u + s^*, s^*)|\, du \le -\alpha \quad \text{for all} \quad t\,,$$

then the conditions of the corollary hold.

Proof. Define a Liapunov functional by

$$V(t, z(\cdot)) = |z(t)| + \int_s^t \int_{t-u}^\infty |B(v + s^*, s^*)|\, dv\, |z(u)|\, du$$

so that

$$V'(t, z(\cdot)) \le -A(t)|z(t)| + \int_s^t |B(t,u)|\, |z(u)|\, du$$

$$+ \int_0^\infty |B(v + s^*, s^*)|\, dv\, |z(t)|$$

$$- \int_s^t |B(t - u + s^*, s^*)|\, |z(u)|\, du$$

$$\le -\alpha|z(t)|\,.$$

Hence, the single column of R satisfies

$$0 \leq V(t, z(\cdot)) \leq V(s, e_1) - \alpha \int_s^t |z(u, s, e_1)| \, du$$

or

$$\int_s^\infty |z(u, s, e_1)| \, du \leq \frac{V(s, e_1)}{\alpha} = |e_1|/\alpha = 1/\alpha \,.$$

This completes the proof.

Perhaps a more transparent way of doing such an example is to ask for a function $C(t - s)$ with $C(t - s) \geq |B(t, s)|$ and with

$$-A(t) + \int_0^\infty C(s) \, ds \leq -\alpha \,.$$

Then V would be defined by

$$V(t, z(\cdot)) = |z(t)| + \int_s^t \int_{t-u}^\infty |C(v)| \, dv \, |z(u)| \, du \,.$$

Here is one of the main applications of Theorem 7.2.1. It is known that there are examples of (7.2.11) which do have periodic solutions; indeed, when $B = 0$ they are very common. But for a general B they are rare. The reason for that is that the right-hand-side of (7.2.11) is generally not periodic even when y is periodic. But (7.2.11) can have solutions which are asymptotically periodic in the sense described below.

Under the conditions of Theorem 7.2.1, the following is shown in Burton (1985; p. 102). Suppose that

$$\lim_{n \to \infty} \int_{-nT}^t |B(t, s)| \, ds = \int_{-\infty}^t |B(t, s)| \, ds \qquad (7.2.23)$$

is bounded and continuous in t, that

$$\int_0^t |R(t, s)| \, ds \leq M \quad \text{for} \quad t \geq 0 \,, \qquad (7.2.24)$$

that $R(t, 0) \to 0$ as $t \to \infty$, and that $\mathbf{f}(t + T) = \mathbf{f}(t)$. Then there exists a sequence of positive integers $\{n_j\}$ such that the function $\mathbf{y}$ defined in (7.2.12) satisfies

$$\mathbf{y}(t + n_j T, 0, \mathbf{y}_0) \to \int_{-\infty}^t R(t, s) \mathbf{f}(s) \, ds := \mathbf{x}(t) \,, \quad j \to \infty \,, \qquad (7.2.25)$$

where $x(t)$ is a $T-$periodic solution of

$$\frac{d\mathbf{x}(t)}{dt} = A(t)\mathbf{x} + \int_{-\infty}^{t} B(t,s)x(s)\,ds + \mathbf{f}(t) \qquad (7.2.25)$$

on $(-\infty, \infty)$.

7.3   UAS and Integrability of the Resolvent

The basic perturbation result in the previous chapter, Theorem 6.4.5, concerns the equation

$$\mathbf{x}' = A\mathbf{x} + \int_0^t C(t,s)\mathbf{x}(s)\,ds + \int_0^t D(t,s)\mathbf{r}(\mathbf{x}(s))\,ds$$
$$+ \mathbf{q}(t,\mathbf{x}) + \mathbf{H}(t,\mathbf{x}(\cdot)) + \mathbf{p}(t,\mathbf{x})\,. \qquad (6.4.11)$$

It depends on the Liapunov functional

$$V(t,\mathbf{x}(\cdot)) = \left[\mathbf{x}^T B\mathbf{x}\right]^{1/2} + \bar{K}\int_0^t \int_t^\infty |C(u,s)|\,du|\mathbf{x}(s)|\,ds$$

for the system

$$\mathbf{x}' = A\mathbf{x} + \int_0^t C(t,s)\mathbf{x}(s)\,ds\,,$$

which is required to have nice properties. Other perturbation results are found in Chapters 2 and 5.

In this section we present and discuss the work of Hino and Murakami (1996) and Zhang (1997) who consider a system

$$\mathbf{x}'(t) = A(t)\mathbf{x}(t) + \int_0^t B(t,s)\mathbf{x}(s)\,ds \qquad (7.3.1)$$

where A, B are $n \times n$ matrix functions, $A(t)$ is continuous on $[0, +\infty)$, $B(t,s)$ is continuous for $0 \leq s \leq t < \infty$. The classical resolvent equation of (7.3.1) is

$$\frac{\partial R(t,s)}{\partial s} = -R(t,s)A(s) - \int_s^t R(t,u)B(u,s)\,du, \quad R(t,t) = I, \quad (7.3.2)$$

for $t \geq s \geq 0$, while Becker's resolvent is

$$\frac{\partial R(t,s)}{\partial t} = A(t)R(t,s) + \int_s^t B(t,u)R(u,s)\,du \quad R(s,s) = I, \qquad (7.3.3)$$

for $t \geq s \geq 0$, where I is the $n \times n$ identity matrix. When A is a constant matrix and $B(t, s) = B(t - s)$ is of convolution type, equation (7.3.1) becomes

$$\mathbf{x}'(t) = A\mathbf{x}(t) + \int_0^t B(t - s)\mathbf{x}(s)\, ds \qquad (7.3.4)$$

and the resolvent equation (7.3.3) is

$$Z'(t) = AZ(t) + \int_0^t B(t - s)Z(s)\, ds, \quad Z(0) = I.$$

If we apply the standard variation of parameters formula to

$$\mathbf{x}'(t) = A(t)\mathbf{x}(t) + \int_0^t B(t, s)\mathbf{x}(s)\, ds + \mathbf{p}(t)$$

then a solution $\mathbf{x}(t) = \mathbf{x}(t, t_0, \boldsymbol{\phi})$ of the perturbed equation may be expressed as

$$\mathbf{x}(t, t_0, \boldsymbol{\phi}) = R(t, t_0)\boldsymbol{\phi}(t_0) + \int_{t_0}^t R(t, s) \int_0^{t_0} B(s, u)\boldsymbol{\phi}(u)\, du\, ds$$
$$+ \int_{t_0}^t R(t, s)\mathbf{p}(s)\, ds.$$

Moreover, $\mathbf{p}$ may depend on $\mathbf{x}$; thus, we already have useful tools for dealing with perturbations of (7.3.1) when the resolvent is integrable. In the convolution case with A constant, Miller (1971b) showed that for $B \in L^1[0, \infty)$, the zero solution of (7.3.4) is uniformly asymptotically stable if and only if the resolvent $Z \in L^1[0, \infty)$. We now follow the work of Hino and Murakami (1996) and Zhang (1997) to prove an extension of Miller's result to (7.1.1). We require that

(H$_1$) $\sup\limits_{t \geq 0} \left\{ |A(t)| + \int_0^t |B(t, s)|\, ds \right\} < \infty$,

(H$_2$) for any $\sigma > 0$, there exists an $S = S(\sigma) > 0$ such that

$$\int_0^{t-S} |B(t, u)|\, du < \sigma \quad \text{for all} \quad t \geq S,$$

(H$_3$) $A(t)$ and $B(t, t + s)$ are bounded and uniformly continuous in

$$(t, s) \in \left\{ (t, s) \in [0, \infty) \times K \mid -t \leq s \leq 0 \right\}$$

$$\text{for any compact set } K \subset (-\infty, 0].$$

Let $R = (-\infty, \infty)$, $R^+ = [0, \infty)$, and $R^- = (-\infty, 0]$ respectively. For $\mathbf{x} \in R^n$, $|\cdot|$ denotes the Euclidean norm of x. For any $n \times n$ matrix A, define the norm $|A|$ of A by $|A| = \sup \{|Ax| : |x| \leq 1\}$. For any interval $J \subset R$, we denote by $C(J)$ the set of continuous functions $\phi : J \to R^n$, and set $|\phi|_J = \sup \{|\phi(s)| : s \in J\}$. For each $t_0 \in R^+$ and $\phi \in C([0, t_0])$, there is a unique function $x : R^+ \to R^n$ which satisfies (7.3.1) on $[t_0, +\infty)$ with $\mathbf{x}(s) = \phi(s)$ for $0 \leq s \leq t_0$ (see Section 2.1). Such a function $\mathbf{x}(t)$ is called a solution of (7.3.1) through (t_0, ϕ), and is denoted by $\mathbf{x}(t, t_0, \phi)$.

Definition 7.3.1. *The zero solution of (7.3.1) is uniformly stable (US) if for any $\varepsilon > 0$, there exists a $\delta = \delta(\varepsilon) > 0$ such that $[t_0 \geq 0,\ \phi \in C([0, t_0]),\ |\phi|_{[0,t_0]} < \delta]$ imply $|\mathbf{x}(t, t_0, \phi)| < \varepsilon$ for $t \geq t_0$.*

Definition 7.3.2. *The zero solution of (7.3.1) is uniformly asymptotically stable (UAS) if it is US and there exists a $\delta_0 > 0$ with the property that for each $\varepsilon > 0$ there exists $T = T(\varepsilon)$ such that $[t_0 \geq 0,\ \phi \in C([0, t_0]),\ |\phi|_{[0,t_0]} < \delta_0,\ t \geq t_0 + T]$ imply $|\mathbf{x}(t, t_0, \phi)| < \varepsilon$.*

Definition 7.3.3. *The zero solution of (7.3.1) is totally stable (TS) if for any $\varepsilon > 0$, there exists a $\delta = \delta(\varepsilon) > 0$ such that $[t_0 \geq 0,\ \phi \in C([0, t_0]),\ \mathbf{p} \in C([t_0, \infty)),\ |\phi|_{[0,t_0])} < \delta,\ |\mathbf{p}|_{[t_0, \infty)} < \delta]$ imply $|\mathbf{x}(t, t_0, \phi, \mathbf{p})| < \varepsilon$, where $\mathbf{x}(t) = \mathbf{x}(t, t_0, \phi, \mathbf{p})$ is a solution of*

$$\mathbf{x}'(t) = A(t)\mathbf{x}(t) + \int_0^t B(t, s)\mathbf{x}(s)\, ds + \mathbf{p}(t) \tag{7.3.5}$$

such that $\mathbf{x}(s) = \phi(s)$ for $s \in [0, t_0]$.

Theorem 7.3.1. Zhang (1997)   *Under (H_1), (H_2), and (H_3), the zero solution of (7.3.1) is UAS if and only if*

$$\sup_{t \geq 0} \int_0^t |R(t, s)|\, ds < \infty. \tag{7.3.6}$$

The proof of this theorem is based on a series of results of Hino and Murakami (1996) and Zhang (1997) on uniform asymptotic stability and total stability of (7.3.1).

Lemma 7.3.1. Zhang (1997)   *If (H_1) and (7.3.6) hold, then there exists a constant K such that $|R(t, s)| \leq K$ for all $0 \leq s \leq t < \infty$.*

Proof. Since $R(t, s)$ is a solution of (7.3.2), we obtain

$$R(t, s) = I + \int_s^t R(t, u)A(u)\, du + \int_s^t \int_v^t R(t, u)B(u, v)\, du\, dv\,.$$

Interchange the order of integration in the last term to obtain

$$R(t, s) = I + \int_s^t R(t, u)A(u)\, du + \int_s^t R(t, u) \int_s^u B(u, v)\, dv\, du$$

and

$$|R(t, s)| \le 1 + \int_s^t |R(t, u)|\,|A(u)|\, du$$

$$+ \int_s^t |R(t, u)| \int_s^u |B(u, v)|\, dv\, du\,. \qquad (7.3.7)$$

By (H_1) and (7.3.6), there are positive constants M and L such that

$$\sup_{t \ge 0} \left\{ |A(t)| + \int_0^t |B(t, s)|\, ds \right\} < M$$

and

$$\sup_{t \ge 0} \int_0^t |R(t, s)|\, ds < L\,.$$

It then follows from (7.3.7) that $|R(t, s)| \le 1 + ML =: K$. This completes the proof.

Lemma 7.3.2. Zhang (1997)   *The matrix functions $A(t)$ and $B(t, t+s)$ can be continuously extended to $(t, s) \in R \times R^-$ with $\bar{A}(t) = A(t)$ and $\bar{B}(t, t + s) = B(t, t + s)$ on $\Omega = \{(t, s) \in R^+ \times R^- \mid -t \le s \le 0\}$. Moreover, if (H_1)–(H_3) hold for $A(t)$ and $B(t, t + s)$, then the extensions $\bar{A}(t)$ and $\bar{B}(t, t + s)$ satisfy the following conditions:*

$(\bar{H}_1)$ $\displaystyle \sup_{t \in R} \left\{ |\bar{A}(t)| + \int_{-\infty}^t |\bar{B}(t, s)|\, ds \right\} =: M < \infty,$

$(\bar{H}_2)$ *for any $\sigma > 0$, there exists an $S = S(\sigma) > 0$ such that*

$$\int_{-\infty}^{t-S} |\bar{B}(t, u)|\, du < \sigma \quad \text{for all} \quad t \in R\,,$$

$(\bar{H}_3)$ $\bar{A}(t)$ *and $\bar{B}(t, t + s)$ are bounded and uniformly continuous in $(t, s) \in R \times K$ for any compact set $K \subset R^-$.*

We omit the proof of Lemma 7.3.2 here and refer the reader to Zhang (1997) for detailed construction of these extensions.

Theorem 7.3.2.   Hino and Murakami (1996)   *Suppose that* (H_1) *and* (H_2) *hold. If the zero solution of* (7.3.1) *is TS, then it is UAS.*

Proof. First, notice by definition that when the zero solution of (7.3.1) is TS then it is US. Let $t_0 \in R^+$ and $\phi \in C([0, t_0])$ with $\|\phi\| < \delta(1)$, $\delta(\cdot)$ is the one given for the TS of the zero solution of (7.1.1). Then $|\mathbf{x}(t, t_0, \phi)| < 1$ for all $t \geq t_0$. Now for any $\varepsilon > 0$ $(0 < \varepsilon < 1)$, $\alpha > 0$, we set

$$u(t) = u(t, \alpha, \varepsilon) = \begin{cases} \dfrac{1 + 2\alpha t}{1 + \varepsilon \alpha t} & \text{if } t \geq 0 \\ 1 & \text{if } t < 0, \end{cases}$$

define $\mathbf{y}(t)$ by

$$\mathbf{y}(t) = u(t - t_0)x(t), \quad t \in R^+,$$

and $\mathbf{p}(t)$ by

$$\mathbf{p}(t) = u'(t - t_0)\mathbf{x}(t) + \int_0^t B(t, s)\mathbf{x}(s)\big[u(t - t_0) - u(s - t_0)\big]\, ds \quad (7.3.8)$$

for $t \geq t_0$. One may verify that $\mathbf{y}(t)$ satisfies (7.3.5) for $t \geq t_0$ with $\mathbf{p}(t)$ defined in (7.3.8). Notice also that for $t \geq 0$

$$u(t) = \frac{2}{\varepsilon}\left(1 - \frac{2 - \varepsilon}{2 + 2\varepsilon\alpha t}\right)$$

and

$$u'(t) = \frac{\alpha(2 - \varepsilon)}{(1 + \alpha\varepsilon t)^2}.$$

This yields $1 \leq u(t) \leq 2/\varepsilon$, $|u(t) - u(s)| \leq 2\alpha|t - s|$ for $t, s \in R$. It follows from (H_2) that for any $\eta > 0$, there exists an $S = S(\eta) > 0$ such that

$$\int_0^{t-S} |B(t, u)|\, du < \eta$$

for all $t \geq S(\eta)$. By (H_1), there exists a constant $M^* > 0$ such that

$$\sup_{t \geq 0} \int_0^t |B(t, s)|\, ds < M^*.$$

Let $t \geq t_0$. Without loss of generality, we may assume that $t_0 \geq S(\eta)$. By (7.3.8) we have

$$
|\mathbf{p}(t)| \leq 2\alpha + \int_{t-S(\eta)}^{t} |B(t,s)| \, |u(t-t_0) - u(s-t_0)| \, ds
$$

$$
+ \int_{0}^{t-S(\eta)} |B(t,s)| \, |u(t-t_0) - u(s-t_0)| \, ds
$$

$$
\leq 2\alpha + 2\alpha M^* S(\eta) + 4\eta/\varepsilon
$$

for all $\eta > 0$. Thus, we may choose $\eta > 0$ and $\alpha = \alpha(\varepsilon)$ so small that $|\mathbf{p}(t)| < \delta(1)$. Since the zero solution of (7.3.1) is TS, we obtain $|\mathbf{y}(t)| < 1$ for all $t \geq t_0$. Hence if $t \geq t_0 + (1 - \varepsilon)/(\varepsilon\alpha)$, then

$$
|\mathbf{x}(t, t_0, \boldsymbol{\phi})| = |\mathbf{y}(t)/u(t-t_0)| < \big[1 + \varepsilon\alpha(t-t_0)\big]/\big[1 + 2\alpha(t-t_0)\big] \leq \varepsilon
$$

which proves the theorem.

Next we shall discuss the converse of the above theorem. To do this, we assume that $(\bar{H}_1)$–$(\bar{H}_3)$ holds and study the limiting equation of (7.3.1) with $A(t)$ and $B(t,s)$ replaced by $\bar{A}(t)$ and $\bar{B}(t,s)$.

By the Ascoli-Arzela's theorem, $(\bar{H}_3)$ implies that for any sequence $\{t'_k\}$ with $t'_k \to \infty$ as $k \to \infty$, there exists a subsequence $\{t_k\}$ of $\{t'_k\}$ and functions $D(t)$ and $E(t,s)$ such that $\bar{A}(t+t_k) \to D(t)$ and $\bar{B}(t+t_k, t+t_k+s) \to E(t, t+s)$ as $k \to \infty$ uniformly on $J \times K$ for any compact sets $J \subset R$ and $K \subset R^-$. We denote by $\Gamma(A, B)$ the set all pairs (D, E) which satisfy the above situation for some sequence $\{t_k\}$ with $t_k \to \infty$ as $k \to \infty$. We can easily see that each $(D, E) \in \Gamma(A, B)$ also satisfies $(\bar{H}_1)$–$(\bar{H}_3)$ with the same number M and $S(\sigma)$. In particular, $(A, B) \in \Gamma(A, B)$ whenever $A(t)$ and $B(t, t+s)$ are almost periodic in $t \in R$ uniformly for $s \in R^-$ [see Hino and Murakami (1991)]. If $(D, E) \in \Gamma(A, B)$, then the equation

$$
\mathbf{x}'(t) = D(t)\mathbf{x}(t) + \int_{-\infty}^{t} E(t,s)\mathbf{x}(s) \, ds \tag{L_∞}
$$

is called a limiting equation of (7.3.1). [See (7.2.23).]

In a similar way, one can also define the stability of the zero solution of (L_∞) by taking $|\phi|_{(-\infty,\tau]}$ in the place of $|\phi|_{[0,\tau]}$. It is also known (see Hino and Murakami (1991b)) that if the zero solution of (7.3.1) is UAS, then the zero solution of each limiting equation (L_∞) is also UAS with a common triple $(\delta_0, \delta(\cdot), T(\cdot))$.

Theorem 7.3.3. Hino and Murakami (1996)   *Suppose that* (H_1)–(H_3) *hold. Then the zero solution of (7.3.1) is UAS if and only if it is TS.*

Proof. The "if" part has been shown in Theorem 7.3.2. We shall prove the "only if" part by assuming that the zero solution of (7.3.1) is UAS. By Lemma 7.3.2 and the remark above, we may also assume that $(\bar{H}_1)$–$(\bar{H}_3)$ hold and the zero solution of each limiting equation (L_∞) is UAS with the same common triple $(\delta_0, \delta(\cdot), T(\cdot))$.

We now claim that for any $\varepsilon > 0$, there exists an $\alpha(\varepsilon) \geq 0$ and $\tilde{\delta}(\varepsilon) > 0$ such that $\tau \geq \alpha(\varepsilon)$, $\phi \in C([0, \tau])$, $\mathbf{p} \in C([\tau, \infty))$ with $|\phi|_{[0,\tau]} < \tilde{\delta}(\varepsilon)$, and $|\mathbf{p}|_{[\tau,\infty)} < \tilde{\delta}(\varepsilon)$ imply $|x(t, \tau, \phi, p)| < \varepsilon$ for all $t \geq \tau$. Then the zero solution of (7.3.1) is TS since it is unique. We prove the claim by the method of contradiction. Suppose there exists an ε, $0 < \varepsilon < \delta_0/2$, and sequences $\{\tau_k\} \in R^+$, $\tau_k \to \infty$ as $k \to \infty$, $\{r_k\}, r_k > 0$, and functions $\phi^k \in C([0, \tau_k])$, $\mathbf{p}^k \in C([\tau_k, \infty))$ and the solution $\mathbf{x}^k(t) = x(t, \tau_k, \phi^k, \mathbf{p}^k)$ of (7.3.5) with $\mathbf{p} = \mathbf{p}^k$ through (τ_k, ϕ^k) such that

$$|\phi^k|_{[0,\tau_k]} < \frac{1}{k}, \quad |\mathbf{p}^k|_{[\tau_k,\infty)} < \frac{1}{k}, \quad |\mathbf{x}^k(\tau_k + r_k)| = \varepsilon$$

$$\text{and} \quad |\mathbf{x}^k(t)| < \varepsilon \quad \text{for} \quad t \in [0, \tau_k + r_k). \tag{7.3.9}$$

Let $T = T(\varepsilon)$ be given in Definition 7.3.2 for UAS. We first consider the case in which the sequence $\{r_k\}$ is unbounded. Without loss of generality, we assume that

$$A(t + \tau_k + r_k - T) \to D(t)$$

and

$$B(t + \tau_k + r_k - T, t + s + \tau_k + r_k - T) \to E(t, t + s)$$

as $k \to \infty$ uniformly on any compact set in $R \times R^-$ for some $(D, E) \in \Gamma(A, B)$. Define $\mathbf{y}^k(t) = \mathbf{x}^k(t + \tau_k + r_k - T)$ for $t \geq T - \tau_k - r_k$. Then $\mathbf{y}^k(t)$ satisfies

$$\frac{d}{dt}\mathbf{y}^k(t) = A(t + \tau_k + r_k - T)\mathbf{y}^k(t)$$

$$+ \int_{T-r_k}^{t} B(t + \tau_k + r_k - T, u + \tau_k + r_k - T)\mathbf{y}^k(u)\,du$$

$$+ \int_{-\tau_k+T-r_k}^{T-r_k} B(t + \tau_k + r_k - T, u + \tau_k + r_k - T)$$

$$\cdot \phi^k(u + \tau_k + r_k - T)\,du$$

$$+ \mathbf{p}^k(t + \tau_k + r_k - T)$$

for $t \geq T - r_k$. In this case we may assume that $\{\mathbf{y}^k\}$ converges to a function y uniformly on any compact set in $(-\infty, T]$. Moreover, $\mathbf{y}$ is a solution of

(L_∞) on $[0, T]$. Letting $k \to \infty$ in (7.3.9), we have $|\mathbf{y}(t)| \leq \varepsilon$ on $(-\infty, T]$ and $|\mathbf{y}(T)| = \varepsilon$. This is a contradiction since $|\mathbf{y}|_{(-\infty,0]} \leq \varepsilon < \delta_0$ implies $|\mathbf{y}(T)| < \varepsilon$. Therefore, the sequence $\{r_k\}$ must be bounded. So, we may assume that $r_k \to r$ as $k \to \infty$ for some $r \in R^+$ and set $\tilde{\mathbf{x}}^k(t) = \mathbf{x}^k(t + \tau_k)$ for $t \geq -\tau_k$. Then $\tilde{\mathbf{x}}^k(t)$ satisfies

$$\frac{d}{dt}\tilde{\mathbf{x}}^k(t) = A(t + \tau_k)\tilde{\mathbf{x}}^k(t) + \int_0^t B(t + \tau_k, u + \tau_k)\tilde{\mathbf{x}}^k(u)\,du$$

$$+ \int_{-\tau_k}^0 B(t + \tau_k, u + \tau_k)\boldsymbol{\phi}^k(u + \tau_k)\,du + \mathbf{p}^k(t + \tau_k)$$

for $t \geq 0$. Again, we may assume that the sequence $\{\tilde{\mathbf{x}}^k\}$ converges to a function $\tilde{\mathbf{x}}$ uniformly on any compact subset $(-\infty, r]$. By the same reasoning as for $\mathbf{y}$, we see that $\tilde{\mathbf{x}}$ is a solution of some limiting equation of (7.3.1). On the other hand, it follows from (7.3.9) that $\tilde{\mathbf{x}}(t) = 0$ on R^- and $\tilde{\mathbf{x}}(r) = \varepsilon$. This is again a contradiction since we must have $\tilde{\mathbf{x}}(t) = 0$ on R by the uniqueness of solutions of (L_∞) with respect to initial functions. This shows that the zero solution of (7.3.1) is TS if it is UAS.

We are now ready to prove Theorem 7.3.1 by applying Perron's theorem [Perron (1930)] and using the properties of the resolvent $R(t, s)$ defined in (7.3.2). It is also verified in Hino and Murakami (1996) that resolvent equations (7.3.2) and Becker's resolvent (7.3.3) are equivalent.

Proof of Theorem 7.3.1. First we suppose that the zero solution of (7.3.1) is UAS. By Theorem 7.3.3, it is TS. Let $\mathbf{p} \in C(R^+)$ be bounded and $\mathbf{x}_p \in C(R^+)$ satisfy

$$\mathbf{x}'(t) = A(t)\mathbf{x}(t) + \int_0^t B(t, s)\mathbf{x}(s)\,ds + \mathbf{p}(t)$$

for $t \geq 0$ with $\mathbf{x}_p(0) = 0$. By the variation of parameters formula, we obtain

$$\mathbf{x}_p(t) = \int_0^t R(t, s)\mathbf{p}(s)\,ds \quad \text{for} \quad t \geq 0.$$

Since the zero solution of (7.3.1) is TS, we see that $\mathbf{x}_p$ is bounded on R^+. This implies that $\int_0^t R(t, s)\mathbf{p}(s)\,ds$ is bounded on R^+ whenever $\mathbf{p} \in C(R^+)$ is bounded. Applying Perron's theorem, we obtain that $\sup_{t \in R^+} \int_0^t |R(t, s)|\,ds < \infty$, and hence (7.3.6) holds.

Conversely, suppose that (7.3.6) holds with $\sup_{t \in R^+} \int_0^t |R(t, s)|\,ds = L$ for some $L > 0$. By Lemma 7.3.1, there exists a constant $K > 0$ such that

$|R(t, s)| \leq K$ for all $t \geq s \geq 0$. Let $\mathbf{x}(t) = \mathbf{x}(t, t_0, \phi, \mathbf{p})$ be a solution of (7.3.5). By the variation of parameters formula again, we obtain

$$\mathbf{x}(t, t_0, \phi) = R(t, t_0)\phi(t_0) + \int_{t_0}^{t} R(t, s) \int_{0}^{t_0} B(s, u)\phi(u)\, du\, ds$$
$$+ \int_{t_0}^{t} R(t, s)\mathbf{p}(s)\, ds\,.$$

This implies that

$$|\mathbf{x}(t)| \leq |R(t, t_0)|\,|\phi(t_0)| + \int_{0}^{t} |R(t, s)| \int_{0}^{s} |B(s, u)|\, du\, ds\, |\phi|_{[0, t_0]}$$
$$+ \int_{t_0}^{t} |R(t, s)|\, ds\, |\mathbf{p}|_{[t_0, \infty)}$$
$$\leq (K + LM^*)|\phi|_{[0, t_0]} + L|\mathbf{p}|_{[t_0, \infty)}\,.$$

Here we have used condition (H_1) with $\sup_{t \geq 0} \int_{0}^{t} |B(t, s)|\, ds < M^*$. For any $\varepsilon > 0$, choose $\delta > 0$ such that $(K + LM^* + L)\delta < \varepsilon$. If $|\phi|_{[0, t_0]} < \delta$ and $|\mathbf{p}|_{[t_0, \infty)} < \delta$, then $|\mathbf{x}(t)| < \varepsilon$ for all $t \geq t_0$. Therefore, the zero solution of (7.3.1) is TS. By Theorem 7.3.2, it is UAS. This completes the proof.

Remark 7.3.1. The integral condition (7.3.6) on the resolvent seems to be very difficult to verify directly. When (7.3.1) is "periodic", Section 7.2 provides an alternative condition which is rather easy to check. Indeed, under the assumptions that $A(t + \omega) = A(t)$, $B(t + \omega, s + \omega) = B(t, s)$ for some $\omega > 0$ and $\sup_{s \geq 0} \int_{s}^{\infty} |B(t, s)|dt < \infty$, condition (7.3.6) follows from

$$\sup_{s \geq 0} \int_{s}^{\infty} |R(t, s)|\, dt < \infty\,. \tag{7.3.10}$$

which is satisfied if there exists a Liapunov functional $V(t, \phi)$ such that

$$V'_{(7.3.1)}(t, \mathbf{x}_t) \leq -c|\mathbf{x}(t)|\,, \quad c > 0$$

along a solution $\mathbf{x}(t) = \mathbf{x}(t, t_0, \phi)$ of (7.3.1).

The work of Sections 7.1 and 7.2 continues in Burton (2005) by means of fixed point theory. Many simple examples of stability by fixed point theory are found in Seiji (1989), Burton-Furumochi (2001), Burton (2003), and for difference equations in Serban (2001). When we study stability by means of Liapunov's direct method, the central problem is to construct a Liapunov functional which is positive definite with a derivative which is at least negative semi-definite. But when we study stability by means of fixed point theory, then the central problem is to define a mapping into a space of functions which would be acceptable stable solutions; one must then show that the mapping has a fixed point and that the fixed point satisfies the differential equation. It turns out that the union of the two methods is far better than either alone.

Chapter 8

Functional Differential Equations

8.0 Introduction

This chapter contains a survey of results concerning problems in general functional differential equations that we encountered in previous chapters for integral and integro-differential equations. Those problems were extensively discussed earlier, so we suppose the reader to be familiar with the background and, hence, primarily just state and prove theorems.

Sections 8.1 and 8.2 deal with existence, uniqueness, continuation, stability, and asymptotic stability for a very general functional differential equation. The reader should consult Chapters 2, 3, and 6 for general facts, problems, and insights concerning these questions, as well as their relations to corresponding problems in ordinary differential equations. The view throughout is that the subject may be developed using Liapunov functionals and Razumikhin techniques.

Sections 8.3, 8.5, and 8.6 are concerned primarily with a functional differential equation

$$\mathbf{x}' = \mathbf{F}(t, \mathbf{x}_t),$$

where $\mathbf{x}_t$ is defined on $[t - h, t]$ for some $h > 0$. In particular, 8.3 deals with boundedness and stability; 8.5 concerns limit sets of autonomous systems; and 8.6 studies the existence of periodic solutions.

Section 8.7 considers limit sets of nonautonomous systems, usually with an unbounded delay.

We concentrate on the following problems.

(a)   What conditions on a Liapunov functional are needed for uniform asymptotic stability?

(b)   What are alternatives to the condition $\mathbf{x}'(t)$ bounded for $\mathbf{x}(\cdot)$ bounded in proving stability and boundedness?

(c)   If a Liapunov functional satisfies $V' \leq 0$ for $|\mathbf{x}(t)| \geq M > 0$, then what more is needed to conclude boundedness, uniform boundedness, or uniform ultimate boundedness?

8.1   Existence and Uniqueness

We consider a system of Volterra functional differential equations $x_i'(t) = f_i(t, x_1(s), \ldots, x_n(s))$; $\alpha \leq s \leq t)$ for $t \geq t_0$, $\alpha \geq -\infty$, $\alpha \leq t_0$, and $i = 1, \ldots, n$. These equations are written as

$$\mathbf{x}'(t) = \mathbf{F}(t, \mathbf{x}(\cdot)), \quad t > t_0, \tag{8.1.1}$$

where $\mathbf{x}(\cdot)$ represents the function $\mathbf{x}$ on the interval $[\alpha, t]$ with the value of t always determined by the first coordinate of $\mathbf{F}$ in (8.1.1). Thus, (8.1.1) is a delay differential equation.

This section and part of the next will closely follow the excellent paper by Driver (1962), which remains the leading authority on the subject of fundamental theory for (8.1.1). As Driver notes, much of his material is found elsewhere in varying forms; in particular, the early work is from Krasovskii, El'sgol'ts, Myshkis, Corduneanu, Lakshmikantham, and Razumikhin. But important formulations, corrections, and general synthesis are by Driver.

Notation.

(a) If $\mathbf{x} \in R^n$, then $|\mathbf{x}| = \max_{i=1,\ldots,n} |x_i|$.

(b) If $\boldsymbol{\psi} : [a, b] \to R^n$, then

$$\|\boldsymbol{\psi}\|^{[a,b]} = \sup_{a \leq s \leq b} \|\boldsymbol{\psi}(s)\|.$$

(c) For any interval $[a, b]$ and any $D \subset R^n$, then $C([a, b] \to D)$ denotes the class of continuous functions $\boldsymbol{\psi} : [a, b] \to D$.

Because α can be $-\infty$, one accepts the following.

Convention. If $\alpha = -\infty$, then intervals $[\alpha, t]$ and $[\alpha, \gamma)$ mean $(-\infty, t]$ and $(-\infty, \gamma)$, respectively, and $\boldsymbol{\psi} \in C([\alpha, t] \to D)$ means that there is a compact set $L_{\boldsymbol{\psi}} \subset D$ such that $\boldsymbol{\psi} \in C((-\infty, t] \to L_{\boldsymbol{\psi}})$. This implies that $\boldsymbol{\psi} \in C([\alpha, t] \to D)$ with $t_0 \leq t < \gamma$ means $\boldsymbol{\psi} \in C([\alpha, t] \to L_{\boldsymbol{\psi}})$ for some compact set $L_{\boldsymbol{\psi}} \subset D$, regardless of whether α is finite or not.

Definition 8.1.1. *The functional* $\mathbf{F}(t, \mathbf{x}(\cdot))$ *will be called*

(a) *continuous in* t *if* $\mathbf{F}(t, \mathbf{x}(\cdot))$ *is a continuous function of* t *for* $t_0 \le t < \gamma$ *whenever* $\mathbf{x} \in C\big([\alpha, \gamma) \to D\big)$,

(b) *locally Lipschitz with respect to* $\mathbf{x}$ *if, for every* $\bar{\gamma} \in [t_0, \gamma)$ *and every compact set* $L \subset D$, *there exists a constant* $K_{\bar{\gamma}, L}$ *such that*

$$\big|\mathbf{F}(t, \mathbf{x}(\cdot)) - \mathbf{F}(t, \mathbf{y}(\cdot))\big| \le K_{\bar{\gamma}, L} \|\mathbf{x} - \mathbf{y}\|^{[\alpha, t]},$$

whenever $t \in [t_0, \bar{\gamma}]$ *and* $\mathbf{x}, \mathbf{y} \in C\big([\alpha, t] \to L\big)$.

Definition 8.1.2. *Given an initial function* $\phi \in C\big([\alpha, t_0] \to D\big)$, *a solution is a function* $\mathbf{x} \in C\big([\alpha, \beta) \to D\big)$, *where* $t_0 < \beta \le \gamma$, *such that* $\mathbf{x}(t) = \phi(t)$ *on* $[\alpha, t_0]$ *and* $\mathbf{x}$ *satisfies* (8.1.1) *for* $t_0 \le t < \beta$. *We write* $\mathbf{x}(t, t_0, \phi)$. *A solution is unique if every other solution* $\mathbf{y}(t, t_0, \phi)$ *agrees with* $\mathbf{x}(t, t_0, \phi)$ *as long as both are defined.*

Theorem 8.1.1, Theorem 8.1.5, and the first paragraph of Theorem 8.1.6 are the basic results on nondifferentiable Liapunov functions and functionals.

Theorem 8.1.1. Driver (1962) *Let* $\omega(t, r)$ *be a continuous, nonnegative function of* t *and* r *for* $t_0 \le t < \beta$, $r \ge 0$. *Let* $v(t)$ *be any continuous, nonnegative function for* $\alpha \le t < \beta$ *such that*

$$\limsup_{\Delta t \to 0^+} \frac{v(t + \Delta t) - v(t)}{\Delta t} \le \omega(t, v(t))$$

at those $t \in [t_0, \beta)$ *at which*

$$v(s) \le v(t) \quad \text{for all} \quad s \in [\alpha, t].$$

Let $r_0 \ge \max_{\alpha \le s \le t_0} v(s)$ *be given, and suppose that the maximal continuous solution* $r(t)$, *of* $r'(t) = \omega(t, r(t))$ *for* $t \ge t_0$ *with* $r(t_0) = r_0$ *exists for* $t_0 \le t < \beta$. *Then*

$$v(t) \le r(t) \quad \text{for} \quad t_0 \le t < \beta.$$

Proof. Choose any $\bar{\beta} \in (t_0, \beta)$. Then $r(t)$ can be represented as

$$r(t) = \lim_{\varepsilon \to 0^+} r(t, \varepsilon) \quad \text{for} \quad t_0 \le t < \bar{\beta},$$

where, for each fixed sufficiently small $\varepsilon > 0$, $r(t, \varepsilon)$ is any solution of

$$r'(t, \varepsilon) = \omega(t, r(t, \varepsilon)) + \varepsilon \quad \text{for} \quad t_0 \le t < \bar{\beta},$$

with $r(t_0, \varepsilon) = r_0$. [See Kamke (1930, p. 83).]

By way of contradiction, suppose that, for some such $\varepsilon > 0$, there exists a $t \in (t_0, \bar{\beta})$ such that $v(t) > r(t, \varepsilon)$. Let $t_1 = \sup \{t \in [t_0, \bar{\beta}) : v(s) \le r(s, \varepsilon) \text{ for all } s \in [t_0, t]\}$. It follows that $t_0 \le t_1 < \bar{\beta}$, and by continuity of $v(t)$ and $r(t, \varepsilon)$, $v(t_1) = r(t_1, \varepsilon)$. Because $r(t, \varepsilon)$ has a positive derivative, $v(s) \le r(t_1, \varepsilon) = v(t_1)$ for all $s \in [\alpha, t_1]$, and therefore

$$\limsup_{\Delta t \to 0^+} \frac{v(t_1 + \Delta t) - v(t_1)}{\Delta t} \le \omega(t_1, v(t_1)).$$

But, $v(t_1 + \Delta t) > r(t_1 + \Delta t, \varepsilon)$ for certain arbitrarily small $\Delta t > 0$. Hence,

$$\limsup_{\Delta t \to 0^+} \frac{v(t_1 + \Delta t) - v(t_1)}{\Delta t} \ge r'(t_1, \varepsilon)$$
$$= \omega(t_1, r(t_1, \varepsilon)) + \varepsilon$$
$$= \omega(t_1, v(t_1)) + \varepsilon$$

is a contradiction.

We then have $v(t) \le r(t, \varepsilon)$ for all sufficiently small $\varepsilon > 0$ and all $t \in [t_0, \bar{\beta})$, so $v(t) \le r(t)$ for all $t \in [t_0, \beta)$. This completes the proof.

Theorem 8.1.2. Driver (1962) *Let the functional $\mathbf{F}(t, \mathbf{x}(\cdot))$ be continuous in t and locally Lipschitz in $\mathbf{x}$. Let $\mathbf{x}(t) = \mathbf{x}(t, t_0, \phi)$ and $\widetilde{\mathbf{x}}(t) = \mathbf{x}(t, t_0, \widetilde{\phi})$ be solutions of (8.1.1) for $\alpha \le t < \beta$ with initial functions ϕ and $\widetilde{\phi} \in C([\alpha, t_0] \to D)$. Then, for any $\bar{\beta} \in (t_0, \beta)$, $\mathbf{x}(t)$ and $\widetilde{\mathbf{x}}(t)$ both map $[\alpha, \bar{\beta}]$ into some compact set $H \subset D$, and*

$$|\mathbf{x}(t) - \widetilde{\mathbf{x}}(t)| \le \|\phi - \widetilde{\phi}\|^{[\alpha, t_0]} \exp[K_{\bar{\beta}, L}(t - t_0)]$$

for $t_0 \le t \le \bar{\beta}$.

Proof. Let G be a compact subset of D such that ϕ and $\widetilde{\phi} \in C([\alpha, t_0] \to G)$. Because $\mathbf{x}(t)$ and $\widetilde{x}(t)$ are continuous on the compact set $[t_0, \bar{\beta}]$, it follows that $\mathbf{x}, \widetilde{\mathbf{x}} \in C([t_0, \bar{\beta}] \to F_1)$, where F_1 is a compact subset of D. Let $H = G \cup F_1$.

For $t_0 \le t < \bar{\beta}$, we have

$$\limsup_{\Delta t \to 0^+} \frac{\left|\mathbf{x}(t + \Delta t) - \widetilde{\mathbf{x}}(t + \Delta t)\right| - \left|\mathbf{x}(t) - \widetilde{\mathbf{x}}(t)\right|}{\Delta t}$$
$$\le \limsup_{\Delta t \to 0^+} \frac{\left|\mathbf{x}(t + \Delta t) - \mathbf{x}(t) - \widetilde{\mathbf{x}}(t + \Delta t) + \widetilde{\mathbf{x}}(t)\right|}{\Delta t}$$
$$= \left|\mathbf{F}(t, \mathbf{x}(\cdot)) - \mathbf{F}(t, \widetilde{x}(\cdot))\right|$$
$$\le K_{\bar{\beta}, H} \|\mathbf{x}(\cdot) - \widetilde{\mathbf{x}}(\cdot)\|^{[\alpha, t]}.$$

The result now follows from Theorem 8.1.1 by taking $v(t) = |\mathbf{x}(t) - \widetilde{\mathbf{x}}(t)|$, $\omega(t, r) = K_{\bar{\beta}, H} r$, and $r_0 = \|\boldsymbol{\phi} - \widetilde{\boldsymbol{\phi}}\|^{[\alpha, t_0]}$.

Remark 8.1.1. This result is the primary one on continual dependence of solutions on initial conditions. A more detailed set of information may be found in Driver (1963).

Theorem 8.1.3. Driver (1962)   *Let* $\mathbf{F}(t, \mathbf{x}(\cdot))$ *be continuous in* t *and locally Lipschitz in* $\mathbf{x}$ *and let* $\boldsymbol{\phi} \in C([\alpha, t_0] \to D)$. *Then there exists an* $h > 0$ *such that a unique solution* $\mathbf{x}(t) = \mathbf{x}(t, t_0, \boldsymbol{\phi})$ *exists for* $\alpha \leq t < t_0 + h$.

Proof. Uniqueness is immediate from Theorem 8.1.2.

Existence is obtained by Picard's approximations. Let G be a compact subset of D such that $\boldsymbol{\phi} \in C([\alpha, t_0] \to G)$. Let $\bar{\gamma} \in (t_0, \gamma)$. Define

$$\bar{\phi}(t) = \begin{cases} \phi(t) & \text{for } \alpha \leq t \leq t_0 \\ \phi(t_0) & \text{for } t_0 < t \leq \bar{\gamma}. \end{cases}$$

Then $\mathbf{F}(t, \bar{\phi}(\cdot))$ is a continuous function of t on the compact set $[t_0, \bar{\gamma}]$, and hence, $|\mathbf{F}(t, \bar{\phi}(\cdot))| \leq M_1$ there.

Next, we show that there exist constants $b > 0$ and $M \geq M_1$ such that $|\mathbf{F}(t, \boldsymbol{\psi}(\cdot))| \leq M$ whenever $t \in [t_0, \bar{\gamma}]$, $\boldsymbol{\psi} \in C([\alpha, \bar{\gamma}] \to D)$, and $\|\psi - \bar{\phi}\|^{[\alpha, \bar{\gamma}]} < b$. By way of contradiction, we suppose this to be false. Then, for each $i = 1, 2, \ldots$, there would exist $t_i \in [t_0, \bar{\gamma}]$ and $\boldsymbol{\psi}_{(i)} \in C([\alpha, \bar{\gamma}] \to D)$ such that $\|\boldsymbol{\psi}_{(i)} - \bar{\phi}\|^{[\alpha, \bar{\gamma}]} < 1/i$, and yet $\left| F(t_i, \boldsymbol{\psi}_{(i)}(\cdot)) - F(t_i, \bar{\phi}(\cdot)) \right| \geq i$. We now choose a subsequence such that $\lim_{k \to \infty} t_{i_k}$ exists, contradicting the continuity properties of F.

Now, suppose b is so small that the rectangle

$$R = \left\{ \mathbf{x} \in R^n : |\mathbf{x} - \boldsymbol{\phi}(t_0)| \leq b \right\} \subset D,$$

and let $H = G \cup R$.

Choose $h > 0$ with

$$h \leq \min(\bar{\gamma} - t_0, b/M).$$

Define a function space

$$S = \{ \boldsymbol{\psi} \in C([\alpha, t_0 + h] \to R^n) : \boldsymbol{\psi}(t) = \boldsymbol{\phi}(t) \quad \text{for } \alpha \leq t \leq t_0, \quad \text{and}$$
$$|\boldsymbol{\psi}(t) - \boldsymbol{\psi}(\hat{t})| \leq M|t - \hat{t}| \quad \text{when } t, \hat{t} \in [t_0, t_0 + h] \}.$$

Now $\|\boldsymbol{\psi} - \boldsymbol{\eta}\|^{[\alpha, t_0 + h]}$ is a metric and S is a complete metric space.

For every $\psi \in S$ and $t \in [t_0, t_0 + h]$, $|\psi(t) - \phi(t_0)| \leq M|t - t_0| \leq b$, so $\psi \in R \subset D$. Thus, $\mathbf{F}(t, \psi(\cdot))$ is continuous in t, Lipschitz with constant $K = K_{\bar{\gamma}, H}$ with respect to ψ, and bounded with bound M. Thus the mapping

$$(T\psi)(t) = \begin{cases} \phi(t) & \text{for } \alpha \leq t \leq t_0 \\ \psi(t_0) + \int_{t_0}^{t} \mathbf{F}(s, \psi(\cdot))\, ds & \text{for } t_0 \leq t \leq t_0 + h \end{cases}$$

is well defined for all $\psi \in S$. Moreover, $T : S \to S$.

Let $\mathbf{x}_{(0)} \in S$ and define a sequence $\{\mathbf{x}_{(i)}\}$ by $\mathbf{x}_{(i)} = T\mathbf{x}_{(i-1)}$ for $i = 1, 2, \ldots$. Then $\|\mathbf{x}_{(i+1)} - \mathbf{x}_{(i)}\|^{[\alpha,t]} \leq BK^i(t - t_0)/i!$ for $t_0 \leq t \leq t_0 + h$, $i = 1, 2, \ldots$, where $B = \|\mathbf{x}_{(1)} - \mathbf{x}_{(0)}\|^{[\alpha,t_0+h]}$. For $t = t_0 + h$, we see that $\{\mathbf{x}_{(i)}\}$ is a Cauchy sequence in S. Thus, $\lim_{i\to\infty} \mathbf{x}_{(i)} = \mathbf{x}$ exists and $\mathbf{x} \in S$. For $i \geq 0$, we find that

$$\|T\mathbf{x} - \mathbf{x}\|^{[\alpha,t_0+h]} \leq \|T\mathbf{x} - T\mathbf{x}_{(i)}\|^{[\alpha,t_0+h]} + \|\mathbf{x}_{(i+1)} - \mathbf{x}\|^{[\alpha,t_0+h]}$$
$$\leq Kh\|\mathbf{x} - \mathbf{x}_{(i)}\|^{[\alpha,t_0+h]} + \|\mathbf{x}_{(i+1)} - \mathbf{x}\|^{[\alpha,t_0+h]},$$

which tends to zero as $i \to \infty$. Thus, $T(\mathbf{x}) = \mathbf{x}$ or

$$\mathbf{x}(t) = \begin{cases} \phi(t) & \text{for } \alpha \leq t \leq t_0 \\ \phi(t_0) + \int_{t_0}^{t} \mathbf{F}(s, \mathbf{x}(\cdot))\, ds & \text{for } t_0 \leq t \leq t_0 + h. \end{cases}$$

This completes the proof.

The next result indicates what must happen if a solution cannot be continued beyond some value of t. Again we see the contrast between ordinary differential equations and functional differential equations, as we saw in Section 3.3.

Theorem 8.1.4. Driver (1962)   *Let* $\mathbf{F}(t, \mathbf{x}(\cdot))$ *be continuous in* t *and locally Lipschitz in* $\mathbf{x}$ *and let* $\phi \in C([\alpha, t_0] \to D)$. *Then there exists a unique solution* $\mathbf{x}(t) = \mathbf{x}(t, t_0, \phi)$ *on* $[\alpha, \beta)$ *where* $t_0 < \beta \leq \gamma$, *and if* $\beta < \gamma$ *and* β *cannot be increased, then for any compact set* $H \subset D$ *there is a sequence of numbers* $t_0 < t_1 < \cdots \to \beta$ *such that* $\mathbf{x}(t_k) \in D - H$ *for* $k = 1, 2, \ldots$.

Proof. Uniqueness has already been proved. Now, suppose $\mathbf{x}(t)$ is a solution for $\alpha \leq t < \beta$, where $t_0 < \beta < \gamma$. Let G be a compact subset of D with $\phi \in C([\alpha, t_0] \to G)$.

Suppose there is a compact set $H \subset D$ such that $\mathbf{x}(t) \in H$ for $t_0 \leq t < \beta$. Let $G_1 = G \cup H$. Then, as in the proof of Theorem 8.1.3, $|\mathbf{F}(t, \bar{\phi}(\cdot))| \leq M_1$ for $t_0 \leq t \leq \beta$. Now by the Lipschitz condition

$$|\mathbf{F}(t, \mathbf{x}(\cdot))| \leq M_{\beta, G_1} \equiv M_1 + 2K_{\beta, G_1} \sup_{x \in G_1} |\mathbf{x}| \,.$$

Hence,

$$|\mathbf{x}(t) - \mathbf{x}(\hat{t})| \leq M_{\beta, G_1} |t - \hat{t}|$$

for $t, \hat{t} \in [t_0, \beta]$. By the Cauchy criterion $\lim_{t \to \beta^-} \mathbf{x}(t)$ exists. Thus $\mathbf{x}(t)$ is extended continuously to $\alpha \leq t \leq \beta$.

Now consider a new initial-data problem in which t_0 is replaced by β, with $\hat{\phi}(t) \equiv \mathbf{x}(t)$ for $\alpha \leq t \leq \beta$. Because $\hat{\phi} \in C([\alpha, \beta] \to G_1)$, Theorem 8.1.3 yields a solution $\mathbf{x}(t, \beta, \hat{\phi})$ on $\alpha \leq t < \beta + h$, some $h > 0$. This completes the proof.

Remark 8.1.2. Stability definitions for (8.1.1) are identical to those for Volterra integro-differential equations. To speak of the zero solution, we must assume that $\gamma = +\infty$, that $D = B_H = \{\mathbf{x} \in R^n : |\mathbf{x}| < H, \, 0 < H \leq \infty\}$, and that $\mathbf{F}(t, \mathbf{0})$ is zero. Moreover, one refers to stability to the right of some fixed t_0, which we shall always take to be zero.

Definition 8.1.3. *Let $V(t, \psi(\cdot))$ be a scalar-valued functional defined for $t \geq 0$ and $\psi \in C([\alpha, t] \to B_H)$. Then the derivative of V with respect to (8.1.1) is*

$$V'_{(8.1.1)}(t, \psi(\cdot)) = \limsup_{\Delta t \to 0^+} \frac{V(t + \Delta t, \psi^*(\cdot)) - V(t, \psi(\cdot))}{\Delta t},$$

where

$$\psi^*(s) = \begin{cases} \psi(s) & \text{for } \alpha \leq s \leq t \\ \psi(t) + \mathbf{F}(t, \psi(\cdot))(s - t) & \text{for } t \leq s \leq t + \Delta t. \end{cases}$$

Remark 8.1.3. When $V(t, \psi(\cdot)) = V(t, \psi(t))$, a function depending only on t and $\psi(t)$, then V' is still a functional if $\mathbf{F}(t, \mathbf{x}(\cdot))$ is a functional; moreover, if $V(t, \psi(t)) \in C^1$, then V' obtained by the chain rule, is

$$V'_{(8.1.1)}(t, \psi(\cdot)) = \frac{\partial V}{\partial t}(t, \mathbf{x}) \bigg|_{\mathbf{x} = \psi(t)} + \sum_{i=1}^{n} \frac{\partial V(t, \mathbf{x})}{\partial x_i} \bigg|_{\mathbf{x} = \psi(t)} F_i(t, \mathbf{x}(\cdot)),$$

which is the form virtually always used in applications, but not in the general theory. In the applications V is usually locally Lipschitz in ψ and, actually, C^1 except for a few "corners" about which we do need to worry.

Notice that the derivative of V in Definition 8.1.3 is a quantity that we obtain without knowledge of the solution, except for its existence. The next theorem shows that this derivative is the same as the derivative along the actual solution.

Theorem 8.1.5. Driver (1962)   Let $V(t, \boldsymbol{\psi}(\cdot))$ be defined and continuous for $t \geq 0$ and $\boldsymbol{\psi} \in C\big([\alpha, t] \to B_H\big)$ and let V be locally Lipschitz in $\boldsymbol{\psi}$. Then for each $t \geq 0$ and every $\boldsymbol{\psi} \in C\big([\alpha, t] \to B_H\big)$ we have

$$V'_{(8.1.1)}(t, \boldsymbol{\psi}(\cdot)) = \limsup_{\Delta t \to 0^+} \frac{V(t + \Delta t, \mathbf{x}(\cdot, t, \boldsymbol{\psi})) - V(t, \boldsymbol{\psi}(\cdot))}{\Delta t},$$

where $\mathbf{x}(s, t, \boldsymbol{\psi})$ is the unique solution of (8.1.1) with initial conditions $t, \boldsymbol{\psi}$.

Proof. Given $t, \boldsymbol{\psi}$, there is a unique solution on $[\alpha, t + h]$ for some $h > 0$. It will suffice to show that

$$V(t + \Delta t, \mathbf{x}(\cdot, t, \boldsymbol{\psi})) - V(t + \Delta t, \boldsymbol{\psi}^*(\cdot)) = o(\Delta t) \quad \text{as} \quad \Delta t \to 0^+ .$$

Choose $h_1 \in (0, h)$ so small that both $\mathbf{x}(s, t, \boldsymbol{\psi})$ and $\boldsymbol{\psi}^*(s) \in Q$, some compact subset of B_H, for $\alpha \leq s \leq t + h_1$. Let K be the Lipschitz constant for $V(t, \boldsymbol{\psi}(\cdot))$ associated with $t + h_1$ and Q. Then for $0 < \Delta t \leq h_1$, we have

$$\begin{aligned}
&\big|V(t + \Delta t, \mathbf{x}(\cdot, t, \boldsymbol{\psi})) - V(t + \Delta t, \boldsymbol{\psi}^*(\cdot))\big| \\
&\quad \leq K \sup_{t \leq s \leq t + \Delta t} \big|\mathbf{x}(s, t, \boldsymbol{\psi}) - \boldsymbol{\psi}(t) - \mathbf{F}(t, \boldsymbol{\psi}(\cdot))(s - t)\big| \\
&\quad \leq K \sup_{t \leq s \leq t + \Delta t} \big|\mathbf{F}(\bar{t}, \mathbf{x}(\cdot, t, \boldsymbol{\psi})) - \mathbf{F}(t, \boldsymbol{\psi}(\cdot))\big| \Delta t ,
\end{aligned}$$

where $t < \bar{t} < s$. Owing to the continuity of $\mathbf{F}(t, \mathbf{x}(\cdot))$, this quantity is $o(\Delta t)$. This completes the proof.

Notation.   In the following, $\omega(t, r)$ is any continuous function on $[0, \infty) \times [0, \infty) \to R$ with

$$\omega(t, 0) = 0$$

and the zero solution of

$$r' = \omega(t, r)$$

is stable. Naturally, we mean $t_0 \geq 0$ and $r_0 \geq 0$.

We are now ready to state a stability result, and in preparation, we offer an outline as a rough summary.

(i) Theorem 8.1.1 says that if

$$\limsup_{\Delta t \to 0^+} \frac{v(t + \Delta t) - v(t)}{\Delta t} \leq \omega(t, v(t)),$$

then $v(t) \leq r(t)$, where $r' = \omega(t, r)$.

(ii) Definition 8.1.3 allows us to take a derivative of V, say $V'_{(8.1.1)}$, along a solution of (8.1.1) without knowing the solution.

(iii) Theorem 8.1.5 says that if V is continuous in t and locally Lipschitz in $\mathbf{x}$, then the derivative in (ii) really is the derivative

$$\limsup_{\Delta t \to 0^+} \frac{V(t + \Delta t, \mathbf{x}(\cdot, t, \boldsymbol{\psi})) - V(t, \boldsymbol{\psi}(\cdot))}{\Delta t}.$$

(iv) Theorem 8.1.6 will tell us to let $v(t) = V(t, x(\cdot))$, apply (iii), and accept the conclusion of (i).

Theorem 8.1.6.   Driver (1962)   *If* $V(t, \boldsymbol{\psi}(\cdot))$ *is defined for* $t \geq 0$ *and* $\boldsymbol{\psi} \in C([\alpha, t] \to B_H)$ *with*

(a) $V(t, \mathbf{0}) \equiv 0$,

(b) V *continuous in* t *and Lipschitz in* $\boldsymbol{\psi}$,

(c) $V(t, \boldsymbol{\psi}(\cdot)) \geq W(|\boldsymbol{\psi}(t)|)$, W *a wedge, and if*

(d) $V'_{(8.1.1)}(t, \boldsymbol{\psi}(\cdot)) \leq \omega(t, V(t, \boldsymbol{\psi}(\cdot)))$,

then the zero solution of (8.1.1) *is stable.*

If the zero solution of $r' = \omega(t, r)$ *is asymptotically stable, then the zero solution of* (8.1.1) *is asymptotically stable.*

Proof. Let $t_0 \geq 0$ be given, $\boldsymbol{\phi} \in C([\alpha, t] \to B_H)$, and consider $\mathbf{x}(t) = \mathbf{x}(t, t_0, \boldsymbol{\phi})$ for $\alpha \leq t < \beta$. Define $v(t) = V(t, \mathbf{x}(\cdot))$ for $t_0 \leq t < \beta$. Then by (d) and Theorem 8.1.5 we have

$$\limsup_{\Delta t \to 0^+} \frac{v(t + \Delta t) - v(t)}{\Delta t} \leq \omega(t, v(t))$$

for $t_0 \leq t < \beta$. Choose $r_0 \geq v(t_0) = V(t_0, \boldsymbol{\phi}(\cdot))$. The result of Theorem 8.1.1 with $\alpha = t_0$ now holds [the nonnegative property of $\omega(t, r)$ was not needed for this part]. Thus, letting $r(t)$ be the maximal solution of $r'(t) = \omega(t, r(t))$ with $r(t_0) = r_0$, we have

$$V(t, \mathbf{x}(\cdot)) = v(t) \leq r(t) \tag{8.1.2}$$

for all $t \geq t_0$ at which both the solutions $x(t)$ and $r(t)$ exist.

Now, let $\varepsilon \in (0, H)$ be given. Choose $r_0 = r_0(\varepsilon, t_0) > 0$ so that

$$r(t) < W(\varepsilon) \quad \text{for} \quad t \geq t_0 ,$$

where $r(t)$ is the maximal solution of $r' = \omega(t, r)$ for $t \geq t_0$ with $r(t_0) = r_0$. Choose $\delta = \delta(\varepsilon, t_0) \in (0, \varepsilon)$ with $\sup \left\{ V(t_0, \boldsymbol{\psi}(\cdot)) : \boldsymbol{\psi} \in C\left([\alpha, t_0] \to B_\delta\right) \right\} \leq r_0(\varepsilon, t_0)$. That is, $\delta \leq \min(\varepsilon, r_0/K)$, where K is the Lipschitz constant for $V(t, \boldsymbol{\psi}(\cdot))$ associated with t_0 and the closure of B_ε.

Let $\mathbf{x}(t) = \mathbf{x}(t, t_0, \boldsymbol{\phi})$ with $\boldsymbol{\phi} \in C([\alpha, t_0] \to B_\delta)$. Then by (8.1.2) we have

$$W(|\mathbf{x}(t)|) \leq V(t, \mathbf{x}(\cdot)) \leq r(t) < W(\varepsilon) ,$$

for all $t \geq t_0$ at which $\mathbf{x}(t)$ exists. Because $|\mathbf{x}(t)| \leq \varepsilon < H$, the solution exists for all $t \geq \alpha$ and $|\mathbf{x}(t)| < \varepsilon$. This completes the proof.

Remark 8.1.4. Frequently, the function $\omega(t, r)$ is identically zero. Driver (1962, p. 416, para. 2) states that the part concerning asymptotic stability is not very useful because the condition can only be satisfied in extremely special cases. Since Driver's paper, investigators have constructed many nontrivial examples with ω nonzero. We have seen (e.g., Theorem 2.5.3 and Section 7.2) that it is a rather natural and quite fundamental condition for integro-differential equations and perturbations. Thus, we see an example of a mathematical idea that initially seems too complicated to be useful and later becomes quite central. The book by Lakshmikantham and Leela (1969) treats such differential inequalities for ordinary differential equations in great and interesting detail.

The next result is the fundamental stability theorem using Liapunov functions instead of functionals. This is Driver's formulation of Razumikhin's result.

Theorem 8.1.7. *Razumikhin, [see Driver (1962)]   If $\omega(t, r) \geq 0$ and if there is a function $V : [\alpha, \infty) \times B_H \to [0, \infty)$ with*

(a) *$V(t, \mathbf{0}) \equiv 0$,*
(b) *V continuous in t and Lipschitz in $\mathbf{x}$,*
(c) *$V(t, \mathbf{x}) \geq W(|\mathbf{x}|)$, W a wedge, and*
(d) *for $Z(t, \boldsymbol{\psi}(\cdot)) = V(t, \boldsymbol{\psi}(t))$,*

we have $Z'_{(8.1.1)}(t, \boldsymbol{\psi}(\cdot)) \leq \omega(t, V(t, \boldsymbol{\psi}(t))$ whenever $t \geq 0$, $\boldsymbol{\psi} \in C([\alpha, t] \to B_H)$, and $V(s, \boldsymbol{\psi}(s)) \leq V(t, \boldsymbol{\psi}(t))$ for all $s \in [\alpha, t]$, then the zero solution of (8.1.1) is stable.

Proof. Define the functional $\mathbf{\Omega}(t, \boldsymbol{\psi}(\cdot)) = \sup_{\alpha \leq s \leq t} \leq V(s, \boldsymbol{\psi}(s))$ for $t \geq 0$, $\boldsymbol{\psi} \in C([\alpha, t] \to B_H)$. Now $\mathbf{\Omega}$ satisfies parts (a), (b), and (c) of Theorem 8.1.6.

There are two cases to be considered to show that $\mathbf{\Omega}(t, \boldsymbol{\psi}(\cdot))$ satisfies (d) of Theorem 8.1.6. Let $t \geq 0$ and $\boldsymbol{\psi} \in C([\alpha, t] \to B_H)$.

If $V(t, \boldsymbol{\psi}(t)) < \mathbf{\Omega}(t, \boldsymbol{\psi}(\cdot))$, then the continuity of $V(s, \boldsymbol{\psi}^*(s))$ assures us that $V(t + \xi, \boldsymbol{\psi}^*(t + \xi)) < \mathbf{\Omega}(t, \boldsymbol{\psi}(\cdot))$ for all sufficiently small $\xi > 0$. Thus, $\mathbf{\Omega}(t + \Delta t, \boldsymbol{\psi}^*(\cdot)) = \mathbf{\Omega}(t, \boldsymbol{\psi}(\cdot))$ for all sufficiently small $\Delta t > 0$, and hence, $\mathbf{\Omega}'_{(8.1.1)}(t, \boldsymbol{\psi}(\cdot)) = 0 \leq w(t, \mathbf{\Omega}(t, \boldsymbol{\psi}(\cdot)))$.

If $V(t, \boldsymbol{\psi}(t)) = \mathbf{\Omega}(t, \boldsymbol{\psi}(\cdot))$, we have $V(s, \boldsymbol{\psi}(s)) \leq V(t, \boldsymbol{\psi}(t))$ for all $s \in [\alpha, t]$. Thus, by (d) of this result

$$V(t + \xi, \boldsymbol{\psi}^*(t + \xi)) - V(t, \boldsymbol{\psi}(t)) \leq \omega(t, V(t, \boldsymbol{\psi}(t)))\xi + \varepsilon(\xi)\xi$$

for all sufficiently small $\xi > 0$, where $\varepsilon(\xi)$ is a positive function that tends to zero monotonically as $\xi \to 0$. Using the fact that $V(t, \boldsymbol{\psi}(t)) = \mathbf{\Omega}(t, \boldsymbol{\psi}(\cdot))$ and letting ξ range over $(0, \Delta t]$, with $\Delta t > 0$ and sufficiently small, then

$$\mathbf{\Omega}(t + \Delta t, \boldsymbol{\psi}^*(\cdot)) - \mathbf{\Omega}(t, \boldsymbol{\psi}(\cdot)) \leq \omega(t, \mathbf{\Omega}(t, \boldsymbol{\psi}(\cdot)))\Delta t + \varepsilon(\Delta t)\Delta t \,.$$

Hence, (d) of Theorem 8.1.6 also holds. This completes the proof.

Example 8.1.1. The trivial solution of

$$x'(t) = -x(t) + ax(t - 1)$$

is stable if $|a| < 1$.

Proof. Take $V(t, x(t)) = x^2(t)/2$ so that

$$V'(t, x(\cdot)) \leq -x^2(t) + |a|\,|x(t)|\,|x(t - 1)| \leq 0$$

if $x^2(s)/2 \leq x^2(t)/2$ for $1 \leq s \leq t$. The result may also be proved using the functional

$$V(t, x(\cdot)) = x^2 + \int_{t-1}^{t} x^2(s)\, ds \,,$$

which yields

$$\begin{aligned}
V'(t, x(\cdot)) &= -2x^2(t) + 2ax(t)x(t - 1) + x^2(t) - x^2(t - 1) \\
&\leq -(x^2(t) + x^2(t - 1)) + |a|\,(x^2(t) + x^2(t - 1)) \\
&\leq 0 \,.
\end{aligned}$$

In either case, the proof is complete.

8.2   Asymptotic Stability

In Chapter 6 we discussed the problem of asymptotic stability for ordinary differential equations in some detail. If we make the Liapunov function decrescent, then we obtain uniform asymptotic stability. (Of course, other conditions were also needed.) Our problem with functional differential equations is that we do not really know the proper formulation for a "decrescent" Liapunov functional. In Theorem 2.5.1 we gave one possible formulation; other possible formulations were given in Theorems 6.4.2 and 6.4.4 as well as in Definition 6.4.3 and 6.4.4.

The prevalent method is the Marachkov technique (see Theorem 6.1.3) of asking $x'(t)$ bounded for $x(t)$ bounded. But that method has very serious defects. To study the asymptotic stability of the scalar equation

$$x' = -x + \int_0^t C(t-s)x(s)\,ds$$

requires that $C(t) \in L^1[0, \infty)$. And we would certainly like to study

$$x' = -x + \int_0^t [\delta + C(t-s)]x(s)\,ds$$

with $\delta \neq 0$ and $C(t) \in L^1[0, \infty)$.

The first result here is Driver's formulation of a Krasovskii theorem. It is considered to be the standard asymptotic stability result for functional differential equations with unbounded delay.

Theorem 8.2.1. *Krasovskii, [see Driver (1962)]   For some $H_1 \in (0, H)$ let there exist a constant M with $|\mathbf{F}(t, \boldsymbol{\psi}(\cdot))| \leq M$ when $t \geq 0$ and $\boldsymbol{\psi} \in C([\alpha, t] \to B_{H_1})$. If there is a functional $V(t, \boldsymbol{\psi}(\cdot))$ for $t \geq t_0$ and $\boldsymbol{\psi} \in C([\alpha, t] \to B_H)$, with*

(a) $V(t, \mathbf{0}) \equiv 0$,
(b) $V(t, \boldsymbol{\psi}(\cdot))$ *continuous in t and locally Lipschitz in* $\boldsymbol{\psi}$,
(c) $V(t, \boldsymbol{\psi}(\cdot)) \geq W(|\boldsymbol{\psi}(t)|)$, W *a wedge*,
(d) $V'_{(8.1.1)}(t, \mathbf{x}(\cdot)) \leq -W_1(|\mathbf{x}(t)|)$, W_1 *a wedge*,

then the zero solution of (8.1.1) is asymptotically stable.

Proof. Stability follows from Theorem 8.1.6.

The asymptotic stability is almost identical to Marachkov's theorem (Theorem 6.1.3).

Let $\mathbf{x}(t)$ by any solution of (8.1.1) on an interval $[t_0, \infty)$ with $|\mathbf{x}(t)| < H_1$. If $\mathbf{x}(t) \not\to 0$, then there is an $\varepsilon > 0$ and a sequence $\{t_n\} \to \infty$ with

$|\mathbf{x}(t_n)| \geq \varepsilon$. Because $|\mathbf{x}'(t)| \leq M$, there is a $T > 0$ with $|\mathbf{x}(t)| \geq \varepsilon/2$ for $t_n \leq t \leq t_n + T$. Thus, $V'_{(8.1.1)}(t, \mathbf{x}(\cdot)) \leq -W_1(\varepsilon/2)$ for $t_n \leq t \leq t_n + T$. By picking a subsequence, if necessary, we may suppose the intervals $[t_n, t_n + T]$ disjoint. Thus,

$$0 \leq V(t, \mathbf{x}(\cdot)) \leq V(t_0, \boldsymbol{\phi}(\cdot)) - W_1(\varepsilon/2)nT,$$

where $t > t_n + T$, a contradiction for large n. This completes the proof.

The classical Razumikhin-type result for asymptotic stability placed strong conditions on the delay; in fact, it cannot apply to Volterra equations of the type we study. However, in 1975 Grimmer and Seifert extended it to equations of type (8.1.1).

In the classical result it is supposed that $\mathbf{F}(t, \mathbf{x}(\cdot))$ depends only on $t \geq 0$ and on $\mathbf{x}(s)$ for $g(t) \leq s \leq t$, where $\alpha \leq g(t) \leq t$. In this case we write (8.1.1) (with all its conditions) as

$$\mathbf{x}'(t) = \mathbf{F}(t, \mathbf{x}(\cdot), g(t)). \tag{8.2.1}$$

Theorem 8.2.2. Krasovskii, [see Driver (1962)] *Consider* (8.2.1) *with* $g(t) \to \infty$ *as* $t \to \infty$. *If there is a function* $V : [\alpha, \infty) \times B_H \to [0, \infty)$ *such that*

(a) $V(t, \mathbf{x}) \leq W(|\mathbf{x}|)$, *W a wedge,*
(b) *V continuous in t and locally Lipschitz in* $\mathbf{x}$,
(c) $V(t, \mathbf{x}) \geq W_1(|\mathbf{x}|)$, W_1 *a wedge, and*
(d) *there is a continuous, nondecreasing function* $f : [0, \infty) \to [0, \infty)$ *with* $f(r) > r$ *for all* $r > 0$ *and a wedge* W_2 *with*

$$V'_{(8.2.1)}(t, \mathbf{x}(\cdot), g(t)) \leq -W_2(|\mathbf{x}(t)|),$$

whenever $t \geq t_0$, $x \in C([\alpha, t] \to B_H)$, *and* $V(s, \mathbf{x}(s)) < f(V(t, \mathbf{x}(t)))$ *for all* $s \in [g(t), t]$,

then the zero solution of (8.2.1) *is uniformly stable and asymptotically stable. If* $g(t) \geq t - h$ *for* $t \geq 0$, *where* $h \geq 0$ *is a constant, then the zero solution is U.A.S.*

Proof. The proof is broken into three parts.

(i) *Uniform stability and definition of* δ_1. Let $\varepsilon \in (0, H)$ be given. Find $\delta \in (0, \varepsilon)$ with $W(\delta) < W_1(\varepsilon)$. Then for any $t_0 \geq 0$ and $\boldsymbol{\phi} \in C([\alpha, t] \to B_\delta)$, we have $V'_{(8.2.1)}(t, \mathbf{x}(\cdot, t_0, \boldsymbol{\phi}), g(t)) < 0$, whenever $t \geq t_0$, and $V(s, \mathbf{x}(s)) \leq V(t, \mathbf{x}(t))$ for $\alpha \leq s \leq t$. Thus, by Theorem 8.1.1 we have $V(t, \mathbf{x}(t)) \leq$

$\sup_{\alpha \le s \le t_0} V(s, \phi(s)) \le W(\delta) < W_1(\varepsilon)$, so that $|\mathbf{x}(t)| < \varepsilon$ if $t \ge \alpha$. This is uniform stability.

For the remainder of the proof let $H_1 \in (0, H)$ be fixed, and let $\delta_1 = \delta(H_1)$ be that of uniform stability so that $t_0 \ge \alpha$ and $\phi \in C([\alpha, t_0] \to B_{\delta_1})$ imply $|\mathbf{x}(t, t_0, \phi)| < H_1$.

(ii) *Construction of a number $T(t_0, \eta) > t_0$ for every $t_0 \ge \alpha$ and $\eta > 0$.* Let $t_0 \ge \alpha$ and $\eta \in (0, \delta_1)$ be given. Then $W_1(\eta) \le W(\delta_1)$. Let $a = a(\eta)$ be any positive number with $f(r) - r > a$ for $W_1(\eta) \le r \le W(\delta_1)$ and let $N = N(\eta)$ be any positive integer with

$$W_1(\eta) + Na \ge W(\delta_1).$$

If $\mathbf{x}(t)$ is a solution and if $V(t, \mathbf{x}(t)) \ge W_1(\eta)$ at some $t \ge t_0$, then hypothesis (a) yields $H_1 > |\mathbf{x}(t)| \ge \delta_2(\eta)$ for some $\delta_2(\eta) > 0$. This implies that $W_2(|\mathbf{x}(t)|) \ge \delta_3(\eta)$ for some $\delta_3(\eta) > 0$.

Now construct $N + 1$ numbers $t_k(t_0, \eta)$ for $k = 0, 1, \ldots, N$ by setting $t_0(t_0, \eta) = t_0$, and for $0 \le k \le N - 1$, choose $t_{k+1}(t_0, \eta)$ to be any number such that $g(t) \ge t_k(t_0, \eta)$ for all $t \ge t_{k+1}(t_0, \eta) - a(\eta)/\delta_3(\eta)$. Note that

$$t_{k+1}(t_0, \eta) \ge t_k(t_0, \eta) + a(\eta)/\delta_3(\eta).$$

Finally, let $T(t_0, \eta) = t_N(t_0, \eta)$.

Observe that if $g(t) \ge t - h$, then

$$t_k(t_0, \eta) = t_0 + k[a(\eta)/\delta_3(\eta) + h]$$

for $k = 0, 1, \ldots, N$ would suffice. In that case, $T(t_0, \eta) = t_0 + T(\eta)$, where $T(\eta) = N(\eta)[a(\eta)/\delta_3(\eta) + h]$ is independent of t_0.

(iii) *Proof that $|\mathbf{x}(t, t_0, \phi)| \le \eta$ for all $t \ge T(t_0, \eta)$.* It is sufficient to show that $V(t, \mathbf{x}(t)) \le W_1(\eta) + (N-k)a$ for all $t \ge t_k(t_0, \eta)$, $k = 0, 1, \ldots, N$. The proof is by induction.

For $k = 0$ this result follows from (i) of this proof and the definition of N. Assume the result true for some $k < N$. Thus, if for some $t \ge t_{k+1} - a/\delta_3$ we have

$$V(t, \mathbf{x}(t)) \ge W_1(\eta) + (N - k - 1)a,$$

then

$$f(V(t, \mathbf{x}(t))) \ge W_1(\eta) + (N - k)a \ge V(s, \mathbf{x}(s))$$

for $g(t) \le s \le t$, and hence,

$$V'_{(8.2.1)}(t, \mathbf{x}(\cdot), g(t)) \le -\delta_3 < 0.$$

This shows that if $V(t, \mathbf{x}(t)) \le W_1(\eta) + (N-k-1)a$ for some $t \ge t_{k+1} - a/\delta_3$, then the same inequality will hold in the future.

Now, if

$$V(t, \mathbf{x}(t)) \geq W_1(\eta) + (N - k - 1)a$$

for $t_{k+1} - a/\delta_3 \leq t \leq \bar{t}$ for some $\bar{t}$, then we have

$$V(\bar{t}, \mathbf{x}(\bar{t})) \leq W_1(\eta) + (N - k)a - \delta_3(\bar{t} - t_{k+1} + a/\delta_3)$$
$$\leq V(\bar{t}, \mathbf{x}(\bar{t})) - \delta_3(\bar{t} - t_{k+1}).$$

This shows that $\bar{t} \leq t_{k+1}$. That is, $V(t, \mathbf{x}(t)) \leq W_1(\eta) + (N - k - 1)a$ for all $t \geq t_{k+1}(t_0, \eta)$. This completes the proof.

Example 8.2.1. Driver (1962). The zero solution of

$$x'(t) = -a(t)x(t) - b(t)x(g(t))$$

is asymptotically stable if, for $t \geq 0$, then $a(t)$, $b(t)$, and $g(t)$ are continuous with $\alpha \leq g(t) \leq t$, $a(t) \geq a$, $a(t)$ bounded, and $|b(t)| \leq qa(t)$, where $a > 0$ and $q \in (0, 1)$ are constants and $g(t) \to \infty$ as $t \to \infty$.

To prove the assertion, let

$$V(t, x) = x^2, \quad f(r) = r/q, \quad \text{and} \quad W_2(|x|) = 2a(1 - \sqrt{q})x^2.$$

The preceding theorem is a fundamental one and its primary weakness is the requirement that $g(t) \to \infty$. This asks that F have a very strongly "fading memory." Much has been written about fading memory and we intuitively understand that it is a natural physical concept. Although we agree that in most problems there is heredity, we also understand that there is a certain duration of heredity. Volterra, in his predator-prey formulations reasoned that one might want to consider $\int_{t-T}^{t}$, where T was the duration of heredity. This, however, is generally too drastic. A system should remember its past, but the memory should grow dim with passing time. For example, our Theorem 2.5.1(d) illustrates a far more reasonable type of fading memory for Volterra equations.

The upcoming Grimmer-Seifert result will offer one remedy to this situation, but it is a complicated one. We suggest that, with a little care, one could extend Theorem 8.2.2 by dropping the requirement that $g(t) \to \infty$ as $t \to \infty$ and by asking for a gradual fading of the memory. A properly constructed result could greatly enhance the Razumikhin method.

In this connection, Seifert has constructed a very interesting example concerning functional differential equations, the Razumikhin technique, and a nonfading memory.

Example 8.2.2. Seifert (1973).   Consider the scalar equation

$$x'(t) = -2x(t) + x(0),$$

with solutions

$$x(t, x_0) = (1 + e^{-2t})x_0/2.$$

Take

$$V(t) = x^2/2$$

and find

$$V'(x(t)) = -2x^2(t) + x(t)x(0).$$

Let $f(r) = 2r$ and note that for any solution $x(t)$ such that

$$f(V(x(t))) > V(x(s)), \quad 0 \le s \le t$$

(which means that $\sqrt{2}\,|x(t)| > |x(s)|, 0 \le s \le t$), we then have

$$V'(x(t)) \le (-2 + \sqrt{2})\,x^2(t)$$

because we must have $\sqrt{2}\,|x(t)| > |x(0)|$ for such a solution.

If we review Theorem 8.2.2 we see that (a)–(d) are easily satisfied. However, $g(t) \equiv 0$ and solutions do not tend to zero.

Seifert then formulates an indirect fading memory principle that yields asymptotic stability. He considers a system

$$\mathbf{x}'(t) = \mathbf{F}(t, \mathbf{x}(s); \ 0 \le s \le t), \tag{8.2.2}$$

which we again write as

$$\mathbf{x}' = \mathbf{F}(t, \mathbf{x}(\cdot)),$$

and ask that $\mathbf{F}$ satisfy the conditions of (8.1.1) with $\gamma = \infty$, $D = R^n$, and $\mathbf{F}(t, \mathbf{0}) = \mathbf{0}$ as in Remark 8.1.2. Although Seifert asks that $\alpha = 0$, we have observed no problem with asking that $\alpha \ge -\infty$. Moreover, Seifert seems to start all solutions at $t = 0$; that is, his initial condition is always $\mathbf{x}(0) = \mathbf{x}_0$, but we detect no problem considering $\mathbf{x}(t, t_0, \boldsymbol{\phi})$ for $t_0 \ge 0$ and $\boldsymbol{\phi}$ continuous.

Beware.   Stability definitions for ordinary differential equations look similar to those for functional differential equations. Moreover, functional differential equations can frequently be converted to ordinary differential equations; however, the ordinary differential equation may be stable, but the functional differential equation unstable.

Exercise 8.2.1. Verify that

$$x' = x - \int_0^t \alpha e^{-\alpha(t-s)} x(s) \, ds \qquad \text{(a)}$$

may be differentiated to obtain

$$x'' + (\alpha - 1)x' = 0 \,,$$

which is stable if $\alpha > 1$ and has solutions

$$x(t) = c_1 + c_2 e^{(1-\alpha)t} \,. \qquad \text{(b)}$$

See Chapter 2 to learn how to solve (a) with an initial function using (b), a translation $y(t) = x(t + t_0)$, and the variation of parameters formula. Conclude that (a) is not stable for $\alpha > 1$. Then carefully check the details of Theorem 8.2.3 to see if Seifert was correct in insisting that only initial points, instead of initial functions, be used. In other words, can this result be extended to include initial functions?

Theorem 8.2.3. Seifert (1974) *Suppose there is a continuous function $V : [0, \infty) \times R^n \to [0, \infty)$ that is locally Lipschitz in $\mathbf{x}$ and satisfies the following conditions:*

(a) *$W_1(|\mathbf{x}|) \leq V(t, \mathbf{x}) \leq W_2(|\mathbf{x}|)$ for $t \geq 0$, $\mathbf{x} \in R^n$, W_i wedges.*
(b) *There is a continuous function $f : [0, \infty) \to [0, \infty)$ with $f(s) > s$ for $s > 0$. For each fixed solution $\mathbf{x}(t)$ of (8.2.2) on $[0, T)$, $T \leq \infty$, there is a number $r > 0$ and a wedge W_3 such that:*

    (b_1) *$V(s, \mathbf{x}(s)) < f(V(t, \mathbf{x}(t)))$ for $s \in [t_0, t]$, $t > 0$, where $t_0 = \max[0, t - r]$, implies*
    (b_2) *$V'_{(8.2.2)}(t, \mathbf{x}(t)) \leq -W_3(|\mathbf{x}(t)|)$.*

Under these conditions, if $\mathbf{x}(t)$ is a bounded solution of (8.2.2) on $[0, \infty)$, then $\mathbf{x}(t) \to 0$ as $t \to \infty$.

Proof. Let $\mathbf{x}(t)$ be a solution of (8.2.2) on $[0, \infty)$ with

$$\sup_{t \geq 0} |\mathbf{x}(t)| = M < \infty \,.$$

Let $\varepsilon > 0$ be given with $W_1(\varepsilon) < W_2(M)$. Then there exists $a = a(\varepsilon) > 0$ such that $f(s) - s > a$ for $s \in [W_1(\varepsilon), W_2(M)]$. Let $N = N(\varepsilon) > 0$ be the smallest integer such that $W_2(M) \leq W_1(\varepsilon) + Na$, and define $\varepsilon_j = W_1(\varepsilon) + (N - j)a$, $j = 0, 1, 2, \ldots, N$. Note that $V(t, \mathbf{x}(t)) \leq \varepsilon_0$ for $t \geq 0$.

Suppose $V(t, \mathbf{x}(t)) \geq \varepsilon_1$ for all $t \geq r$. Then for any such t we have $W_2(|\mathbf{x}(t)|) \geq \varepsilon_1$, and hence, $|\mathbf{x}(t)| \geq W_2^{-1}(\varepsilon_1) > 0$. Also, for such t we have $W_1(\varepsilon) \leq V(t, \mathbf{x}(t)) \leq W_2(M)$, so

$$f(V(t, \mathbf{x}(t))) > V(t, \mathbf{x}(t)) + a$$
$$\geq W_1(\varepsilon) + (N - 1)a + a$$
$$= W_1(\varepsilon) + Na.$$

But $V(s, \mathbf{x}(s)) \leq W_1(\varepsilon) + Na$ for all $s \geq 0$, and thus also for $s \in [t - r, t]$, $t \geq r$. Using (b) with $j = 0$, we conclude that

$$V'(t, \mathbf{x}(t)) \leq -W_3(|\mathbf{x}(t)|), \quad t \geq r.$$

Define $\rho_1 = W_2^{-1}(\varepsilon_1)$ and $\gamma_1 = W_3(\rho_1)$. It follows from $V' \leq -W_3$ that

$$0 \leq V(t, \mathbf{x}(t)) \leq V(r, \mathbf{x}(r)) - \gamma_1(t - r)$$
$$\leq \varepsilon_0 - \gamma_1(t - r), \quad \text{for all } t \geq r.$$

This is a contradiction for large t.

Thus, there exists $t_1 > r$ such that $V(t_1, \mathbf{x}(t_1)) < \varepsilon_1$. If $V(\widetilde{t}_1, \mathbf{x}(\widetilde{t}_1)) = \varepsilon_1$ for some $\widetilde{t}_1 > t_1$, we suppose $\widetilde{t}_1$ chosen so that $V(t, \mathbf{x}(t)) < \varepsilon_1$ for $t \in [t_1, \widetilde{t}_1)$, and it follows that

$$V'(\widetilde{t}_1, \mathbf{x}(\widetilde{t}_1)) \geq 0.$$

However,

$$f(\varepsilon_1) = f(V(\widetilde{t}_1, \mathbf{x}(\widetilde{t}_1))) > V(\widetilde{t}_1, \mathbf{x}(\widetilde{t}_1)) + a$$
$$= \varepsilon_1 + a = \varepsilon_0.$$

We also have $V(s, \mathbf{x}(s)) \leq \varepsilon_0$ for $s \in [\widetilde{t}_1 - r, \widetilde{t}_1]$, so it follows from (b) that $V'(\widetilde{t}_1, \mathbf{x}(\widetilde{t}_1)) \leq -W_3(|\mathbf{x}(\widetilde{t}_1)|) < 0$, a contradiction. Hence, we conclude that $V(t, \mathbf{x}(t)) < \varepsilon_1$ for all $t \geq t_1$.

Suppose $V(t, \mathbf{x}(t)) \geq \varepsilon_2$ for all $t \geq t_1$. Then for $t \geq t_1 + r$ we have $W_2(|\mathbf{x}(t)|) \geq \varepsilon_2$, and so $|\mathbf{x}(t)| \geq W_2^{-1}(\varepsilon_2) \overset{\text{def}}{=} \rho_2$. Because $\varepsilon_2 \leq V(t, \mathbf{x}(t)) \leq \varepsilon_1$ for $t \geq t_1 + r$, for such t we have

$$f(V(t, \mathbf{x}(t))) > V(t, \mathbf{x}(t)) + a$$
$$\geq W_1(\varepsilon) + (N - 1)a$$
$$= \varepsilon_1 \geq V(s, \mathbf{x}(s))$$

for $s \in [t - r, t]$. Thus, by (b) we have

$$V'(t, \mathbf{x}(t)) \leq -W_3(|\mathbf{x}(t)|), \quad t \geq t_1 + r.$$

If $\gamma_2 = W_3(\rho_2)$, then $\gamma_2 > 0$, and from $V' \leq -W_3$ we have

$$0 \leq V(t, \mathbf{x}(t))$$
$$\leq V(t_1 + r, \mathbf{x}(t_1 + r)) - \gamma_2(t - t_1 - r)$$
$$\leq \varepsilon_1 + \gamma_2(t - t_1 - r),$$

a contradiction for $t \geq t_1 + r$. Thus, there exists $t_2 \geq t_1 + r$ with $V(t_2, \mathbf{x}(t_2)) < \varepsilon_2$.

Suppose for some $\widetilde{t}_2 > t_2$, $V(\widetilde{t}_2, \mathbf{x}(\widetilde{t}_2)) = \varepsilon_2$, when $V(t, \mathbf{x}(t)) < \varepsilon_2$ for $t \in [t_2, \widetilde{t}_2]$. Then $V'(\widetilde{t}_2, \mathbf{x}(\widetilde{t}_2)) \geq 0$. However,

$$f(\varepsilon_2) = f(V(\widetilde{t}_2, \mathbf{x}(\widetilde{t}_2)))$$
$$> V(\widetilde{t}_2, \mathbf{x}(\widetilde{t}_2)) + a$$
$$= \varepsilon_2 + a = \varepsilon_1.$$

But $V(s, \mathbf{x}(s)) \leq \varepsilon_1$ for $s \in [\widetilde{t}_2 - r, \widetilde{t}_2]$ when $V(t, \mathbf{x}(t)) < \varepsilon_1$ for all $t \geq t_1$, and $s \geq t_2 - r \geq t_1$. Thus, $f(V(\widetilde{t}_2, \mathbf{x}(\widetilde{t}_2))) > V(s, \mathbf{x}(s))$ for $s \in [\widetilde{t}_2 - r, \widetilde{t}_2]$, and by (b) we have $V'(\widetilde{t}_2, \mathbf{x}(\widetilde{t}_2)) < 0$, a contradiction. Thus, $V(t, \mathbf{x}(t)) < \varepsilon_2$ for $t \geq t_2$.

Continuing in this way we get, for $j = 0, 1, \ldots, N$, that there exists t_j such that $V(t, \mathbf{x}(t)) < \varepsilon_j$ for $t \geq t_j$, where $t_j \geq t_{j-1} + r$, and $t_0 = 0$. But $\varepsilon_N = W_1(\varepsilon)$, so $W_1(|\mathbf{x}(t)|) \leq V(t, \mathbf{x}(t)) \leq W_1(\varepsilon)$ for $t \geq t_N$ yields $|\mathbf{x}(t)| < \varepsilon$ for $t \geq t_N$. This completes the proof.

Grimmer and Seifert obtain a similar result yielding uniform asymptotic stability for a system of functional differential equations containing a parameter

$$\mathbf{x}'(t) = \mathbf{F}(t, \mathbf{x}(s), \mu; \ -\infty < s \leq t), \quad t > 0, \tag{8.2.3}$$

where μ is an arbitrary parameter set and, for each fixed μ, then $\mathbf{F} : [0, \infty) \times CB \to R^n$ is continuous, with CB being bounded, continuous vector functions $\phi : (-\infty, t] \to R^n$ with the sup norm.

Theorem 8.2.4. Grimmer-Seifert (1975)   *Suppose $\mathbf{F}(t, \mathbf{0}, \mu) \equiv \mathbf{0}$ and the zero solution of (8.2.3) is uniformly stable. Let $V : (-\infty, \infty) \times R^n \to [0, \infty)$ be continuous and satisfy a local Lipschitz condition in $\mathbf{x}$. If*

(a) *$W_1(|\mathbf{x}|) \leq V(t, \mathbf{x}) \leq W_2(|\mathbf{x}|)$ on $(-\infty, \infty) \times R^n$ for W_1 and W_2 wedges, and*

(b) *there exists $M > 0$ and a monotonic sequence of positive reals $\{r_j\}$ and $\{u_j\}$ with $r_j \to \infty$ and $u_j \to 0$ as $j \to \infty$, and continuous functions $w_j(s)$ which are positive on $u_j/2 \leq s \leq M$, so that given a* solution $\mathbf{x}(s) = \mathbf{x}(s, t_0, \phi, \mu)$ of (8.2.3) with $|\mathbf{x}(s)| \leq M$ for all s, if for some $t \geq r_j + t_0$ one has $u_j \leq |\mathbf{x}(t)|$ and $V(s, \mathbf{x}(s)) < f(V(t, \mathbf{x}(t)))$ for $t - r_j \leq s \leq t$, then $V'_{(8.2.3)}(t, \mathbf{x}(t)) \leq -w_j(|\mathbf{x}(t)|)$, where $f(s) > s$ for $s > 0$ as before.

Then $\mathbf{x} = \mathbf{0}$ is U.A.S.

The proof is similar to that given for Theorem 8.2.3.

Grimmer and Seifert (1975) state without proof a result (Theorem 8.2.5) based on work of Seifert (1973) and on the Razumikhin-Driver result (Theorem 8.1.7). That result is then used to obtain an exceptionally strong perturbation result (Theorem 8.2.6).

Definition 8.2.1. *Solutions of (8.1.1) are uniform bounded if, for each $K > 0$, there exists $B > 0$ such that*

$$t_0 \geq 0, \quad t \geq t_0, \quad \|\phi\|^{[\alpha, t_0]} \leq K$$

imply $|\mathbf{x}(t, t_0, \phi)| \leq B$.

Definition 8.2.2. *Solutions of (8.1.1) are uniform ultimate bounded for bound B if there exists $B > 0$ such that for each $\gamma > 0$ there exists $T = T(\gamma) > 0$ such that*

$$t_0 \geq 0, \quad \|\phi\|^{[\alpha, t_0]} < \gamma, \quad t \geq t_0 + T$$

imply $|\mathbf{x}(t, t_0, \phi)| < B$.

When we speak of these global properties it is assumed that the differential equation under discussion is continuous on the whole space.

Theorem 8.2.5. Grimmer-Seifert (1975)   *Again consider (8.2.3) and suppose there is a continuous function $V : (-\infty, \infty) \times R^n \to [0, \infty)$ that is locally Lipschitz in $\mathbf{x}$ and satisfies*

(a) $W_1(|\mathbf{x}|) \leq V(t, \mathbf{x}) \leq W_2(|\mathbf{x}|)$, W_i *wedges,*
(b) *there exists $M \geq 0$, so that if $\mathbf{x}(t)$ is a solution of (8.2.3) with $|\mathbf{x}(t)| \geq M$ for some $t \geq 0$ and $V(s, \mathbf{x}(s)) < f(V(t, \mathbf{x}(t)))$ for $s \leq t$ and $f(r) > r$ as before, then $V'_{(8.2.3)}(t, \mathbf{x}(t)) \leq 0$.*

Then solutions of (8.2.3) are uniform bounded.

Grimmer and Seifert then consider

$$\mathbf{x}' = A\mathbf{x} + \int_0^t C(t, s)\mathbf{x}(s)\, ds + \mathbf{g}(t), \tag{8.2.4}$$

where A and C are $n \times n$ matrices, A constant, all characteristic roots of A have negative real parts, C continuous for $0 \leq s \leq t < \infty$ (actually, they ask less than continuity), and $\mathbf{g} : [0, \infty) \to R^n$ is continuous. Select $B = B^T$ with

$$A^T B + BA = -I$$

and let α^2 and β^2 be the smallest and largest (respectively) characteristic roots of B.

Theorem 8.2.6. Grimmer-Seifert (1975) *Let the above stated conditions hold and suppose there is an $M > 0$ with*

$$\int_0^t |BC(t,s)| \, ds \leq M, \quad t \geq 0,$$

where $2\beta M/\alpha < 1$. If, in addition, g is bounded, then all solutions of (8.2.4) are bounded.

Proof. Define $V(t, \mathbf{x}) = \mathbf{x}^T B \mathbf{x}$, so that $\alpha^2 |\mathbf{x}|^2 \leq V(t, \mathbf{x}) \leq \beta^2 |\mathbf{x}|^2$, yielding (a) of Theorem 8.2.5. Now

$$V'_{(8.2.4)}(t, \mathbf{x}(\cdot)) = -|\mathbf{x}|^2 + 2\mathbf{x}^T(t) \int_0^t BC(t,s)\mathbf{x}(s) \, ds + 2\mathbf{x}^T(t) B\mathbf{g}(t)$$

$$\leq -|\mathbf{x}|^2 + 2|\mathbf{x}| \int_0^t |BC(t,s)| \, |\mathbf{x}(s)| \, ds + 2|\mathbf{x}| \, |B| \, \|\mathbf{g}\|^{[0,\infty)}.$$

Now, if $h^2 V(t, \mathbf{x}(t)) > V(s, \mathbf{x}(s))$ for $s \leq t$, where $h > 1$ is a constant to be determined, then

$$h^2 \beta^2 |\mathbf{x}(t)|^2 \geq h^2 V(t, \mathbf{x}(t)) \geq V(s, \mathbf{x}(s))$$
$$\geq \alpha^2 |\mathbf{x}(s)|^2$$

and

$$(h\beta/\alpha)|\mathbf{x}(t)| \geq |\mathbf{x}(s)|, \quad s \leq t.$$

Thus,

$$V'_{(8.2.4)}(t, \mathbf{x}(\cdot))$$

$$\leq -|\mathbf{x}|^2 + (2h\beta/\alpha)|\mathbf{x}|^2 \int_0^t |BC(t,s)| \, ds + 2|\mathbf{x}| \, |B| \, \|\mathbf{g}\|^{[0,\infty)}$$

and, because $2\beta M/\alpha < 1$, h may be chosen so that $h > 1$ and $2h\beta M/\alpha < 1$ yielding

$$V'_{(8.2.4)}(t, \mathbf{x}(\cdot)) \leq \left[(2h\beta M/\alpha) - 1 \right] |\mathbf{x}|^2 + 2|B| \, \|\mathbf{g}\|^{[0,\infty)} |\mathbf{x}|$$
$$\leq 0$$

if $|\mathbf{x}| \geq 2|B| \, \|\mathbf{g}\|^{[0,\infty)} \left[1 - (2h\beta M/\alpha) \right]$. The result now follows from Theorem 8.2.5.

If we review the perturbation results in Chapters 2, 6, and 7 in the nonconvolution case and in cases in which a differential inequality

$$V' \leq -\alpha V + K|g(t)|, \quad \alpha > 0,$$

was not obtained, then one sees that Theorem 8.2.6 is a strong result. Of course, in the two exceptional cases mentioned the desired conclusion is easily obtained; in fact, **g** need not even be bounded.

Remark 8.2.1.   While we have noted earlier that some investigators claim that the method of Liapunov functions is better than that of Liapunov functionals, the careful investigator will notice that they are not always comparable. In Theorem 8.2.6 the authors use a Liapunov functional which integrates the second coordinate of C, while in Theorem 2.5.1a,b a Liapunov functional is used which integrates the first coordinate of C. In Theorem 2.5.1c,d, the Liapunov functional integrates both coordinates of C. They are independent tools and the investigator who uses both will certainly be the winner. Hale (1971, p. 58) proves a theorem about Liapunov functions by converting the Liapunov function to a Liapunov functional.

8.3   Equations with Bounded Delay

We consider the system

$$\mathbf{x}'(t) = \mathbf{F}(t, \mathbf{x}_t), \tag{8.3.1}$$

where $\mathbf{x}_t$ is that segment of $\mathbf{x}(s)$ on $[t - h, t]$ shifted to $[-h, 0]$, where $h > 0$ is a fixed constant. This is, of course, an example of (8.1.1). The literature on (8.3.1) is enormous and we will repeat little of it here. The classical treatment is found in Yoshizawa (1966, pp. 183–213).

There is fairly standard notation concerning (8.3.1) throughout the literature and, because it differs in substantial detail from what we have used, it is worth noting here. The interested reader may then consult pertinent literature without pondering over details of notation.

Notation.   For $x \in R^n$, $|\mathbf{x}| = \max |x_i|$. For $h > 0$, C denotes the space of continuous functions mapping $[-h, 0]$ into R^n, and for $\phi \in C$, $\|\phi\| = \sup_{-h \leq \theta \leq 0} |\phi(\theta)|$. C_H denotes the set of $\phi \in C$ with $\|\phi\| \leq H$. If **x** is a continuous function of u defined on $-h \leq u < A$, $A > 0$, and if t is a fixed number satisfying $0 \leq t < A$, then $\mathbf{x}_t$ denotes the restriction of **x** to the interval $[t-h, t]$ so that $\mathbf{x}_t$ is an element of C defined by $\mathbf{x}_t(\theta) = \mathbf{x}(t+\theta)$ for $-h \leq \theta \leq 0$.

In (8.3.1), $\mathbf{x}'(t)$ denotes the right hand derivative of **x** at t and $\mathbf{F}(t, \phi) \in R^n$ is continuous on $[0, \infty) \times C_H$.

We denote by $\mathbf{x}(t_0, \boldsymbol{\phi})$ a solution of (8.3.1) with initial condition $\boldsymbol{\phi} \in C_H$, where $\mathbf{x}_{t_0}(t_0, \boldsymbol{\phi}) = \boldsymbol{\phi}$ and by $\mathbf{x}(t, t_0, \boldsymbol{\phi})$ the value of $\mathbf{x}(t_0, \boldsymbol{\phi})$ at t.

It is supposed that $\mathbf{F}(t, \mathbf{0}) = \mathbf{0}$, so that the zero function is a solution. The stability definitions are as before, but in this notation we say that the zero solution is uniformly stable if for each $\varepsilon > 0$ there exists $\delta > 0$ such that

$$t_0 \geq 0, \quad t \geq t_0, \quad \text{and} \quad \|\boldsymbol{\phi}\| < \delta$$

imply $|\mathbf{x}(t, t_0, \boldsymbol{\phi})| < \varepsilon$. If, in addition, there exists $\delta > 0$ such that for any $\varepsilon > 0$ there exists $T > 0$ such that

$$t_0 \geq 0, \quad t \geq t_0 + T, \quad \text{and} \quad \|\boldsymbol{\phi}\| < \delta$$

imply $|\mathbf{x}(t, t_0, \boldsymbol{\phi})| < \varepsilon$, then $\mathbf{x} = \mathbf{0}$ is uniformly asymptotically stable.

It is important now to review Section 6.1, particularly the definition of a wedge, Definition 6.1.2, the Marachkov idea in Theorem 6.1.3, and the annulus argument in Definition 6.1.5. The annulus argument has been central in stability theory and Hatvani (1997b) has an exhaustive study of such arguments and references.

In the 1950s Krasovskii (1963, pp. 143–151) reviewed what was essentially Theorem 6.1.1 and constructed a parallel set of theorems for functional differential equations with a finite delay. His idea was simple in the extreme. Everywhere there was an R^n norm, he replaced it by a supremum norm. The astonishing part was that the resulting theorems were true and they even had converses. Here are three of his main results.

Theorem 8.3.1. Krasovskii (1963) *Suppose there is a continuous and locally Lipschitz functional $V : [0, \infty) \times C_H \to [0, \infty)$ and wedges W_i such that:*

(i) *If $[W_1(\|\boldsymbol{\phi}\|) \leq V(t, \boldsymbol{\phi}), V(t, 0) = 0, V'_{(8.3.1)}(t, \boldsymbol{\phi}) \leq 0]$, then the zero solution of (8.3.1) is stable. If, in addition, $V(t, \boldsymbol{\phi}) \leq W_2(\|\boldsymbol{\phi}\|)$, then the zero solution of (8.3.1) is uniformly stable.*

(ii) *If $[W_1(\|\boldsymbol{\phi}\|) \leq V(t, \boldsymbol{\phi}) \leq W_2(\|\boldsymbol{\phi}\|)$ and $V'_{(8.3.1)}(t, \mathbf{x}_t) \leq -W_3(\|\boldsymbol{\phi}\|)]$, then the zero solution of (8.3.1) is uniformly asymptotically stable.*

Almost immediately Krasovskii realized that such functionals could only rarely be found. The search began for alternatives and very early Krasovskii (1963, p. 155) proved a result on asymptotic stability for which investigators were very successful in constructing the required Liapunov functionals. It had the form of Theorem 8.3.2 below except that its conclusion was simple asymptotic stability and it had the Marachkov boundedness condition embedded in a Lipschitz condition. Burton (1978) showed that the conclusion of Krasovskii's theorem was actually uniform asymptotic stability and that the Marachkov condition was not necessary. Here, $||| \cdot |||$ denotes the L^2-norm on C_H and has become the standard notation.

Theorem 8.3.2.   Burton (1978)   *If there is a continuous functional* $V : [0, \infty) \times C_H \to [0, \infty)$, *locally Lipschitz in* ϕ, *and wedges* W_i *such that*

   (i) $W(|\phi(0)|) \leq V(t, \phi) \leq W_1(|\phi(0)|) + W_2(|||\phi|||)$

and

   (ii) $V'_{(8.3.1)}(t, \phi) \leq -W_3(|\phi(0)|)$

then the zero solution of (8.3.1) *is uniformly asymptotically stable.*

   Remark 8.3.1.   This result is a consequence of a recent one of Zhang (1995), Theorem 8.3.6, as well as one by Hatvani (2002), Theorem 8.3.14, so it will not be proved here. But its method of proof is worth a remark because it initiates a very different point of view. In the Burton (1978) proof of UAS we "play the function W_2 against the function W_3." It seems to be an entirely new idea in stability theory and we will see it used throughout this section. In the annulus argument we integrate $-W_3(|x(t)|)$ when the solution is in the annulus and rely on the speed of the solution to tell us the value of the integral. In the present arguments, the upper bound on the Liapunov functional actually performs the integration for us and it tells us the value of the integral; the speed of the solution and the location within an annular ring are of no importance at all.

   Theorem 8.3.2 was not proved until 1978 and in the meantime Krasovskii had shown that there was a counterpart to the Marachkov result, Theorem 6.1.3.

Theorem 8.3.3.   *Suppose there is a continuous functional* $V : [0, \infty) \times C_H \to [0, \infty)$ *which is locally Lipschitz in* ϕ *and satisfies*

   (i) $W_1(|\phi(0)|) \leq V(t, \phi) \leq W_2(\|\phi\|)$

and

   (ii) $V'_{(8.3.1)}(t, \phi) \leq -W_3(|\phi(0)|)$.

   If, in addition, $|F(t, \phi)|$ *is bounded on* $[0, \infty) \times C_H$, *then the zero solution of* (8.3.1) *is uniformly asymptotically stable.*

One of the first Liapunov functionals which Krasovskii (1963, p. 170) constructed was very simple, it is of continuing interest, and it paved the way for a result very much like Liapunov's theorem for ordinary differential equations.

Example 8.3.1. Consider the scalar equation

$$x'(t) = -a(t)x(t) + b(t)x(t - h)$$

for $h > 0$, a and b continuous, and $K \geq 1$ with

$$a(t) \geq K|b(t + h)|.$$

Define

$$V(t, x_t) = |x(t)| + K \int_{t-h}^{t} |b(s + h)x(s)| \, ds$$

so that along a solution we have

$$\begin{aligned} V'(t, x_t) &\leq -a(t)|x(t)| + |b(t)x(t - h)| \\ &\quad + K|b(t + h)x(t)| - K|b(t)x(t - h)| \\ &= \left[-a(t) + K|b(t + h)| \right] |x(t)| - (K - 1)|b(t)x(t - h)|. \end{aligned}$$

Many conclusions can be drawn from this. For example, if $b(t)$ is bounded and if $-a(t) + K|b(t + h)| \leq -\alpha < 0$, then the conditions of Theorem 8.3.2 are satisfied and the zero solution is uniformly asymptotically stable. But the interested reader will be able to derive numerous other results when later theorems are presented. In particular, if $K > 1$ then we can argue that $V'(t, x_t) \leq -k \left[|x'(t)| + |x(t)| \right]$ for some $k > 0$, a condition we have seen before in Section 2.5.

It is well recognized that most elementary stability applications of non-linear ordinary differential equations are studied by means of a result of Liapunov, often referred to as Perron's theorem, as shown in Section 6.2 starting with Equation (6.2.27). Consider an equation

$$\mathbf{x}' = A\mathbf{x} + \mathbf{H}(t, \mathbf{x})$$

where A is an $n \times n$ constant matrix, all of whose characteristic roots have negative real parts, $\mathbf{H}(t, \mathbf{x})$ continuous, and

$$\lim_{|\mathbf{x}| \to 0} |\mathbf{H}(t, \mathbf{x})| / |\mathbf{x}| = 0$$

uniformly for $0 \leq t < \infty$. By means of a Liapunov functional or Gronwall's inequality, it can be shown that the zero solution is uniformly asymptotically stable. In effect, the function H can be ignored for small initial conditions. We can treat delay terms in the same way. In this result we can easily see how $A x$ dominates $H(t, x)$, but it is a great surprise that Krasovskii's Liapunov functional shows us that Ax also dominates a function $g(t, x(t-h))$. There are many nonlinear extensions which we will not consider here.

Theorem 8.3.4. Krasovskii-Liapunov   *Let A be an $n \times n$ real constant matrix, all of whose characteristic roots have negative real parts. Suppose that D is an open neighborhood of $\mathbf{x} = \mathbf{0}$ in R^n and that $\mathbf{f}, \mathbf{g} : [0, \infty) \times D \to R^n$ are continuous. If*

$$\lim_{|\mathbf{x}| \to 0} \left(\frac{|f(t, \mathbf{x})|}{|\mathbf{x}|} + \frac{\mathbf{g}(t+h, \mathbf{x})}{|\mathbf{x}|} \right) = 0$$

uniformly for $0 \le t < \infty$, then the zero solution of

$$\mathbf{x}' = A\mathbf{x} + \mathbf{f}(t, \mathbf{x}) + \mathbf{g}(t, \mathbf{x}(t-h))$$

is uniformly asymptotically stable.

Proof. We can find a positive definite and symmetric matrix B with $A^T B + BA = -I$, as in Section 5.1. Also, there are positive constants α, β with $\alpha^2 |\mathbf{x}|^2 \le \mathbf{x}^T B \mathbf{x} \le \beta^2 |\mathbf{x}|^2$. Then define a Liapunov functional by

$$V(t, \mathbf{x}_t) = \mathbf{x}^T B \mathbf{x} + k \int_{t-h}^{t} |\mathbf{g}(s+h, \mathbf{x}(s))|^2 \, ds$$

where $k = 4|B|^2$. If we take the derivative of V along a solution of our equation we have

$$\begin{aligned}
V'(t, x_t) &\le \left[A\mathbf{x} + \mathbf{f}(t, \mathbf{x}) + \mathbf{g}(t, \mathbf{x}(t-h) \right]^T B \mathbf{x} \\
&\quad + \mathbf{x}^T B \left[A\mathbf{x} + \mathbf{f}(t, \mathbf{x}) + \mathbf{g}(t, \mathbf{x}(t-h)) \right] \\
&\quad + k|\mathbf{g}(t+h, \mathbf{x}(t))|^2 - k|\mathbf{g}(t, \mathbf{x}(t-h))|^2 \\
&\le -|\mathbf{x}|^2 + 2\,|B|\,|\mathbf{x}|\,|\mathbf{f}(t, \mathbf{x})| + 2\,|B|\,|\mathbf{x}|\,|\mathbf{g}(t, \mathbf{x}(t-h)| \\
&\quad + k|\mathbf{g}(t+h, \mathbf{x})|^2 - k|\mathbf{g}(t, \mathbf{x}(t-h))|^2 \, .
\end{aligned}$$

By our limit assumption we can find $\gamma > 0$ such that

(i) $|\mathbf{x}| \le \gamma$ and $t \in [0, \infty)$

imply

(ii) $2\,|B|\,|\mathbf{f}(t, \mathbf{x})| \le |\mathbf{x}|/4$ and $4\,|B|^2 \mathbf{g}^2(t+h, \mathbf{x}) \le |\mathbf{x}|^2/4$.

Now, for $|\mathbf{x}| \leq \gamma$ we have

$$V'(t, \mathbf{x}) \leq -|\mathbf{x}|^2 + (1/4)|\mathbf{x}|^2 + \left((1/4)|\mathbf{x}|^2 + 4\,|B|^2|\mathbf{g}(t, \mathbf{x}(t-h))|^2\right)$$
$$+ 4\,|B|^2|\mathbf{g}(t+h, \mathbf{x})|^2 - 4\,|B|^2|\mathbf{g}(t, \mathbf{x}(t-h))|^2$$
$$\leq -(1/4)|\mathbf{x}|^2\,.$$

For $|\mathbf{x}| \leq \gamma$ and $t \in [0, \infty)$ then

$$\alpha^2|\mathbf{x}(t)|^2 \leq V(t, \mathbf{x}_t) \leq \beta^2|\mathbf{x}(t)|^2 + (1/4)\int_{t-h}^{t} |\mathbf{x}(s)|^2\,ds$$

and

$$V'(t, \mathbf{x}_t) \leq -(1/4)|\mathbf{x}(t)|^2\,.$$

By Theorem 8.3.2 the zero solution is uniformly asymptotically stable.

The foremost problem in stability theory of functional differential equations from 1950 to 1992 was to prove or give a counterexample to the conjecture that the Marachkov boundedness condition in Theorem 8.3.3 could be removed. We formally state it as follows.

Conjecture 8.3.1. *If there is a continuous and locally Lipschitz functional $V : [0, \infty) \times C_H \to [0, \infty)$ and wedges W_i such that*

(i) $W_1(|\boldsymbol{\phi}(0)|) \leq V(t, \boldsymbol{\phi}) \leq W_2(\|\boldsymbol{\phi}\|)$

and

(ii) $V'_{(8.3.1)}(t, \boldsymbol{\phi}) \leq -W_3(|\boldsymbol{\phi}(0)|)$

then the zero solution of (8.3.1) is uniformly asymptotically stable.

Makay (1991) gave a counterexample to the version of this conjecture in which the conditions are required only along the solutions of (8.3.1). Using Makay's idea, in 1992 Junji Kato (1996) gave a sophisticated counterexample to Conjecture 8.3.1. Then Makay (1994) simplified this example.

In view of the counterexample, investigators took the view that condition (i) of Theorem 8.3.2 asks too much, while condition (i) of Conjecture 8.3.1 asks too little. Most examples which have been constructed turn out to satisfy condition (i) of Theorem 8.3.2 as often as they satisfy condition (i) of Conjecture 8.3.2; however, the single supremum norm is certainly simpler. Moreover, converse theorems yield condition (i) of Conjecture 8.3.2, while no one seems to have obtained converse theorems having condition (i) of Theorem 8.3.2. Wang (1992) shows that condition

(i) of Theorem 8.3.2 is far more general than it appears. Many advances have been made in weakening the conditions of Theorem 8.3.2 and we will be looking at a number of them. Zhang (1995) added just a small part of (i) in Theorem 8.3.2 to (i) in Conjecture 8.3.1, while weakening (ii) in the conjecture to obtain a valid result which was very near what the original conjecture claimed. Hatvani (2002) kept (i), but weakened (ii) to allow V' to vanish on long intervals.

Remark 8.3.2. A major reason for wanting to prove Conjecture 8.3.1 in view of the entirely satisfactory Theorem 8.3.2 is that once we leave the small and secure world of the supremum norm in Theorem 8.3.1, we see that there are simply too many possibilities. Once we have proved Theorem 8.3.2 using the L^2-norm, must we state new versions with L^p−norms and other measures which are, perhaps, not norms at all? Wang (1992) studied that question in depth and shows that we really need only consider the R^n norms, $|\phi(0)|$ and integral expressions of the form $\int_{-h}^{0} W(|\phi(s)|)ds$, as we have seen in Theorem 8.3.2. After choosing a variant of the upper bound given in Theorem 8.3.2, our main choice will be to ask either

$$V'(t, \phi) \leq -W(|\phi(0)|)$$

or

$$V'(t, \phi) \leq -W_1\left(\int_{-h}^{0} W_2(|\phi(s)|)\, ds \right).$$

Here is a typical result from Wang (1992). The L^p-norm will be denoted by $\| \cdot \|_p$.

Theorem 8.3.5. Wang (1992, p. 143) *Let* $V : [0, \infty) \times C_H \to [0, \infty)$ *be continuous. Then for each* $\phi \in C_H$ *and each solution* $x(t) = x(t, t_0, \phi)$ *of* (8.3.1), V *satisfies*

(i) $W_1(|\mathbf{x}(t)|) \leq V(t, \mathbf{x}_t) \leq W_2(|\mathbf{x}(t)|) + W_3(\|\mathbf{x}_t\|_2),$

and

(ii) $V'_{(8.3.1)}(t, \mathbf{x}_t) \leq -\eta(t)W_4(|\mathbf{x}(t)|),\ \eta(t) \geq 0,$

if and only if there exist wedges $\bar{W}_3, W_5$ *and a constant* $\delta > 0$ *such that for each* $\phi \in C_\delta$ *and each solution* $x(t) = x(t, t_0, \phi)$ *of* (8.3.1), V *satisfies*

(iii) $W_1(|\mathbf{x}(t)|) \leq V(t, \mathbf{x}_t) \leq W_2(|\mathbf{x}(t)|) + \bar{W}_3\left(\int_{t-h}^{t} W_5(|\mathbf{x}(u)|)\, du \right),$

and

(iv) $V'_{(8.3.1)}(t, \mathbf{x}_t) \leq -\eta(t)W_4(|\mathbf{x}(t)|).$

It turns out that most of the proofs are greatly enhanced by use of Jensen's inequality.

Definition 8.3.1. *A scalar function G defined on a closed interval $[a, b]$ is said to be* convex downward *if*

$$G([x + y]/2) \leq [G(x) + G(y)]/2$$

for any $x, y \in [a, b]$.

A proof of the following result is found in Natanson (1960).

Lemma 8.3.1. *Jensen's inequality   Let $G : R \to R$ be continuous and convex downward. If f and p are continuous on $[a, b]$ with $p(t) \geq 0$ and $\int_a^b p(t)\, dt > 0$, then*

$$G\left[\frac{\int_a^b f(t)p(t)\, dt}{\int_a^b p(t)\, dt}\right] \leq \frac{\int_a^b G(f(t))p(t)\, dt}{\int_a^b p(t)\, dt}.$$

If we have an expression

$$V'(t, \mathbf{x}_t) \leq -W(|\mathbf{x}(t)|)$$

where W is a wedge and we are using it for $0 \leq r \leq 1$, then we can create a new wedge

$$W^*(r) := \int_0^r W(s)\, ds.$$

Notice three things. First, W^* is a wedge. Because W is increasing we have

$$W^*(r) = \int_0^r W(s)\, ds \leq rW(r) \leq W(r).$$

Hence, we can say that

$$V'(t, \mathbf{x}_t) \leq -W^*(|\mathbf{x}(t)|).$$

Finally, one can show that W^* is convex downward. In conclusion, we may always assume that this wedge associated with V' in the above described way is convex downward.

We are going to use the L^1-norm in this theorem for conceptual reasons. However, it will be easy to see that we can change that to the L^2-norm, if desired, by using the Schwarz inequality. Moreover, Wang (1992) has shown how we can move freely in a given theorem from one kind of norm to another and, in fact, to general wedges of functions. Those results will be discussed later. But the reader should keep in mind that a given theorem can be, thereby, made to fit a wide variety of problems.

We begin with an important property of the integral of the solution. It has its roots in Zhang (1995) who uses the uniform continuity of separate solutions.

Proposition 8.3.1. *Let* $m = \{\phi : [0, \infty) \to R^n : \phi \in C, |\phi(t)| \le H\}$ *and let* $M = \{\Phi : [h, \infty) \to [0, \infty) : \Phi(t) = \int_{t-h}^{t} |\phi(s)| \, ds, \; \phi \in m\}$. *Then* M *is a uniformly bounded and equicontinuous set. For a given* $\varepsilon > 0$, *the* δ *for equicontinuity is* $\varepsilon/2H$.

Proof. For an arbitrary $\Phi \in M$ we find the associated $\phi \in m$ so that $\Phi(t) = \int_{t-h}^{t} |\phi(s)| \, ds$. Then $|\Phi(t_1) - \Phi(t_2)| \le 2H|t_1 - t_2|$, from which the result follows.

The result means that if we have the L^1-norm of a solution in the upper bound on V or in the derivative of V and if that integral is large at one point, it remains large in a neighborhood of the point.

The following result was obtained by combining two of the theorems in Zhang (1995) and adding an upper wedge on V. The only reason we add the upper wedge in (i) is to ensure uniform stability. One may drop W_2 and ask that the zero solution be uniformly stable. Frequently, independent proofs of uniform stability are simple to supply, often using a Razumikhin argument. There is both a theorem and an example of this kind in Wang (1994a).

Theorem 8.3.6. Zhang (1995)   *Suppose there are wedges* W_i *so that*

  (i) $W_1(|\phi(0)|) \le V(t, \phi) \le W_2(\|\phi\|)$

and

  (ii) $V'_{(8.3.1)}(t, \phi) \le 0$.

Suppose also that there are positive constants d *and* L, *together with a monotone increasing sequence* $\{t_n\}$ *tending to infinity, and satisfying* $h \le t_{n+1} - t_n \le L$, *such that*

  (iii) $V(t, \phi) \le W_3(|\phi(0)|) + W_4\left(\int_{-h}^{0} |\phi(s)| \, ds\right)$ *for* $t_n - d \le t \le t_n$

and

  (iv) $V'_{(8.3.1)}(t, \phi) \le -W_5(|\phi(0)|)$ *for* $t_n - h \le t \le t_n$.

Then the zero solution of (8.3.1) *is uniformly asymptotically stable.*

Proof. The zero solution is uniformly stable. For the $H > 0$ find B of uniform stability so that $t_0 \geq 0$, $\boldsymbol{\psi} \in C_B$, $t \geq t_0$ imply that any solution $\mathbf{x}(t) := \mathbf{x}(t, t_0, \boldsymbol{\psi})$ satisfies $|\mathbf{x}(t)| < H/2$. The solution can then be continued for all future time. To prove uniform asymptotic stability, let $\gamma > 0$ be given. We will find $T > 0$ such that $|\mathbf{x}(t)| < \gamma$ for $t \geq t_0 + T$, independently of t_0 and $\boldsymbol{\psi} \in C_B$.

For any $t_0 \geq 0$ which is selected, let $[t_N - d, t_N]$ be the next element of our sequence to the right of t_0. *We come now to the central idea mentioned in Remark 8.3.1.* So long as $V(t, \mathbf{x}_t) \geq W_1(\gamma)$, on $[t_N - d, t_N]$ (and on any subsequent member of that sequence) either

$$W_3(|\mathbf{x}(t)|) \geq W_1(\gamma)/2 \quad \text{for all} \quad t \in [t_N - d, t_N]$$

or

$$\text{there is a} \quad t^* \in [t_N - d, t_N] \quad \text{with} \quad W_4\left(\int_{t^*-h}^{t^*} |\mathbf{x}(s)|\, ds\right) \geq W_1(\gamma)/2.$$

Applying the first case to (iv), we have

$$V'_{(8.3.1)}(t, \mathbf{x}_t) \leq -W_5\big(W_3^{-1}(W_1(\gamma)/2)\big) \quad \text{on} \quad [t_N - d, t_N].$$

For brevity we denote $V(t) = V(t, \mathbf{x}_t)$ and integrate that last inequality to obtain

$$V(t_N) - V(t_N - d) \leq -dW_5\big(W_3^{-1}(W_1(\gamma)/2)\big) =: -d_1,$$

where d_1 depends only on γ and not on the particular solution or t_0.

In the other case, when t^* exists, we can assume without loss of generality that W_5 is convex downward since such a wedge can be constructed under W_5. Now by the equicontinuity proved in Proposition 8.3.1 we can assume that $d < h$ and that d is so small that

$$\int_{t_N-h}^{t_N} |\mathbf{x}(s)|\, ds \geq W_4^{-1}(W_1(\gamma)/2)/2.$$

We then integrate (iv), apply Jensen's inequality, and obtain

$$V(t_N) - V(t_N - h) \leq -hW_5\left(\int_{t_N-h}^{t_N} |\mathbf{x}(s)|\, ds\, /\, h\right)$$

$$\leq -hW_5\big(W_4^{-1}(W_1(\gamma)/2)/2h)\big) =: -d_2.$$

Let $d^* = \min[d_1, d_2]$. As $V(t_0, \psi) \leq W_2(B)$ and $V' \leq 0$, there is an integer k with

$$V(t, \mathbf{x}_t) \leq W_2(B) - kd^* < W_1(\gamma)$$

if $t \geq t_0 + kL \geq t_{N+k-1}$. We select $T = kL$, to complete the proof.

There is a simple way in which one can obtain asymptotic stability results from Theorem 8.3.6. First, replace W_2 with the statement that $V(t, 0) = 0$. This will allow us to prove that the zero solution is stable. Next, drop the constant L bounding $t_{n+1} - t_n$ and change d to d_n where $\sum_{n=1}^{\infty} d_n = \infty$. The main details of the proof just given will show that $V(t, x_t)$ can be driven to zero, but not in a prescribed amount of time.

Theorem 8.3.7. *Suppose there are wedges W_i so that*

   (i) $W_1(|\boldsymbol{\phi}(0)|) \leq V(t, \boldsymbol{\phi}), \quad V(t, \mathbf{0}) = 0$

and

   (ii) $V'_{(8.3.1)}(t, \boldsymbol{\phi}) \leq 0.$

Suppose also that there is a sequence of positive constants $\{d_n\}$ with $\sum_{n=1}^{\infty} d_n = \infty$, together with a monotone increasing sequence $\{t_n\}$ tending to infinity and satisfying $h \leq t_{n+1} - t_n$ such that

   (iii) $V(t, \boldsymbol{\phi}) \leq W_3(|\boldsymbol{\phi}(0)|) + W_4\left(\int_{-h}^{0} |\boldsymbol{\phi}(s)|\, ds\right)$ *for* $t_n - d_n \leq t \leq t_n$

and

   (iv) $V'_{(8.3.1)}(t, \boldsymbol{\phi}) \leq -W_5(|\boldsymbol{\phi}(0)|)$ *for* $t_n - h \leq t \leq t_n$.

Then the zero solution of (8.3.1) *is asymptotically stable.*

Proof. The zero solution is stable. For $t_0 \geq 0$, find δ of stability for $\varepsilon_0 = H/2$ so that $\boldsymbol{\psi} \in C_\delta$, $t \geq t_0$ imply that any solution $\mathbf{x}(t) := \mathbf{x}(t, t_0, \boldsymbol{\psi})$ satisfies $|\mathbf{x}(t)| < H/2$. The solution can then be continued for all future time. We will show that $\mathbf{x}(t, t_0, \boldsymbol{\psi}) \to 0$ as $t \to \infty$.

For any $t_0 \geq 0$ which is selected, let $[t_N - d_N, t_N]$ be the next element of our sequence to the right of t_0. Again, so long as $V(t, \mathbf{x}_t) \geq W_1(\gamma)$ on $[t_N - d_N, t_N]$ (and on any subsequent member of that sequence) we have either

$$W_3(|\mathbf{x}(t)|) \geq W_1(\gamma)/2 \quad \text{for all} \quad t \in [t_N - d_N, t_N]$$

or

there is a $\quad t^* \in [t_N - d_N, t_N] \quad$ with $\quad W_4\left(\int_{t^*-h}^{t^*} |\mathbf{x}(s)|\, ds\right) \geq W_1(\gamma)/2.$

Now by the equicontinuity proved in Proposition 8.3.1 we can choose $d_0 > 0$ independent of solutions such that

$$\int_{s-h}^{s} |\mathbf{x}(u)|\, du \geq W_4^{-1}(W_1(\gamma)/2)/2$$

whenever $|t^* - s| < d_0$.

We will assume that $d_j < d_0$ for all $j \geq N$. Applying the first case to (iv), we have

$$V'_{(8.3.1)}(t, \mathbf{x}_t) \leq -W_5\big[W_3^{-1}(W_1(\gamma)/2)\big] \quad \text{on} \quad [t_N - d_N, t_N].$$

For brevity we denote $V(t) = V(t, \mathbf{x}_t)$ and integrate that last inequality to obtain

$$V(t_N) - V(t_N - d_N) \leq -d_N W_5\big[W_3^{-1}(W_1(\gamma)/2)\big] =: -d_N^*.$$

In the other case, when t^* exists, we can assume without loss of generality that W_5 is convex downward since such a wedge can be constructed under W_5. Since $t^* - t_N < d_0$, by the definition d_0, we have

$$\int_{t_N - h}^{t_N} |\mathbf{x}(s)|\, ds \geq W_4^{-1}(W_1(\gamma)/2)/2.$$

We then integrate (iv), apply Jensen's inequality, and obtain

$$V(t_N) - V(t_N - h) \leq -hW_5\left(\int_{t_N - h}^{t_N} |\mathbf{x}(s)|\, ds \,/\, h\right)$$

$$\leq -hW_5\big[W_4^{-1}(W_1(\gamma)/2)/2h\big] =: -\ell.$$

We may assume $d_0 W_5(W_3^{-1}(W_1(\gamma)/2)) < \ell$. Thus, $d_j^* < \ell$ for all $j \geq N$. As $\sum_{j=1}^{\infty} d_j^* = \infty$ and $V' \leq 0$, there is an integer k with

$$W_1(|\mathbf{x}(t)|) \leq V(t, \mathbf{x}_t) \leq V(t_0, \boldsymbol{\psi}) - \sum_{j=0}^{k} d_{N+j}^* < W_1(\gamma)$$

if $t > t_{N+k}$. This implies $\mathbf{x}(t, t_0, \boldsymbol{\phi}) \to 0$ as $t \to \infty$, and hence the zero solution of (8.3.1) is asymptotically stable.

Example 8.3.2. Zhang (1995).   Consider the scalar equation

$$x'(t) = -a(t)x(t) + b(t)x(t-1)$$

where

$$a(t) = 3\big(|\sin(\pi t/2)| - \sin(\pi t/2)\big)^2 g^2(t),$$

$$b(t) = 2\big(|\sin(\pi t/2)| - \sin(\pi t/2)\big)\big(|\cos(\pi t/2)| + \cos(\pi t/2)\big)g(t)g(t-1)$$

and $g : R \to [0, \infty)$ is any continuous function with $1 \leq g(t)$ for all $t \in \big[4k - \frac{3}{2}, 4k - \frac{1}{2}\big]$ and $g(t) \leq B$ on $\big[4k - \frac{3}{2} - \sigma, 4k - \frac{1}{2}\big]$ for some positive numbers σ and B, where $k = 1, 2, \ldots$. Then the zero solution is asymptotically stable.

To prove that statement, define

$$V(t, \phi) = \frac{1}{2}|\phi(0)|^2 + \frac{1}{3}\int_{-1}^{0} a(t+s)\phi^2(s)s$$

for $(t, \phi) \in [0, \infty) \times C$ and $W_1(r) = r^2/2$. It follows that

$$W_1(|\phi(0)|) \leq V(t, \phi).$$

Along a solution we have

$$V'(t, x_t) = -\frac{2}{3}a(t)x^2(t) + b(t)x(t)x(t-1) - \frac{1}{3}a(t-1)x^2(t-1).$$

Since $a(t-1) = 3(|\cos(\pi t/2)| + \cos(\pi t/2))^2 g^2(t-1)$ and $b(t)x(t)x(t-1) \leq \frac{1}{3}a(t)x^2(t) + \frac{1}{3}a(t-1)x^2(t-1)$ it follows that

$$V'(t, x_t) \leq -\frac{1}{3}a(t)x^2(t)$$

for all $t \geq t_0$ and

$$V'(t, x_t) \leq -2x^2(t) \quad \text{on} \quad [t_k - 1, t_k]$$

where $t_k = 4k - \frac{1}{2}$. Define $W_2(r) = r^2/2$ and $W_3(r) = 4B^2r^2$. Since $g(t) \leq B$ on $[t_k - 1 - \sigma, t_k]$, it follows that

$$V(t, \phi) \leq W_2(|\phi(0)|) + W_3(|\phi|_2) \quad \text{for} \quad t \in [t_k - \sigma, t_k].$$

All the conditions of Theorem 8.3.7 are satisfied and the zero solution is asymptotically stable.

Furthermore, it is clear that $V'(t, x_t) \equiv 0$ on $[4(k-1), 4k-2]$. If we choose $g(4k - \frac{1}{4}) = k$, then $V(t, \phi)$ is not bounded for fixed ϕ.

On the other hand, if g is bounded in this example, then the zero solution is uniformly asymptotically stable.

Our Liapunov functionals are usually imperfect in some way and we are forced to use some devious method in conjunction with the Liapunov functional to prove UAS. But once we have proved UAS, we can often invoke a converse theorem which says that there is a seemingly perfect Liapunov functional which can be used effectively on a perturbed equation. The common converse theorems require that (8.3.1) be linear or satisfy a Lipschitz condition in ϕ uniformly in t. The effect of the last condition is to impose a Marachkov type condition. We will state, without proof, a converse theorem by Krasovskii. A linear version can be found in Yoshizawa (1966, p. 193) in the case of exponential asymptotic stability. Somolinos (1977) illustrates its use in perturbations.

Theorem 8.3.8. Krasovskii (1963, p. 146)   *Suppose that F in (8.3.1) satisfies a Lipschitz condition in ϕ uniformly in t on $[0,\infty) \times C_H$ and that the zero solution is uniformly asymptotically stable. Then there is a continuous functional $V : [0,\infty) \times C_H \to [0,\infty)$, a constant K, and wedges W_i with*

(i) $W_1(\|\phi\|) \leq V(t, \phi) \leq W_2(\|\phi\|)$,

(ii) $|V(t, \phi) - V(t, \xi)| \leq K\|\phi - \xi\|$,

and

(iii) $V'_{(8.3.1)}(t, \phi) \leq -W_3(\|\phi\|)$.

But if, in a given problem, all of the upper bounds we have considered here fail, then we are often forced to look back at the Marachkov result in Theorem 8.3.3. We understand that we need the boundedness on F because of the details in the annulus argument. A solution might be racing back and forth across an annulus on which V' is negative, but it might be racing so fast that when we try to integrate V' we do not get enough to drive V to zero; hence, we invoke the Marachkov condition to slow the solution down. But, do we have to slow it down that much? Investigations show that it may suffice to slow it down on the order of $|F(t, \phi)| \leq q(t)$ where $\int_0^\infty [1/q(t)] \, dt = \infty$. Wang (1994a) obtains such a result in case one can independently prove uniform stability. Hatvani and Krisztin (1995) consider a scalar example and show that when F is separated into stable and unstable parts, such a bound need only hold on the unstable part. This is reasonable since we know that it is not the speed of the solution which causes problems, rather it is the direction.

Krasovskii (1963) had also stated Theorem 8.3.3 with (i) replaced by

(i*) $W(|\phi(0)|) \leq V(t, \phi), \quad V(t, 0) = 0$

and had concluded that the zero solution of (8.3.1) is asymptotically stable. The basic question was this: Can we add just a bit to (i*) and, thereby, rid ourselves of the Marachkov boundedness condition? Burton and Makay (1994) obtain a result along those lines which we state next, without proof. In that result we add to (i*) the condition that $V(t_n, \phi) \leq W_2(\|\phi\|)$ holds only along a discrete sequence $\{t_n\}$ tending to infinity. In exchange for this addition, the derivative condition on V which Krasovskii required is relaxed and we allow F to be unbounded of order $J(t + 1)\ln(t + 2)$. It would be a major *coup* if one could prove Zhang's Theorem 8.3.7 with (iii) holding only along a discrete sequence instead of on intervals.

Theorem 8.3.9. Burton-Makay (1994)   *Suppose there is a continuous and locally Lipschitz functional $V : [0,\infty) \times C_H \to [0,\infty)$, wedges W_i, positive constants K and J, a sequence $\{t_n\} \uparrow \infty$ with $t_n - t_{n-1} \leq K$ such that*

(i) $W_1(|\phi(0)|) \leq V(t,\phi)$, $V(t,0) = 0$, $V'_{(8.3.1)}(t,\phi) \leq 0$,

(ii) $V(t_n,\phi) \leq W_2(\|\phi\|)$,

(iii) $V'_{(8.3.1)}(t,\phi) \leq -W_3(|\phi(0)|)$ *if* $t_n - h \leq t \leq t_n$,

and

(iv) $|F(t,\phi)| \leq J(t+1)\ln(t+2)$.

Then the zero solution of (8.3.1) is asymptotically stable.

By varying the length of the delay in the next example one can see all the types of upper bounds which we have discussed arise for different values of t. We employ the function

$$(|\sin t| + \sin t)^{1/q}$$

where q is a large positive integer. Notice that the function is zero on alternate intervals of length π. On the intervals in between we are taking the q-th root of numbers between 0 and 2. If q is very large, it can represent an elementary function having a graph very nearly that of a square wave.

Example 8.3.3. Consider the scalar equation

$$x'(t) = -a(t)x + t(|\sin t| + \sin t)^{1/p}x(t - \pi)$$

where p is a positive integer and

$$a(t) \geq 2(t + \pi).$$

Define

$$V(t,x_t) = |x(t)| + \int_{t-\pi}^{t} (s+\pi)(|\sin(s+\pi)| + \sin(s+\pi))^{1/p}|x(s)|\,ds$$

so that the derivative along a solution satisfies

$$V'(t,x_t) \leq -a(t)|x| + t(|\sin t| + \sin t)^{1/p}|x(t-\pi)|$$
$$+ (t+\pi)(|\sin(t+\pi)| + \sin(t+\pi))^{1/p}|x|$$
$$- t(|\sin t| + \sin t)^{1/p}|x(t-\pi)|$$
$$= \big[-a(t) + (t+\pi)(|\sin(t+\pi)| + \sin(t+\pi))^{1/p}\big]|x(t)|.$$

Thus, if $p > 1$ then V' is negative semi-definite. Moreover, the integral in V is zero along a discrete sequence; hence the conditions of Theorem 8.3.9 hold. If we shorten the delay, then the integral is zero on intervals. Conditions of Theorem 8.3.7 will then hold. The reader is invited to look at

$$x' = -a(t)x + \left[\cos t + t(|\sin t| + \sin t)^{1/p}\right]x(t-1)$$

and determine t_n and d for Theorem 8.3.5.

For at least seventeen years Hatvani (1990, 1996, 1997a,b, 2000, 2002) has studied the problem of using a norm on C_H in the derivative of the Liapunov functional and having a coefficient which is frequently zero. The problem goes back to Burton-Hatvani(1989) in which we investigated the possibility that Krasovskii had abandoned Theorem 8.3.1 too quickly. We studied a series of classical examples in which Liapunov functionals had been constructed whose derivatives along solutions of (8.3.1) satisfied a condition

$$V'_{(8.3.1)}(t, x_t) \leq -W_1\left(\int_{t-h}^{t} |x(s)|\, ds\right) - W_2(|x(t)|)\,.$$

We noted that the investigators had merely discarded the integral, not realizing its importance. Thus, the L^1-norm occurred naturally, instead of the supremum norm which Krasovskii had sought so diligently.

The significance of this is that, while $x(t)$ can move quickly and cause us problems studied in the annulus argument, the L^1-norm always moves very slowly. Not only does it remove the need for the Marachkov condition, but it allows a coefficient $\eta(t)$ which contrasts with the concept of integral positivity in that $\eta(t)$ is zero on long intervals, yet an integration of $-\eta(t)W_1\left(\int_{t-h}^{t} |x(s)|\, ds\right)$ can drive V to zero. The problem has been studied by numerous investigators including Becker-Burton-Zhang (1989), Burton (1987), Burton-Hatvani (1990), Burton-Casal-Somolinos (1987, 1989), Hatvani (1990, 1996, 1997a, 2000, 2002), Wang (1992, 1994a,b), and Zhang (1995). We will not be able to mention all of the results here, but recommend that the interested reader see those papers.

All of the ideas for the following theorem were borrowed from Zhang (1995).

Theorem 8.3.10.   Zhang (1995)   *Let the zero solution of (8.3.1) be uniformly stable. Suppose there are positive constants d and M, a sequence $\{t_n\} \uparrow \infty$ with $h \leq t_{n+1} - t_n \leq M$, wedges W_i, and a continuous and locally Lipschitz functional $V : [0, \infty) \times C_H \to [0, \infty)$ such that*

(i) $W_1(|\phi(0)|) \leq V(t, \phi)$, $V(t, \mathbf{0}) = 0$, $V'_{(8.3.1)}(t, \phi) \leq 0$,

(ii) $V(t, \phi) \leq W_2(|\phi(0)|) + W_3\left(\int_{-h}^{0} |\phi(s)| \, ds\right)$ *for* $t \in [t_n - d, t_n]$,

and

(iii) $V'_{(8.3.1)}(t, \phi) \leq -W_4\left(\int_{-h}^{0} |\phi(s)| \, ds\right)$ *for* $t_n - d \leq t \leq t_n$.

Then the zero solution of (8.3.1) is uniformly asymptotically stable.

Proof. Since the zero solution is uniformly stable, for the $H > 0$ find the δ of uniform stability and call it H^*. Let $\gamma > 0$ be given. We must find $T > 0$ such that $t_0 \geq 0$, $\psi : [t_0 - h, t_0] \to R^n$, $\|\psi\| < H^*$, ψ continuous, imply that $|\mathbf{x}(t, t_0, \psi)| < \gamma$ for $t \geq t_0 + T$. Denote any such a solution by $\mathbf{x}(t)$. On the interval $[t_0, \infty)$, so long as $V(t, \mathbf{x}_t) \geq W_1(\gamma)$, then at any such $t \in [t_n - d, t_n]$ which is to the right of t_0 either

$$W_2(|\mathbf{x}(t)|) \geq W_1(\gamma)/2 \quad \text{or} \quad W_3\left(\int_{t-h}^{t} |\mathbf{x}(s)| \, ds\right) \geq W_1(\gamma)/2 \,.$$

Let $[t_N - d, t_N]$ be the first element of the sequence to the right of t_0. Either

(a) $W_2(|\mathbf{x}(t)|) \geq W_1(\gamma)/2$ for every $t \in [t_N - d, t_N]$

or

(b) there is a $t^* \in [t_N - d, t_N]$ with $W_3\left(\int_{t^*-h}^{t^*} |\mathbf{x}(s)| \, ds\right) \geq W_1(\gamma)/2$.

If (a) holds, at every $t \in [t_N - d, t_N]$ then (iii) yields

$$V'_{(8.3.1)}(t, \mathbf{x}_t) \leq -W_4\left(dW_2^{-1}(W_1(\gamma)/2)\right)$$

so that if we denote $V(t, \mathbf{x}_t) =: V(t)$ then we have

$$V(t_N) - V(t_N - d) \leq -dW_4\left(W_3^{-1}(W_1(\gamma)/2)\right) =: -d_1 \,.$$

If (b) holds, then by the equicontinuity of the integrals, there is a $k > 0$ which depends only on γ and not on t_0 or the initial function with

$$\int_{t-h}^{t} |\mathbf{x}(s)| \, ds \geq (1/2)W_3^{-1}(W_1(\gamma)/2)$$

for all $t \in [t^* - k, t^* + k]$. It may be assumed that $k < d$. Thus, for all these values of t which also lie in $[t_N - d, t_N]$ then (iii) becomes

$$V'_{(8.3.1)}(t, \mathbf{x}_t) \leq -W_4\big((1/2)W_3^{-1}(W_1(\gamma)/2)\big)$$

which holds on an interval of length $k/2$. Hence V decreases by at least

$$(k/2)W_4\big[(1/2)W_3^{-1}(W_1(\gamma)/2)\big] =: d_2\,.$$

If $d_3 = \min[d_1, d_2]$, then V decreases by at least d_3 on $[t_N - d, t_N]$; in fact, so long as $V(t, \mathbf{x}_t) \geq W_1(\gamma)$, then V decreases by at least d_3 on every $[t_n - d, t_n]$ past t_0. Recall that $V' \leq 0$ so V never increases.

Notice that $V(t_N, x_{t_N}) \leq W_2(H) + W_3(hH)$ so there is a fixed integer p with

$$W_2(H) + W_3(hH) - pd_3 < W_1(\gamma)\,.$$

We can then take $T = pM$. This completes the proof.

We now give, without proof, three important results of Wang (1992). These show us that a given Liapunov functional may be far more general than it appears. More results of this same type can be found in Becker-Burton-Zhang (1989) and in Burton-Casal-Somolinos (1987, 1989) where a norm on C_H is brought into the derivative of V.

Theorem 8.3.11. Wang (1992, p. 142)   *Let $V : [0, \infty) \times C_H \to [0, \infty)$ be continuous. Then for some constant $\delta > 0$ and each solution $\mathbf{x}(t) = \mathbf{x}(t, t_0, \phi)$ of (8.3.1) with $\phi \in C_\delta$, V satisfies*

$$W_1(|\mathbf{x}(t)|) \leq V(t, \mathbf{x}_t) \leq W_2(|\mathbf{x}(t)|) + W_3(\|\mathbf{x}_t\|_2)$$

and

$$V'_{(8.3.1)}(t, \mathbf{x}_t) \leq -\eta(t)W_4(\|\mathbf{x}_t\|_2)$$

if and only if there exist wedges $\bar{W}_i$ and constant $\delta_0 > 0$ such that for each $\phi \in C_{\delta_0}$ and each solution $x(t) = x(t, t_0, \phi)$ of (8.3.1), V satisfies

$$W_1(|\mathbf{x}(t)|) \leq V(t, \mathbf{x}_t) \leq W_2(|\mathbf{x}(t)|) + \bar{W}_3\left(\int_{t-h}^{t} \bar{W}_4(|\mathbf{x}(u)|)\, du\right)$$

and

$$V'_{(8.3.1)}(t, \mathbf{x}_t) \leq -\eta(t)\bar{W}_5\left(\int_{t-h}^{t} \bar{W}_6(|\mathbf{x}(u)|)\, du\right).$$

The next two results are something of a bridge between our Liapunov functionals with V' involving $|\phi(0)|$ and those involving $\int_{-h}^{0} |\phi(s)|\,ds$. But care must be taken to see the intervals on which the integrals are defined.

Theorem 8.3.12. Wang (1992, p. 144)   *Let* $V : [0,\infty) \times C_H \to [0,\infty)$ *and* $D : [0,\infty) \times C_H \to [0,\infty)$ *be continuous along the solutions of* (8.3.1) *with*

(i) $W_1(|\mathbf{x}(t)|) \leq V(t, \mathbf{x}_t) \leq W_2(\|\mathbf{x}_t\|),$

and

(ii) $V'_{(8.3.1)}(t, \mathbf{x}_t) \leq -\eta(t)W_3(D(t, \mathbf{x}_t)),$

where W_3 *is convex downward and* $\eta(t) \geq 0$ *is nonincreasing. Then there is a continuous functional* $U : [0,\infty) \times C_H \to [0,\infty)$ *with*

(iii) $hW_1(|\mathbf{x}(t)|) \leq U(t, \mathbf{x}_t) \leq hW_2(\|\mathbf{x}_{t-h}\|),$

and

(iv) $U'_{(8.3.1)}(t, \mathbf{x}_t) \leq -\eta(t)W_4\big(\int_{t-h}^{t} D(s, \mathbf{x}_s)\,ds\big).$

Corollary.   Wang (1992, p. 145)   *Let* $V : [0,\infty) \times C_H \to [0,\infty)$ *be continuous with*

(i) $W_1(|\mathbf{x}(t)|) \leq V(t, \mathbf{x}_t) \leq W_2(\|\mathbf{x}_t\|)$

and

(ii) $V'_{(8.3.1)}(t, \mathbf{x}_t) \leq -\eta(t)W_3(|\mathbf{x}(t)|),$

where $\eta(t) \geq 0$ *is nonincreasing. Then there is a continuous functional* $U : [0,\infty) \times C_H \to [0,\infty)$ *with*

(iii) $hW_1(|\mathbf{x}(t)|) \leq U(t, \mathbf{x}_t) \leq hW_2(\|\mathbf{x}_{t-h}\|)$

(iv) $U'_{(8.3.1)}(t, \mathbf{x}_t) \leq -\eta(t)W_4\big(\int_{t-h}^{t} |\mathbf{x}(s)|\,ds\big).$

Open Problem 8.3.1. These results of Wang deserve careful attention. The functional U is simply

$$U(t, \mathbf{x}_t) = \int_{t-h}^{t} V(s, \mathbf{x}_s) \, ds \,.$$

When we compare Theorems 8.3.14 and 8.3.15 of Hatvani we see the tremendous advantage of having the integral of $\mathbf{x}$ in the derivative of V. Here are some open problems which should be very accessible and affirmative answers should significantly advance the theory. First, in Wang's proof of Theorem 8.3.12 we do not detect the requirement that h in the definition of U be the same as the h on the delay in (8.3.1); this must be examined with great care. Next, it appears that the technique of Zhang in the proof of Theorem 8.3.6 can be employed so that the upper bound on U and the monotonicity of η need only hold on intervals $[t_n - h, t_n]$. Thus, if we can choose the size of h, then $\eta(t) = \sin^2 t$, for example, would be an acceptable function in $U'(t, \mathbf{x}_t) \leq -\eta(t)W_4\big(\int_{t-h}^{t} D(s, \mathbf{x}_s) \, ds\big)$ since η is monotone on certain intervals. Wang asks for the supremum upper bound on V. We need to examine the proof with care to see if the type of upper bound for V in Theorem 8.3.2 will be preserved for U in a usable form. Finally, we would like to see an imaginative stability conclusion under the conditions of the corollary to Theorem 8.3.12. Some work along that last line can be found in Burton-Hatvani (1989).

The next result is a partial converse of the corollary.

Theorem 8.3.13. Wang (1992, pp. 145–146) $\quad$ *Let $U : [0, \infty) \times C_H \to [0, \infty)$ be continuous with*

(i) $W_1(|\mathbf{x}(t)|) \leq U(t, \mathbf{x}_t) \leq W_2(\|\mathbf{x}_t\|),$

and

(ii) $U'_{(8.3.1)}(t, \mathbf{x}_t) \leq -\eta(t)W_3\big(\int_{t-h}^{t} |\mathbf{x}(s)| \, ds\big),$

where $\eta(t) \geq 0$ is nonincreasing and W_3 is convex upward. Then there exists a continuous functional $V : [0, \infty) \times C_H \to [0, \infty)$ and a wedge W_4 satisfying

(iii) $W_1(|\mathbf{x}(t + h)|) \leq V(t, \mathbf{x}_t) \leq W_4(\|\mathbf{x}_{t+h}\|),$

and

(iv) $V'_{(8.3.1)}(t, \mathbf{x}_t) \leq -\eta(t)W_3(h|\mathbf{x}(t)|).$

The next example does two things. First, it should lead us to extend the Krasovskii-Liapunov theorem to include distributed delays such as

$$\mathbf{x}' = A\mathbf{x} + \mathbf{f}(t, \mathbf{x}) + \mathbf{g}(t, \mathbf{x}(t - h)) + \int_{t-h}^{t} \mathbf{r}(s, \mathbf{x}(s)) \, ds.$$

Those kinds of extensions are left as an exercise. In addition, it shows in a simple way how a norm on C_H occurs in the derivative of V. While the supremum norm desired by Krasovskii may be rare, integral norms are ubiquitous.

Example 8.3.4. Wang (1992).   Consider the scalar equation

$$x'(t) = -a(t)x(t) + b(t) \int_{t-h}^{t} x(u) \, du$$

with $a : [0, \infty) \to [0, \infty)$ and $b : [0, \infty) \to R$, both being continuous. If a and b satisfy, for all $t \geq 0$,

$$-a(t) + k \int_{t}^{t+h} |b(u)| \, du \leq 0 \quad \text{for some} \quad k > 1,$$

and

$$\int_{-h}^{0} \left[\int_{t-h}^{t} |b(u - s)|^q \, du \right]^{1/q} ds \leq B \quad \text{for some} \quad B > 0, \, q > 1,$$

then for

$$V(t, x_t) = |x(t)| + k \int_{-h}^{0} \int_{t+s}^{t} |b(u - s)| \, |x(u)| \, du \, ds$$

and for

$$\eta(t) = (k - 1)|b(t)|,$$

we have

$$|x(t)| \leq V(t, x_t) \leq |x(t)| + kB \left[\int_{t-h}^{t} |x(u)|^p \, du \right]^{1/p}$$

for $1/p + 1/q = 1$ with $p, q > 1$ and

$$V'(t, x_t) \leq -\eta(t) \int_{t-h}^{t} |x(u)|^2 \, du$$

for $|x(t)| \leq 1$.

It is interesting to see how much the conditions can change with small changes in the equation.

Example 8.3.5. Consider the scalar equation

$$x' = -a(t)x + \int_{t-1}^{t} b(s)x(s)\,ds$$

and suppose that there is a constant $K > 1$ with

$$-a(t) + K|b(t)| \leq 0\,.$$

Define

$$V(t, x_t) = |x(t)| + K \int_{-1}^{0} \int_{t+v}^{t} |b(s)x(s)|\,ds\,dv$$

and note that

$$|x(t)| \leq V(t, x_t) \leq |x(t)| + K \int_{t-1}^{t} |b(s)x(s)|\,ds\,.$$

Along a solution we have

$$V'(t, x_t) \leq -a(t)|x| + \int_{t-1}^{t} |b(s)x(s)|\,ds$$
$$+ K \int_{-1}^{0} |b(t)x(t)|\,dv - K \int_{-1}^{0} |b(t+v)x(t+v)|\,dv\,.$$

There are many conclusions which can be drawn. As an exercise, the reader may examine each of our theorems and place appropriate conditions on a and b to satisfy the conditions of the theorem. In particular, deduce that (See Section 2.5 for implications.)

$$V'(t, x_t) \leq -\delta\big[|x(t)| + |x'(t)|\big]\,.$$

Example 8.3.6. Burton-Hatvani (1989, p. 68).   Consider the scalar equation

$$x'(t) = b(t)x(t-h)$$

where $b : [-h, \infty) \to [-1, 0]$ is continuous, $h > 0$,

$$-2 + \int_{t-h}^{t} |b(u)|\,du + h \leq 0\,,$$

$$b(t+h) = b(t) \,, \qquad \int_{t-h}^{t} |b(u)| \, du > 0 \,,$$

and

$$\int_{t-h}^{t} [1 - |b(s)|] \, ds > 0 \,.$$

Define

$$V(t, x_t) = \left[x(t) + \int_{t-h}^{t} b(u)x(u) \, du \right]^2 + \int_{-h}^{0} \int_{t+s}^{0} |b(u)| x^2(u) \, du \, ds$$

and obtain

$$V'(t, x_t) \le |b(t)| \left[-2 + \int_{t-h}^{t} |b(u)| \, du + h \right] x^2 + (|b(t)| - 1) \int_{t-h}^{t} |b(s)| x^2(s) \, ds \,.$$

First, we note that we see both types of terms in V' which we have been discussing.

Next, note that if $b(t)$ is a classical square wave function taking the values 0 and -1 on intervals, but continuous, then the first term in V' is without value in classical theory. But in the present context, the integral term is useful. Depending on the conditions imposed, this example can be made to fit every one of our theorems given here.

In the papers by Hatvani mentioned above there is deep study of functions which will be satisfactory coefficients in the derivative of V. We will see two examples here from Hatvani (2002). The symbol $\| \cdot \|_2$ will denote the L^2-norm on C_H.

Definition 8.3.2. *A subset $L \subset [0, \infty)$ is called h-dense on the interval $[t_0, \infty)$ if there exists a constant $\kappa > 0$ such that*

$$\mu([t, t+h] \cap L) \ge \kappa \quad \text{for all} \quad t \ge t_0 \,,$$

where $\mu(\cdot)$ denotes the Lebesgue measure.

Lemma 8.3.2. Hatvani (2002) *For $\phi \in C_H$ and $\kappa \in (0, H)$, let*

$$P(\phi, \kappa) := \left\{ u \in [-h, 0] : |\phi(u)| \ge \kappa \right\} \,.$$

Then

$$\mu\big(P(\phi, \|\phi\|_2 / \sqrt{2h})\big) \ge \|\phi\|_2^2 / 2H^2 \,.$$

Proof. We have

$$\|\phi\|_2^2 = \int_{-h}^0 |\phi(u)|^2 \, du \le H^2 \mu(P(\phi, \kappa)) + \kappa^2 \big[h - \mu(P(\phi, \kappa)) \big] ,$$

whence

$$\mu(P(\phi, \kappa)) \ge \frac{\|\phi\|_2^2 - \kappa^2 h}{H^2 - \kappa^2} .$$

Choose $\kappa := \|\phi\|_2 / \sqrt{2h}$. Then $\kappa^2 < H^2$, and

$$\mu\big(P(\phi, \|\phi\|_2 / \sqrt{2h})\big) \ge \frac{\|\phi\|_2^2 - \frac{\|\phi\|_2^2 h}{2h}}{H^2} = \frac{\|\phi\|_2^2}{2H^2} .$$

Theorem 8.3.14. Hatvani (2002) *Suppose that there exist a continuous and locally Lipschitz functional $V : [0, \infty) \times C_H \to [0, \infty)$, wedges W_i, and a measurable function $\eta : [0, \infty) \to [0, \infty)$ satisfying the conditions*

(i) $W_1(|\phi(0)|) \le V(t, \phi)$

(ii) $V(t, \phi) \le W_2(|\phi(0)|) + W_3(\|\phi\|_2)$,

(iii) $V'(t, \phi) \le -\eta(t) W(|\phi(0)|)$

along every solution $x : [t_0, \infty) \to R^n$ of (8.3.1) for every $t \ge t_0$.

(A) *If for every $t_0 \in [0, \infty)$ and every h-dense subset $L \subset [t_0, \infty)$ the divergence*

$$\lim_{K \to \infty} \int_{[t_0, t_0 + K] \cap L} \eta(t) dt = \infty$$

holds, then the zero solution of (8.3.1) is asymptotically stable.

(B) *Suppose that the divergence in (A) is uniform with respect to t_0 and L, i.e., for every J there is an $N = N(J)$ such that $K > N(J)$ implies $\int_{[t_0, t_0 + K] \cap L} \eta(t) \, dt > J$ for all $t_0 \in [0, \infty)$ and for all subsets $L \subset [0, \infty)$ which are h-dense on $[t_0, \infty)$. Then the zero solution of (8.3.1) is uniformly asymptotically stable.*

Proof. Conditions (i) and (ii) imply uniform stability so that for $\varepsilon > 0$ there is a $\delta(\varepsilon) > 0$ such that $\phi \in C_\delta$, $t_0 \in [0, \infty)$, and $t \geq t_0$ imply that $|\mathbf{x}(t, t_0, \phi)| < \varepsilon$. Take $\sigma := \delta(1) > 0$, and consider an arbitrary solution $\mathbf{x}(t_0, \phi)$ for $\phi \in C_\sigma$. For this solution we take $v(t) := V(t, \mathbf{x}_t(t_0, \phi))$ which is non-increasing, non-negative, and satisfies

$$V(t_0, \phi) \leq W_2(\sigma) + W_3(\sigma\sqrt{h}) =: c_1, \tag{1}$$

independent of t_0, ϕ.

To prove asymptotic stability, for every $\gamma > 0$, $t_0 \in [0, \infty)$, and $\phi \in C_\sigma$, we have to find $T = T(\gamma, \mathbf{x}(t_0, \phi))$ such that $|\mathbf{x}(t, t_0, \phi)| < \gamma$ for all $t \geq t_0 + T$. To this end it is sufficient to know that $T > h$ and

$$v(t_0 + T - h) = V(t_0 + T - h, \mathbf{x}_{t_0+T-h}) < W_1(\delta(\gamma)) =: c_2(\gamma), \tag{2}$$

is satisfied. In fact, then

$$W_1(|\mathbf{x}(s)|) \leq V(s, \mathbf{x}_s) \leq V(t_0 + T - h, \mathbf{x}_{t_0+T-h}) < W_1(\delta(\gamma)),$$

since $\delta(\gamma)$ is the constant of uniform stability.

For fixed $\gamma > 0$, $t_0 \in [0, \infty)$, $\phi \in C_\sigma$, the fixed solution $\mathbf{x}(t_0, \phi)$, and $K > 0$ define

$$L_0^K = L_0^K(\gamma, \mathbf{x}(t_0, \phi))$$

$$:= \left\{ t \in [t_0, t_0 + K] : |\mathbf{x}(t)| > W_2^{-1}\left(\frac{c_2(\gamma)}{3}\right) =: c_3(\gamma) \right\},$$

$$L_2^K = L_2^K(\gamma, \mathbf{x}(t_0, \phi))$$

$$:= \left\{ t \in [t_0, t_0 + K] : \|\mathbf{x}_t\|_2 > W_3^{-1}\left(\frac{c_2(\gamma)}{3}\right) =: c_4(\gamma) \right\}.$$

If we cannot find $T \leq K + h$ with (2) holding, then L_0^K and L_2^K together cover the interval $[t_0, t_0 + K]$, otherwise we would have $t_* \in [t_0, t_0 + K]$ with $|\mathbf{x}(t_*)| \leq c_3$, $\|\mathbf{x}_{t_*}\|_2 \leq c_4$, and we would obtain

$$c_2 \leq v(t_0 + K) \leq v(t_*) = V(t_*, \mathbf{x}_{t_*}(t_0, \phi))$$

$$\leq W_2(|\mathbf{x}(t_*)|) + W_3(\|\mathbf{x}_{t_*}\|_2) \leq \frac{c_2}{3} + \frac{c_2}{3} < c_2,$$

a contradiction. Therefore, if there is no T in $[t_0, t_0+K+h]$, then $L_0^K \cup L_2^K \supset [t_0, t_0 + K]$.

By Lemma 8.3.2, if $t \in L_2^K$, then there is a set $Q(t) \subset [t - h, t]$ with $\mu(Q(t)) \geq c_4^2(\gamma)/2 =: c_5(\gamma)$ such that

$$|x(s)| \geq c_4/\sqrt{2h} \quad \text{for} \quad s \in Q(t). \tag{3}$$

Let

$$Q^K := \left(\cup_{t \in L_2^K} Q(t) \right) \backslash L_0^K.$$

Now, we can estimate the decrease of V along the solution $\mathbf{x}(t_0, \phi)$. From (iii), the definitions of L_0^K, L_2^K, and inequalities (1) and (3) we get

$$
\begin{aligned}
0 \leq v(t_0 + K + h) &\leq v(t_0) + \int_{t_0}^{t_0+K} v'(t)\, dt \\
&\leq c_1 + \int_{L_0^K} v'(t)\, dt + \int_{Q^K} v'(t)\, dt \\
&\leq c_1 - c_6 \int_{L_0^K} \eta(t)\, dt - c_7 \int_{Q^K} \eta(t)\, dt \\
&\leq c_1 - c_8 \int_{L_0^K \cup Q^K} \eta(t)\, dt, \tag{4}
\end{aligned}
$$

where

$$c_6(\gamma) := W(c_3(\gamma)), \quad c_7(\gamma) := W\left(\frac{c_4(\gamma)}{\sqrt{2h}} \right), \quad c_8(\gamma) := \min\left\{ c_6(\gamma); c_7(\gamma) \right\}.$$

Suppose, by way of contradiction, that the zero solution is not asymptotically stable. Then there are $t_0 \in [0, \infty)$, $\phi \in C_\sigma$, $\mathbf{x}(t_0, \phi)$, and $\gamma > 0$ such that there exists no $T < K + h$ with (2) for all $K > 0$. First we show that the set $L = L(\gamma, \mathbf{x}(t_0, \phi)) := \cup_{K>0}(L_0^K \cup Q^K)$ is h-dense in $[t_0, \infty)$ with the constant $\kappa = \kappa(\gamma) := c_5(\gamma)/2$.

In fact, if $t \geq t_0 + h$, then for some $K > 0$ either $[t - c_5/2, t] \subset L_0^K$, or the interval $[t - c_5/2, t]$ contains a point of L_2^K. In the latter case, the interval $[t - h - c_5/2, t]$ contains a subset of Q^K with measure c_5. In any case we have

$$\mu\big([t - h, t] \cap (L_0^K \cup Q^K)\big) \geq c_5/2 = \kappa,$$

which means that L is h-dense in $[t_0, \infty)$ with the constant κ.

Finally, from part A of the theorem and (4) we obtain

$$0 \leq \lim_{t \to \infty} v(t) \leq v(t_0) + \int_{t_0}^{\infty} v'(t)\, dt$$

$$\leq c_1 - c_8(\gamma) \int_{L(\gamma, x(to, \phi))} \eta(t)\, dt = -\infty. \tag{5}$$

This is a contradiction; hence, for every $\mathbf{x}(t_0, \phi)$ and $\gamma > 0$ there is a $K(\gamma, \mathbf{x}(t_0, \phi))$ such that (2) is fulfilled with some $T(\gamma, \mathbf{x}(t_0, \phi)) < K(\gamma, \mathbf{x}(t_0, \phi)) + h$; i.e., the zero solution is asymptotically stable.

To prove statement B in the theorem we have to observe that the size of K (and, consequently, T) is governed only by $v(t_0)$ and the speed of the divergence in our last display. However, $v(t_0) \leq c_1$, and $c_8(\gamma)$ is independent of the solution; hence, if we require that $\int_L \eta$ diverge uniformly with respect to $t_0 \in [0, \infty)$ and h-dense sets $L \subset [0, \infty)$, then $T = T(\gamma)$ is independent of t_0, ϕ, and the solution; that is, the zero solution is uniformly asymptotically stable.

Example 8.3.7. Hatvani (2002, p. 3565).   Consider the scalar equation

$$x'(t) = -a(t)f(x(t)) + b(t)g(x(t - h))$$

where $a, b : [0, \infty) \to R$ are continuous, $f, g : R \to R$ are continuously differentiable, and $a(t) \geq 0$, $xf(x) > 0$ for all $t \in [0, \infty)$ and $x \neq 0$. Suppose also that there exists a constant $c > 0$ such that $|g(x)| \leq c|f(x)|$ for sufficiently small $|x|$.

Define

$$V(t, \phi) := |\phi(0)| + \int_{t-h}^{t} |b(s + h)|\, |g(\phi(s - t))|\, ds.$$

For every neighborhood of the origin there is a constant c_1 such that

$$V(t, \phi) \leq |\phi(0)| + c_1 \left(\int_{t}^{t+h} b^2(s)\, ds \right)^{1/2} \|\phi\|_2$$

for all ϕ in the neighborhood and for all $t \in [0, \infty)$. The derivative of V with respect to our equation satisfies

$$V'(t, x_t) \leq -\big[a(t) - c|b(t + h)|\big]\, |f(x(t))|.$$

Here is the conclusion. Suppose that

(i) $\eta_1(t, c) := a(t) - c|b(t + h)| \geq 0$ for all $t \in [0, \infty)$, and

(ii) $\int_t^{t+h} b^2(s)\,ds$ is bounded on $[0, \infty)$.

If $\int_L \eta_1(t, c)\,dt = \infty$ for all h-dense subsets $L \subset [0, \infty)$, then the zero solution is asymptotically stable. If, in addition, $\int_{[t_0, \infty) \cap L} \eta_1(t, c)\,dt = \infty$ uniformly with respect to $t_0 \in [0, \infty)$ and h-dense sets $L \subset [0, \infty)$, then the zero solution is uniformly asymptotically stable.

Proposition 8.3.2. *Suppose there are wedges and a continuous and locally Lipschitz functional $V : [0, \infty) \times C_H \to [0, \infty)$ such that*

(i) $W_1(|\boldsymbol{\phi}(0)|) \le V(t, \boldsymbol{\phi}) \le W_2(|\boldsymbol{\phi}(0)|) + W_3\big(\int_{-h}^0 |\boldsymbol{\phi}(s)|\,ds\big)$

and

(ii) $V'_{(8.3.1)}(t, \boldsymbol{\phi}) \le 0$.

For each $\gamma > 0$ there exist $\mu = \mu(\gamma) > 0$ such that if $\mathbf{x}(t)$ is a solution of (8.3.1) on $[t_0, \infty)$ and if $\gamma \le V(t, \mathbf{x}_t)$ on $[t_0, \infty)$, then $\int_{t-h}^t |\mathbf{x}(s)|\,ds \ge \mu$ for $t \ge t_0 + h$.

Proof. The zero solution is uniformly stable, so there exists H such that for every t_0 and for every $\boldsymbol{\phi}$ sufficiently small we have $|\mathbf{x}(t, t_0, \boldsymbol{\phi})| < H$ for all $t \ge t_0$. If $\mathbf{x}(t)$ is such a solution, then for $f(t) := \int_{t-h}^t |\mathbf{x}(s)|\,ds$ we have $|f(t_1) - f(t_2)| \le 2|t_1 - t_2|H$ so f is uniformly continuous and for an $\varepsilon > 0$ the corresponding $\delta = \varepsilon/2H$. Assume $\delta < h$.

Now, for any fixed $t_1 \ge t_0$, if $\gamma \le V(t, \mathbf{x}_t)$ either

$$W_2(|\mathbf{x}(t)|) \ge \gamma/2 \quad \text{or} \quad W_3\bigg(\int_{t-h}^t |\mathbf{x}(s)|\,ds\bigg) \ge \gamma/2\,.$$

Take $\varepsilon = (1/4)W_3^{-1}(\gamma/2)$ and find the corresponding $\delta < h$.

First, suppose that $W_2(|\mathbf{x}(t)|) \ge \gamma/2$ for all $t \in [t_1 - \delta, t_1]$. Then $\int_{t_1-h}^{t_1} |\mathbf{x}(s)|\,ds \ge \delta W_2^{-1}(\gamma/2)$.

Next, if there is a $t^* \in [t_1 - \delta, t_1]$ with $W_3\big(\int_{t^*-h}^{t^*} |\mathbf{x}(s)|\,ds\big) \ge \gamma/2$ then

$$\int_{t-h}^t |\mathbf{x}(s)|\,ds \ge (1/4)W_3^{-1}(\gamma/2) \quad \text{for} \quad t^* \le t \le t^* + \delta$$

and that interval includes t_1. Thus, we take

$$\mu(\gamma) := \min\big[(1/4)W_3^{-1}(\gamma/2), \delta W_2^{-1}(\gamma/2)\big]\,.$$

Part (A) of the next theorem appears in Burton-Hatvani (1990) and in Hatvani (2002). It is interesting because there are three essentially different proofs for it and each of these may be useful in other contexts. The proof in Burton-Hatvani (1990) is used by Zhang (1995) in his proof of Theorem 8.3.7.

Theorem 8.3.15.   Hatvani (2002)   *Let the conditions* (i) *and* (ii) *of Theorem 8.3.14 hold, and*

(iii) $V'(t, \phi) \le -\eta(t)W(\|\phi\|_2)$

along every solution of (8.3.1) *on* $[t_0, \infty)$.

(A)   *If for every* $t_0 \in [0, \infty)$ *the divergence*

$$\lim_{K \to \infty} \int_{t_0}^{t_0 + K} \eta(t)\, dt = \infty$$

holds, then the zero solution of (8.3.1) *is asymptotically stable.*

(B)   *If the divergence in* (A) *is uniform with respect to* $t_0 \in [0, \infty)$, *then the zero solution of* (8.3.1) *is uniformly asymptotically stable.*

The result is a direct consequence of Proposition 8.3.2.

Remark 8.3.2. In view of the difficulty in proving Theorem 8.3.14 and the ease in proving Theorem 8.3.15, if the conditions of Theorem 8.3.14 hold and if η is nonincreasing, one would attempt to apply Wang's corollary to Theorem 8.3.14 and reduce it to a much simpler problem.

Suppose that we are beset by adversity. In fact, suppose that no upper bound on V can be found. Be of good cheer for the most pleasant surprise still awaits us and it is both simple and natural. If we have $V'(t, \mathbf{x}_t) \le -\eta(t)W\big(\int_{t-h}^{t} |\mathbf{x}(s)|\, ds\big)$, then we can let η vanish on intervals of length smaller than h and still prove asymptotic stability. A general result of the following kind appears in Burton-Hatvani (1989), but the following sharper and more streamlined one was privately communicated by Tibor Krisztin and appears in Hatvani (1990, p. 232).

Lemma 8.3.3.   *Let* $\mathbf{x} : [-h, \infty) \to R^n$ *and suppose that* $\|\mathbf{x}_{t_0}\| \ge \varepsilon > 0$ $(t_0 \in [0, \infty))$. *Then there exists* $a \in [t_0 - h, t_0]$ *such that* $\|\mathbf{x}_t\| \ge \varepsilon/2$ *for all* $t \in [a, a + h]$.

Proof. Define

$$f_1(s) := \left(\int_{t_0 - h}^{s} |\mathbf{x}(v)|^2\, dv \right)^{1/2}, \quad f_2(s) := \left(\int_{s}^{t_0} |\mathbf{x}(v)|^2\, dv \right)^{1/2}.$$

Since $f_1(t_0 - h) = f_2(t_0) = 0$, $f_1(t_0) = f_2(t_0 - h) = \|\mathbf{x}_{t_0}\| \ge \varepsilon$, there exists a number $a \in [t_0 - h, t_0]$ such that $f_1(a) = f_2(a) = \varepsilon/2$. Then for

$t \in [a, t_0]$ we have $|||\mathbf{x}_t||| \geq f_1(a) = \varepsilon/2$, and for $t \in [t_0, a + h]$ the inequality $|||\mathbf{x}_t||| \geq f_2(a) = \varepsilon/2$ holds. This completes the proof.

In Section 6.1 we introduced the concept of integral positivity in connection with the annulus argument. If we have $V'(t, \mathbf{x}(t)) \leq -\eta(t)W(|\mathbf{x}(t)|)$ where η is integrally positive, and if the Marachkov boundedness condition holds then we can drive V to zero.

Definition 8.3.3. *A measurable function* $\eta : [0, \infty) \to [0, \infty)$ *is said to be integrally positive with parameter* $\delta > 0$ *(IP(δ)) if whenever* $\{t_n\} \to \infty$ *with* $t_{n+1} - t_n \geq \delta$, *then* $\sum_{n=1}^{\infty} \int_{t_n - \delta}^{t_n} \eta(t)\, dt = \infty$.

Theorem 8.3.16. *Suppose there are wedges and a continuous and locally Lipschitz* $V : [0, \infty) \times C_H \to [0, \infty)$, *together with* $\eta \in IP(h)$ *such that*

(i) $W_1(|\boldsymbol{\phi}(0)|) \leq V(t, \boldsymbol{\phi}), V(t, 0) = 0$,

(ii) $V'_{(8.3.1)}(t, \boldsymbol{\phi}) \leq -\eta(t)W_2\big(\int_{t-h}^{t} |\mathbf{x}(s)|\, ds\big)$,

and

(iii) $\mathbf{F}(t, \boldsymbol{\phi})$ *is bounded for* $\boldsymbol{\phi} \in C_H$.

Then the zero solution of (8.3.1) is asymptotically stable. If, in addition, the zero solution is uniformly stable and there is a positive constant k *with*

(iv) $\int_{t-h}^{t} \eta(s)\, ds \geq k$ *for all* $t \geq h$,

then the zero solution of (8.3.1) is uniformly asymptotically stable.

Proof. The zero solution is stable. Let $\mathbf{x}(t)$ be a fixed solution of (8.3.1) with $|\mathbf{x}(t)| < H/2$ on $[t_0, \infty)$. If $\int_{t-h}^{t} |\mathbf{x}(s)|\, ds =: f(t) \to 0$ as $t \to \infty$, it is readily proved from the boundedness of $\mathbf{x}'(t)$ that $\mathbf{x}(t) \to 0$. Thus, we suppose there is a sequence $\{t_n\} \uparrow \infty$ and an $\varepsilon > 0$ with $f(t_n) \geq \varepsilon$. By Krisztin's lemma there is a sequence $\{a_n\} \uparrow \infty$ with $f(t) \geq \varepsilon/2$ on $[a_n, a_n + h]$. On those intervals $V'_{(8.3.1)}(t, \mathbf{x}_t) \leq -\eta(t)W_2(\varepsilon/2)$. An integration sends V to $-\infty$, a contradiction. The U.A.S. is left as an exercise.

8.4 Boundedness with Unbounded Delay

This section is devoted to the problem of extending Theorem 6.1.2 to equations with unbounded delay. We return now to (8.1.1), which we write as

$$\mathbf{x}'(t) = \mathbf{F}(t, \mathbf{x}(\cdot)), \quad t \geq 0, \tag{8.4.1}$$

where $\alpha \geq -\infty$, $\gamma = +\infty$, and $H = \infty$ as discussed in Remark 8.1.1. Here, $|\mathbf{x}|$ denotes Euclidean length.

Theorem 8.4.1. Burton (1982c)   Let $V(t, \mathbf{x}(\cdot))$ be a continuous scalar functional when $t \geq 0$ and $\mathbf{x} : [\alpha, \infty) \to R^n$ is continuous, and let V be locally Lipschitz in $\mathbf{x}$. Suppose there is a continuous function $Q : [0, \infty) \times R^n \to [0, \infty)$, wedges W_1 and W_2, and positive constants U, L, R_1, and μ such that

$$W_1(|\mathbf{x}(t)|) \leq V(t, \mathbf{x}(\cdot)) \leq Q(t, \mathbf{x}(t)) + W_2(\|\mathbf{x}\|^{[\alpha, t]}), \tag{a}$$

$$Q(t, \mathbf{x}) \leq U \text{ implies } |\mathbf{x}| \leq L, \tag{b}$$

$$V'_{(8.4.1)}(t, \mathbf{x}(\cdot)) \leq -\mu |\mathbf{F}(t, \mathbf{x}(\cdot))| \tag{c}$$

if $Q(t, \mathbf{x}(\cdot)) \geq U$, and if $R \geq R_1$, then

$$U + W_2(L + R) - \mu R < W_1(L + R). \tag{d}$$

Then all solutions of (8.4.1) are bounded.

Proof. Suppose the theorem is false and that $\mathbf{x}(t)$ is an unbounded solution. If $Q(t, \mathbf{x}(t)) \geq U$ on an interval $[a, b]$, then

$$0 \leq V(t, \mathbf{x}(\cdot)) \leq V(a, \mathbf{x}(\cdot)) - \mu \mathbf{x}[a, t],$$

and so we readily argue that there is a sequence $\{s_n\} \to \infty$ with $Q(s_n, \mathbf{x}(s_n)) = U$, as $\mathbf{x}(t)$ is unbounded. See (2.5.12) for the details regarding this notation.

We may therefore find $t_0 \geq 0$ and $R \geq R_1$ with $Q(t_0, \mathbf{x}(t_0)) = U$ and $\|\mathbf{x}\|^{[\alpha, t_0]} < L + R$. Because $|\mathbf{x}(t)|$ is unbounded, there is a first $t_2 > t_0$ with $|\mathbf{x}(t_2)| = L + R$, and therefore, there is a $t_1 \geq t_0$ with $Q(t_1, \mathbf{x}(t_1)) = U$ and $Q(t, \mathbf{x}(t)) > U$ on $(t_1, t_2]$. Now on $[t_1, t_2]$ we have $V'_{(8.4.1)}(t, \mathbf{x}(\cdot)) \leq -\mu |\mathbf{F}(t, \mathbf{x}(\cdot))|$, and so

$$W_1(|\mathbf{x}(t)|) \leq V(t, \mathbf{x}(\cdot)) \leq V(t_1, \mathbf{x}(\cdot)) - \mu \mathbf{x}[t_1, t]$$
$$\leq Q(t_1, \mathbf{x}(\cdot)) + W_2(\|\mathbf{x}\|^{[\alpha, t_1]}) - \mu \mathbf{x}[t_1, t]$$
$$\leq U + W_2(L + R) - \mu \mathbf{x}[t_1, t],$$

so that at $t = t_2$ we have

$$W_1(L + R) \leq U + W_2(L + R) - \mu R,$$

a contradiction in (d). This completes the proof.

Notice that $V = V(t, \mathbf{x}(t), \mathbf{x}(\cdot))$ and (d) asks, essentially, that V depend as much on $\mathbf{x}(t)$ as on $\mathbf{x}(\cdot)$ when $\mathbf{x}(t)$ is large. That is, of course, a type of fading memory condition.

Example 8.4.1. Consider the system

$$\mathbf{x}' = A\mathbf{x} + \int_0^t C(t,s)\mathbf{x}(s)\,ds + \mathbf{F}(t)\,, \tag{8.4.2}$$

in which A is a stable matrix with

$$A^T B + BA = -I\,,$$
$$|B\mathbf{x}| \le K[\mathbf{x}^T B\mathbf{x}]^{1/2}\,,$$
$$|\mathbf{x}| \ge 2k[\mathbf{x}^T B\mathbf{x}]^{1/2}\,,$$

and

$$[\mathbf{x}^T B\mathbf{x}]^{1/2} \ge r|\mathbf{x}|\,,$$

which we have seen in Section 2.5. We ask that $C(t,s)$ be an $n \times n$ matrix continuous for $0 \le s \le t < \infty$ and that $\mathbf{F} : [0,\infty) \to R^n$ be continuous and bounded.

Recall that Theorem 8.2.6 yields boundedness when $\int_0^t |BC(t,s)|\,ds$ is small. Certainly, in the convolution case the variation of parameters formula yields an easy boundedness result. We will use Theorem 8.4.1 to obtain boundedness upon integration of the first coordinate of $C(t,s)$.

We ask that $\int_t^\infty |C(u,s)|\,du$ be continuous for $0 \le s \le t < \infty$, and we choose $\bar{K} > K$ and define

$$V(t,\mathbf{x}(\cdot)) = [\mathbf{x}^T B\mathbf{x}]^{1/2} + \bar{K} \int_0^t \int_t^\infty |C(u,s)|\,du\,|\mathbf{x}(s)|\,ds\,.$$

A calculation yields

$$V'_{(8.4.2)}(t,\mathbf{x}(\cdot)) \le \left[-k + \bar{K} \int_t^\infty |C(u,t)|\,du \right] |\mathbf{x}| + K|\mathbf{F}(t)|$$
$$+ [-\bar{K} + K] \int_0^t |C(t,s)|\,|\mathbf{x}(s)|\,ds\,.$$

Thus, if we ask that there exists $\mu > 0$ such that

$$|\mathbf{x}| \left[k - \bar{K} \int_t^\infty |C(u,t)|\,du \right] \ge \mu[|A\mathbf{x}| + |\mathbf{x}|] \tag{8.4.3}$$

and if we define $\bar{K}$ by

$$\bar{K} - K = \mu\,, \tag{8.4.4}$$

then

$$V'_{(8.4.2)}(t,\mathbf{x}(\cdot)) \le -\mu|\mathbf{x}'(t)| - \mu|\mathbf{x}(t)| + [K+\mu]|\mathbf{F}(t)|\,. \tag{8.4.5}$$

Because $|\mathbf{F}|$ is bounded, there is a $U > 0$ with $V' \le -\mu|\mathbf{x}'|$ if $|\mathbf{x}(t)| \ge 2kU$.

We define

$$[\mathbf{x}^T B \mathbf{x}]^{1/2} = Q(t, \mathbf{x})\,, \tag{8.4.6}$$

so that $Q = U$ implies $\mathbf{x}^T B \mathbf{x} = U^2$ yielding $U^2 = \mathbf{x}^T B \mathbf{x} \geq r^2 |\mathbf{x}|^2$ or

$$|\mathbf{x}| \leq U/r \overset{\text{def}}{=} L\,. \tag{8.4.7}$$

Because $r|\mathbf{x}| \leq [\mathbf{x}^T B \mathbf{x}]^{1/2}$, we have

$$W_1(|\mathbf{x}|) = r|\mathbf{x}|\,. \tag{8.4.8}$$

Also,

$$V(t, \mathbf{x}(\cdot)) \leq Q(t, \mathbf{x}) + \sup_{0 \leq s \leq t} |\mathbf{x}(s)| \int_0^t \int_t^\infty \bar{K} |C(u, s)| \, du \, ds\,,$$

and so if

$$\int_0^t \int_t^\infty \bar{K} |C(u, s)| \, du \, ds \leq D\,, \quad \text{for} \quad D > 0\,, \tag{8.4.9}$$

then

$$W_2(r) = Dr\,. \tag{8.4.10}$$

In summary then, we need

$$U + D(L + R) - \mu R < r(L + R) \tag{8.4.11}$$

for all large R. Because U and L are fixed, this requires

$$D < \mu + r\,. \tag{8.4.12}$$

The conditions of Theorem 8.4.1 are satisfied and we have the following result.

Theorem 8.4.2. *Let* (8.4.3)–(8.4.12) *hold. Then all solutions of* (8.4.2) *are bounded.*

Definition 8.4.1. *A scalar functional* $V(t, \mathbf{x}(\cdot))$ *is uniformly positive definite and strongly decrescent if there are continuous functions* $Q, W :$ $[0, \infty) \times R^n \to [0, \infty)$, *wedges* W_1 *and* W_2, *a continuous scalar function*

$\alpha(t) \geq \alpha$, *a continuous scalar functional* $J(t, W(s, \mathbf{x}(s)); \alpha(t) \leq s \leq t)$, *and a constant* $m < 1$ *with*

$$W(t, \mathbf{x}(t)) \leq V(t, \mathbf{x}(t)) \leq Q(t, \mathbf{x}(t)) + J(t, W(\cdot, \mathbf{x}(\cdot))), \qquad \text{(a)}$$

$$J(t, W(\cdot, \mathbf{x}(\cdot))) \leq m \sup_{\alpha(t) \leq s \leq t} W(s, \mathbf{x}(s)), \qquad \text{(b)}$$

and

$$W_1(|\mathbf{x}|) \leq W(t, \mathbf{x}) \leq Q(t, \mathbf{x}) \leq W_2(|\mathbf{x}|). \qquad \text{(c)}$$

Remark 8.4.1. We introduced Q earlier and its role should now be understood. Also, (c) should be expected. Thus, with the exception of (b), this definition merely organizes relations. In explanation of (b), investigators are aware that the following may occur. Set

$$V(t, \mathbf{x}(\cdot)) = Q(t, \mathbf{x}(t)) + J(t, W)$$

and suppose $V'_{(8.4.1)}(t, \mathbf{x}(\cdot)) \leq 0$ if $Q(t, \mathbf{x}) \geq U$. The solution $\mathbf{x}(t)$ may satisfy $Q(t, \mathbf{x}(t)) < U$ for a time during which V may increase (because V' is possibly positive and *very* large). Then $Q(t, \mathbf{x}(t))$ may exceed U for a time (when V decreases, because $V' \leq 0$), but J is decreasing during this time allowing $Q(t, \mathbf{x}(t))$ to increase and $|\mathbf{x}(t)|$ to become larger than ever before. Then $Q(t, \mathbf{x}(t))$ returns to U for a time during which V uses the large growth of $\mathbf{x}(t)$ to grow even larger than before. This continues and we envision an unbounded solution with

$$\liminf_{t \to \infty} Q(t, \mathbf{x}(t)) < U$$

and

$$\limsup_{t \to \infty} W(t, \mathbf{x}(t)) = +\infty.$$

Part (b) is one way to preclude the above behavior.

Theorem 8.4.3. Burton (1982c) *Let* $V(t, \mathbf{x}(\cdot))$ *be continuous, locally Lipschitz in* $\mathbf{x}$, *uniformly positive definite and strongly decrescent:*

$$W(t, \mathbf{x}(t)) \leq V(t, \mathbf{x}(\cdot))$$
$$\leq Q(t, \mathbf{x}(t)) + m \sup_{\alpha(t) \leq s \leq t} W(s, \mathbf{x}(s))$$

for a continuous function $\alpha(t) \geq \alpha$ *and* $m < 1$. *Suppose there is a positive number* U *with* $V'_{(8.4.1)}(t, \mathbf{x}(\cdot)) \leq 0$ *if* $Q(t, \mathbf{x}(t)) \geq U$. *Then solutions of* (8.4.1) *are uniformly bounded.*

Proof. Let $H > 0$ be given. We must find $D > 0$ such that

$$t_0 \geq 0, \quad t \geq t_0, \quad \|\phi\|^{[\alpha, t_0]} \leq H$$

imply $|\mathbf{x}(t, t_0, \phi)| \leq D$.

For the given $H > 0$, if $\|\phi\|^{[\alpha, t_0]} < H$, then $t \in [\alpha, t_0]$ yields $W(t, \phi(t)) \leq W_2(|\phi(t)|) \leq W_2(H)$ and we can find $M > W_2(H) + U$ with $U + mM < M$. We have $Q(t, \mathbf{x}) \geq W(t, \mathbf{x})$, so that if $W(t, \mathbf{x}) \geq U$, then $V' \leq 0$.

We shall show that solutions are bounded, and so they are continuable. Now, for $\mathbf{x}(t) = \mathbf{x}(t, t_0, \phi)$ either

(a) $W(t, \mathbf{x}(t)) < M$ for all $t \geq t_0$ or

(b) there is a first $t^* > t_0$ with $W(t^*, \mathbf{x}(t^*)) = M$.

In this case, there is an $\bar{M}$ with $Q(t^*, \mathbf{x}(t^*)) < \bar{M}$.

If (b) holds, then either

(b$_1$) $Q(t, \mathbf{x}(t)) > U$ for $t \geq t^*$ or

(b$_2$) there is a first $t_1 > t^*$ with $Q(t_1, \mathbf{x}(t_1)) = U$.

If (b$_2$) holds, there is a $\bar{t} \geq t^*$ (with $t_1 > \bar{t}$) such that $W(\bar{t}, x(\bar{t}))$ is the maximum of $W(t, \mathbf{x}(t))$ on $[\alpha, t_1]$. We claim that $W(\bar{t}, \mathbf{x}(\bar{t}))$ is the maximum of $W(t, \mathbf{x}(t))$ on $[\alpha, \infty)$. If not, then there is a first interval past t_1, say $[t_2, t_3]$ with $W(t, \mathbf{x}(t)) \leq W(\bar{t}, \mathbf{x}(\bar{t}))$ on $[t_2, t_3]$ and with $Q(t, \mathbf{x}(t)) \geq U$ on $[t_2, t_3]$, $Q(t_2, \mathbf{x}(t_2)) = U$, and $W(t_3, \mathbf{x}(t_3)) = W(\bar{t}, \mathbf{x}(\bar{t}))$. This is impossible because $V'_{(8.4.1)}(t, \mathbf{x}(\cdot)) \leq 0$ on $[t_2, t_3]$, and so

$$W(t_3, \mathbf{x}(t_3)) \leq V(t_3, \mathbf{x}(\cdot)) \leq V(t_2, \mathbf{x}(\cdot))$$
$$\leq U + m \sup_{\alpha(t_2) \leq s \leq t_2} W(s, \mathbf{x}(s))$$
$$\leq U + mW(\bar{t}, \mathbf{x}(\bar{t})) < W(\bar{t}, \mathbf{x}(\bar{t}));$$

the last inequality follows from $W(\bar{t}, \mathbf{x}(\bar{t})) > M$.

Next, we find a bound on $W(\bar{t}, \mathbf{x}(\bar{t}))$. We have $V'(t, \mathbf{x}(\cdot)) \leq 0$ on $[t^*, \bar{t}]$, so

$$W(\bar{t}, \mathbf{x}(\bar{t})) \leq V(\bar{t}, \mathbf{x}(\cdot)) \leq V(t^*, \mathbf{x}(\cdot))$$
$$\leq Q(t^*, \mathbf{x}(t^*)) + m \sup_{\alpha(t^*) \leq s \leq t^*} W(s, \mathbf{x}(s))$$
$$\leq \bar{M} + mM$$

and this is the bound on $W(t, \mathbf{x}(t))$ if case (b$_2$) holds.

For (b_1) we have $Q(t, \mathbf{x}(t)) > U$ for $t \geq t^*$, so that $V' \leq 0$ for $t \geq t^*$ implies

$$
\begin{aligned}
W_1(|\mathbf{x}(t)|) \leq W(t, \mathbf{x}(t)) &\leq V(t, \mathbf{x}(\cdot)) \\
&\leq V(t^*, \mathbf{x}(\cdot)) \\
&\leq Q(t^*, \mathbf{x}(t^*)) + m \sup_{\alpha(t^*) \leq s \leq t^*} W(s, \mathbf{x}(s)) \\
&\leq \bar{M} + mM \,.
\end{aligned}
$$

Certainly, if case (a) holds we have $W(t, \mathbf{x}(t)) < M < \bar{M} + mM$. Thus, in all cases,

$$
W_1(|\mathbf{x}(t)|) \leq W(t, \mathbf{x}(t)) \leq \bar{M}(1 + m) \,,
$$

and so

$$
|\mathbf{x}(t)| \leq W_1^{-1}(\bar{M}(1 + m)) \overset{\text{def}}{=} D
$$

and we recall that M was selected by $M > W_2(H) + U$ and $U + mM < M$. Then for $W(t^*, \mathbf{x}(t^*)) = M$, we chose $\bar{M} > Q(t^*, \mathbf{x}(t^*))$. This is uniform boundedness and the proof is complete.

Example 8.4.2. Again consider

$$
\mathbf{x}' = A\mathbf{x} + \int_0^t C(t, s)\mathbf{x}(s)\, ds + \mathbf{F}(t) \tag{8.4.2}
$$

as in Example 8.4.1 with B, K, k, and r as before. Let $|\mathbf{F}(t)| \leq P$ and define

$$
V(t, \mathbf{x}(\cdot)) = [\mathbf{x}^T B \mathbf{x}]^{1/2} + (K/r) \int_0^t \int_t^\infty |C(u, s)|\, du\, [\mathbf{x}^T(s) B \mathbf{x}(s)]^{1/2}\, ds
$$

and obtain

$$
V'_{(8.4.2)}(t, \mathbf{x}(\cdot)) \leq -\left[k - (K/2kr) \int_t^\infty |C(u, t)|\, du \right] |\mathbf{x}| + K|\mathbf{F}(t)| \,.
$$

Hence, if there exists $d > 0$ with

$$
k - (K/2kr) \int_t^\infty |C(u, t)|\, du \geq d \,, \tag{8.4.13}
$$

then

$$
V'_{(8.4.2)}(t, \mathbf{x}(\cdot)) \leq -[2kd][\mathbf{x}^T B \mathbf{x}]^{1/2} + KP \,, \tag{8.4.14}
$$

so that $V' \le 0$ if

$$[\mathbf{x}^T B\mathbf{x}]^{1/2} \ge KP/2kd \overset{\text{def}}{=} U . \tag{8.4.15}$$

We then have $Q(t, \mathbf{x}) = W(t, \mathbf{x}) = [\mathbf{x}^T B\mathbf{x}]^{1/2}$, and

$$J(t, W(\cdot, \mathbf{x}(\cdot))) = (K/r) \int_0^t \int_t^\infty |C(u, s)| \, du \, [\mathbf{x}^T(s) B\mathbf{x}(s)]^{1/2} \, ds .$$

Theorem 8.4.4.   *Suppose that $|\mathbf{F}(t)| \le P$ and (8.4.13) holds. If there exists $m < 1$ with $[K/r] \int_0^t \int_t^\infty |C(u, s)| \, du \, ds \le m$, then solutions of (8.4.2) are uniform bounded.*

Example 8.4.3.   Consider the scalar delay equation

$$x'(t) = -[a + (t \sin t)^2] x(t) + b x(t - r(t)) + F(t) \tag{8.4.16}$$

with $a > 0$ and constant, b constant, $r(t) \le t$, $-M \le r'(t) \le \delta < 1$ for some $M > 0$ and some $\delta > 0$, F continuous on $[0, \infty)$, and $|F(t)| \le P$ for some $P > 0$. Let $a^2(1 - \delta) > b^2$ and define

$$V(t, x(\cdot)) = (1/2)x^2(t) + (a/2) \int_{t-r(t)}^t x^2(s) \, ds$$

so that a calculation will yield

$$V'_{(8.4.16)}(t, x(\cdot)) \le -\mu x^2(t) + P|x(t)|$$

for some $\mu > 0$. This will be negative if $x^2(t)/2 \ge U$ for some $U > 0$.
    Here, $W(t, x) = Q(t, x) = \frac{1}{2}x^2$ and

$$J(t, W(\cdot, x(\cdot))) = a \int_{t-r(t)}^t W(s, x(s)) \, ds$$

$$\le ar(t) \sup_{t-r(t) \le s \le t} W(s, x(s)) .$$

Theorem 8.4.5.   *Let the condition with (8.4.16) hold and suppose that $r(t) \le m/a$ for some $m < 1$. Then solutions of (8.4.16) are uniform bounded.*

    The significance of a functional differential equation is that the behavior of the derivative of a solution at time t depends on much or all of its past history, rather than only on its present position.

In problems of ultimate boundedness the dependence of $\mathbf{x}'(t)$ on the very distant past may gradually fade. An example of this is seen through examination of the V function. Suppose we have the relation

$$W_1(|\mathbf{x}(t)|) \leq V(t, \mathbf{x}(\cdot))$$

and we want to drive $|\mathbf{x}(t)|$ into a certain set by making V decrease. Because V may take into account all of $\mathbf{x}(s)$ for $\alpha \leq s \leq t$, in some fashion V must "forget" early behavior of $\mathbf{x}$. Consider the equations

$$\mathbf{x}'(t) = A\mathbf{x} + \int_0^t C(t, s)\mathbf{x}(s)\, ds\,, \tag{8.4.17}$$

$$V(t, \mathbf{x}(\cdot)) = [\mathbf{x}^T B\mathbf{x}]^{1/2} + \int_0^t \boldsymbol{\Phi}(t, s)|\mathbf{x}(s)|\, ds \tag{8.4.18}$$

with $\boldsymbol{\Phi}(t, s) \geq 0$, and

$$V'(t, \mathbf{x}(\cdot)) \leq -\mu\big[|\mathbf{x}'(t)| + |\mathbf{x}(t)|\big]\,, \tag{8.4.19}$$

under appropriate conditions on A, C, and $\boldsymbol{\Phi}$. We wish to use V' to drive $V(t, \mathbf{x}(\cdot))$ to zero along any solution.

Theorem 8.4.6. $V(t, \mathbf{x}(\cdot))$ *defined by* (8.4.18) *will tend to zero along any solution of* (8.4.17) *only if*

$$\text{for each } T > 0\,, \quad \text{then } \int_0^T \boldsymbol{\Phi}(t, s)\, ds \to 0 \quad \text{as } t \to \infty\,.$$

Proof. If the result is false, then there is a $T > 0$, an $\varepsilon > 0$, and a sequence $\{t_n\} \to \infty$ with $\int_0^T \boldsymbol{\Phi}(t_n, s)\, ds \geq \varepsilon$. Consider a solution $\mathbf{x}(t)$ of (8.4.17) with $|\mathbf{x}(t)| \equiv 1$ on $[0, T]$. By assumption $V(t, \mathbf{x}(\cdot)) \to 0$, so each term of

$$V(t, \mathbf{x}(\cdot)) = [\mathbf{x}^T B\mathbf{x}]^{1/2} + \int_0^T \boldsymbol{\Phi}(t, s)(1)\, ds + \int_T^t \boldsymbol{\Phi}(t, s)|\mathbf{x}(s)|\, ds$$

tends to zero as $t \to \infty$. This contradicts $\int_0^T \boldsymbol{\Phi}(t_n, s)\, ds \geq \varepsilon$, and completes the proof.

The reader may recognize that we have touched on only one small corner of the theory of fading memory spaces in hereditary response. An excellent treatment may be found in the work of Leitman and Mizel (1974), as well as in the extensive references given there. Our use herein, however, is so peripheral that we prefer to use a simple, independent definition that is useful for our purposes and contains independent notation and terminology.

Definition 8.4.2. *Let V satisfy Definition 8.4.1 with W, W_1, W_2, Q, J, $\alpha(t)$, and m as before. Let $U > 0$. V is forgetful if for each $R > 0$ and $T > 0$ there exists $S > 0$ such that $\left[W(t, \mathbf{x}(t)) \leq R$ on $[\alpha, \infty)$, $Q(t_1, \mathbf{x}(t_1)) = U$ for some $t_1 \in [T, t]$, and $t \geq T + S\right]$ imply*

$$J\big(t, W(\cdot, (\mathbf{x}(\cdot)))\big) \leq m \sup_{T \leq s \leq t} W(s, \mathbf{x}(s)).$$

Example 8.4.4. The reader may verify Definition 8.4.2 for

$$V(t, \mathbf{x}(\cdot)) = |x| + \int_0^t K(t - s + 1)^{-2} |x(s)| \, ds$$

for $K < 1$.

Theorem 8.4.7. Burton (1982c)   *Let $V(t, \mathbf{x}(\cdot))$ satisfy the conditions of Theorem 8.4.3 and let V be forgetful. Let $V'_{(8.4.1)}(t, \mathbf{x}(\cdot)) \leq -\delta < 0$ if $Q(t, \mathbf{x}(t)) \geq U$, $M > U$, $m < \bar{m} < 1$, and $U + mM \leq \bar{m}M$. Then any solution of (8.4.1) satisfies $W(t, \mathbf{x}(t)) \leq M/\bar{m}$ for large t.*

Proof. By Theorem 8.4.3 solutions are uniformly bounded. If $\mathbf{x}(t)$ is a solution, because $V'_{(8.4.1)}(t, \mathbf{x}(\cdot)) \leq -\delta$ for $Q(t, \mathbf{x}(t)) \geq U$, we see that $Q(t, \mathbf{x}(t))$ will eventually satisfy $Q(t_1, \mathbf{x}(t_1)) < U$ for some $t_1 \geq t_0$.

Suppose that there is an interval $[t_2, t_3]$ with $Q(t_2, \mathbf{x}(t_2)) = Q(t_3, \mathbf{x}(t_3)) = U$, $Q(t, \mathbf{x}(t)) > U$ on (t_2, t_3), $\sup_{\alpha \leq s \leq t_2} W(s, \mathbf{x}(s)) = \widetilde{M} > M$. Then $V'_{(8.4.1)}(t, \mathbf{x}(\cdot)) \leq 0$ on $[t_2, t_3]$ implies

$$W(t, \mathbf{x}(\cdot)) \leq V(t, \mathbf{x}(\cdot)) \leq V(t_2, \mathbf{x}(\cdot))$$
$$\leq U + m \sup_{\alpha(t_2) \leq s \leq t_2} W(s, \mathbf{x}(s))$$
$$\leq U + m\widetilde{M} < \bar{m}\widetilde{M}.$$

Indeed, if $[t_j, t_{j+1}]$ is any interval with the properties of $[t_2, t_3]$ and $t_j > t_3$, we argue that $W(t, \mathbf{x}(t)) < \bar{m}\widetilde{M}$ on $[t_j, t_{j+1}]$, so that $W(t, \mathbf{x}(t)) < \bar{m}\widetilde{M}$ if $t \geq t_2$ for $W \leq Q$.

In fact, if $T = t_2$ and $R = \widetilde{M}$, then there is an S such that if $t_4 > t_2 + S$ and $Q(t, \mathbf{x}(t)) \geq U$ on $[t_4, t_5]$ with $Q(t_4, \mathbf{x}(t_4)) = U$, then

$$W(t, \mathbf{x}(t)) \leq V(t, \mathbf{x}(\cdot)) \leq V(t_4, \mathbf{x}(\cdot))$$
$$\leq U + m \sup_{t_2 \leq s \leq t_4} W(s, \mathbf{x}(s))$$
$$\leq U + m(\bar{m}\widetilde{M}) \leq \bar{m}(\bar{m}\widetilde{M})$$

as long as $\bar{m}\widetilde{M} > M$.

We continue in this way and conclude that either $Q(t, \mathbf{x}(t))$ ultimately remains smaller than U or $W(t, \mathbf{x}(t)) \leq \bar{m}^k \widetilde{M}$ for k large enough that $\bar{m}^k \widetilde{M} > M$, but $\bar{m}^{k+1} \widetilde{M} < M$. Now $\bar{m}^k \widetilde{M} - M > 0$ and

$$0 < \bar{m}[\bar{m}^k \widetilde{M} - M] = (\bar{m})^{k+1} \widetilde{M} - \bar{m}M < M - \bar{m}M = M(1 - \bar{m}),$$

so that $\bar{m}^k \widetilde{M} - M < M(1 - \bar{m})/\bar{m}$ or $\bar{m}^k \widetilde{M} < M + M(1 - \bar{m})/\bar{m} = M/\bar{m}$ as required.

Definition 8.4.3. *A functional V satisfying Definition 8.4.2 will be called uniformly forgetful if S is independent of T.*

Theorem 8.4.8. Burton (1982c)   *Let the conditions of Theorem 8.4.7 hold with V uniformly forgetful. Then solutions of (8.4.1) are uniform ultimate bounded with $D = M/\bar{m}$.*

Proof. Given $R > 0$, if $W(t, \mathbf{x}(t)) \leq R$ on $[\alpha, \infty)$, then there is an S such that if $Q(t, \mathbf{x}(t)) < U$ on an interval I of minimum length S and if $Q(t, \mathbf{x}(t))$ then exceeds U on some interval $[t_1, t)$ then

$$
\begin{aligned}
W(t, \mathbf{x}(t)) &\leq V(t_1, \mathbf{x}(\cdot)) \\
&\leq Q(t_1, \mathbf{x}(t_1)) + m \sup_{s \in I} W(s, \mathbf{x}(s)) \\
&\leq U + mU < \bar{m}M
\end{aligned}
$$

and it will follow that $W(t, \mathbf{x}(t)) < M$ for all future time.

Because $V'(t, \mathbf{x}(\cdot)) \leq -\delta < 0$ if $Q(t, \mathbf{x}(t)) \geq U$, we see that if $|\mathbf{x}(t)| < R$ on $[\alpha, \infty)$, then there is a number I such that $W(t, \mathbf{x}(t))$ can remain larger than U on an interval of length at most I.

On each interval of length $I + S$, the upper bound on $W(t, \mathbf{x}(t))$ changes from $\bar{m}\widetilde{M}$ to $\bar{m}(\bar{m}\widetilde{M})$. Find k with $\bar{m}^{(k+1)} \widetilde{M} < M$. Then $W(t, \mathbf{x}(t))$ will be bounded by $M/\bar{m}$ by the time $t > t_0 + k(I + s) \stackrel{\text{def}}{=} t_0 + T$. This is unifocrm, ultimate boundedness.

Example 8.4.5. Again consider Example 8.4.2 with U defined so that $V'_{(8.4.16)}(t, \mathbf{x}(\cdot)) \leq -\delta < 0$ if $x^2(t)/2 \geq U$. Because $r(t) \leq m/a$, we have

$$J\big(t, W(\cdot, \mathbf{x}(\cdot))\big) \leq m \sup_{t - m/a \leq s \leq t} W(s, \mathbf{x}(s)),$$

and so $S = m/a$. Also, $W_1(r) = W_2(r) = r^2/2$.

Theorem 8.4.9. *Let the conditions with (8.4.16) hold and let $r(t) \leq m/a$ for some $m < 1$. Then solutions of (8.4.16) are uniform ultimate bounded.*

Remark 8.4.2. Let A be stable and let $\mathbf{P}(t, \mathbf{x})$ be continuous on $[0, \infty) \times R^n \to R^n$ with

$$|\mathbf{P}(t, \mathbf{x})| / |\mathbf{x}| \to 0 \quad \text{as} \quad |\mathbf{x}| \to \infty \quad \text{uniformly in } t. \tag{8.4.20}$$

If B satisfies $A^T B + BA = -I$ and if

$$V(x) = \mathbf{x}^T B \mathbf{x},$$

then it is easy to show that solutions of

$$\mathbf{x}' = A\mathbf{x} + \mathbf{P}(t, \mathbf{x})$$

are uniform ultimate bounded. It is reasonable to believe that a similar result should hold for

$$\mathbf{x}' = A\mathbf{x} + \int_0^t C(t, s)\mathbf{x}(s) \, ds + \mathbf{P}(t, \mathbf{x}) \tag{8.4.21}$$

under certain "smallness" assumptions on $C(t, s)$. Theorem 8.4.8 makes this possible.

Example 8.4.6. Consider (8.4.21) with (8.4.20) holding and with A and C as in Example 8.4.1, $\bar{K} = K$, $A^T B + BA = -I$, $|B\mathbf{x}| \leq K[\mathbf{x}^T B \mathbf{x}]^{1/2}$, $|\mathbf{x}| \geq 2k[\mathbf{x}^T B \mathbf{x}]^{1/2}$, $r|\mathbf{x}| \leq [\mathbf{x}^T B \mathbf{x}]^{1/2}$, $\int_t^\infty |C(u, s)| \, du$ continuous, $k - K \int_t^\infty |C(u, t)| \, du \geq \mu > 0$ for positive constants K, k, r, and μ. Then, for

$$V(t, \mathbf{x}(\cdot)) = [\mathbf{x}^T B \mathbf{x}]^{1/2} + K \int_0^t \int_t^\infty |C(u, s)| \, du \, |\mathbf{x}(s)| \, ds,$$

we shall have

$$V'_{(8.4.1)}(t, \mathbf{x}(\cdot)) \leq -\mu|\mathbf{x}| + K|\mathbf{P}(t, \mathbf{x})| < -\delta$$

if $|\mathbf{x}| \geq U$, where δ and U are appropriate positive constants.
Now

$$K \int_0^t \int_t^\infty |C(u, s)| \, du \, |\mathbf{x}(s)| \, ds$$

$$\leq (K/r) \int_0^t \int_t^\infty |C(u, s)| \, du \, [\mathbf{x}^T(s) B \mathbf{x}(s)]^{1/2} \, ds$$

$$\overset{\text{def}}{=} J(t, W(\mathbf{x}(\cdot))),$$

where $W(t, \mathbf{x}) = W(\mathbf{x}) = Q(t, \mathbf{x}) = [\mathbf{x}^T B \mathbf{x}]^{1/2}$. For convenience we write

$$\mathbf{\Phi}(t, s) = (K/r) \int_t^\infty |C(u, s)| \, du$$

and consult Definitions 8.4.2 and 8.4.3.

We need to show that for each $R > 0$ there is an $S > 0$ such that if $T > 0$ and $W(\mathbf{x}(t)) \leq R$ on $[\alpha, \infty)$, if $Q(\mathbf{x}(t_1)) = U$ for some $t_1 \in [T, t]$, and if $t \geq T + S$, then

$$J(t, W(\mathbf{x}(\cdot))) \leq m \sup_{T \leq s \leq t} W(\mathbf{x}(s)).$$

Because $W(\mathbf{x}(t_1)) = U$ for some $t_1 \in [T, t]$, we have $R \leq (R/U) \times \sup_{T \leq s \leq t} W(\mathbf{x}(s))$, so

$$J(t, W(\mathbf{x}(\cdot)))$$

$$= \int_0^t \mathbf{\Phi}(t, s) W(\mathbf{x}(s))\, ds$$

$$\leq \int_0^T \mathbf{\Phi}(t, s) R\, ds + \int_T^t \mathbf{\Phi}(t, s) W(\mathbf{x}(s))\, ds$$

$$\leq R \int_0^T \mathbf{\Phi}(t, s)\, ds + \int_T^t \mathbf{\Phi}(t, s)\, ds \sup_{T \leq s \leq t} W(\mathbf{x}(s))$$

$$\leq \left[(R/U) \int_0^T \mathbf{\Phi}(t, s)\, ds + \int_T^t \mathbf{\Phi}(t, s)\, ds \right] \sup_{T \leq s \leq t} W(\mathbf{x}(s)).$$

By Theorem 8.4.8 this will yield the following result.

Theorem 8.4.10. *Let B, k, K, r, μ, U, and $\mathbf{\Phi}$ be defined in Example 8.4.6. Suppose there is an $m < 1$ such that for each $R > 0$ there is an $S > 0$ such that if $T \geq 0$ and $t \geq T + S$, then*

$$(R/U) \int_0^T \mathbf{\Phi}(t, s)\, ds + \int_T^t \mathbf{\Phi}(t, s)\, ds \leq m, \tag{8.4.22}$$

then solutions of (8.4.21) are uniform ultimate bounded.

It is of interest to see examples of (8.4.22).

Example 8.4.7. Let $|C(t, s)| \leq h(t - s + 1)^{-3}$ for $h < 2$. Then

$$\mathbf{\Phi}(t, s) \leq \int_t^\infty h(u - s + 1)^{-3}\, du = (h/2)(t - s + 1)^{-2}.$$

In (8.4.22) we have

$$(R/U) \int_0^T (h/2)(t - s + 1)^{-2}\, ds + \int_T^t (h/2)(t - s + 1)^{-2}\, ds$$

$$\leq \left[Rh/2U(t - T + 1) \right] + (h/2).$$

Now $h/2 \overset{\text{def}}{=} m_1 < 1$ and we define $\tilde{m} = (1 - m_1)/2$. Then find S so large that $R m_1/U(t - T + 1) < \tilde{m}$ if $t \geq T + S$. We then have $m = m_1 + \tilde{m} < 1$.

We have sought throughout this discussion to present results without requiring $\mathbf{F}(t, \mathbf{x}(\cdot))$ bounded for $\mathbf{x}(s)$ bounded and to effectively use the upper and lower bounds on V. If $U = 0$, then uniform asymptotic stability will result under reduced relations between these bounds.

We again consider

$$\mathbf{x}'(t) = \mathbf{F}(t, \mathbf{x}(\cdot)), \tag{8.4.1}$$

and because our result is a local one, we only ask F and V continuous and Lipschitz for $0 \leq t < \infty$ and $|\mathbf{x}(s)| \leq 1$.

Theorem 8.4.11. *Let V be continuous and locally Lipschitz in $\mathbf{x}$ and suppose there are wedges W_i and an $m < 1$ with*

(a) $W_1(|\mathbf{x}(t)|) \leq V(t, \mathbf{x}(\cdot)) \leq W_2(|\mathbf{x}(t)|) + W_3(\|\mathbf{x}\|^{[\alpha, t]})$,

(b) $V'_{(8.4.1)}(t, \mathbf{x}(\cdot)) \leq -W_4(|\mathbf{x}(t)|)$ *if $|\mathbf{x}(s)| < 1$, for $\alpha \leq s \leq t$,*

(c) *For each $\gamma > 0$ there exists $S > 0$ such that $|\mathbf{x}(t)| < 1$ on $[\alpha, \infty)$ and $t_1 \geq 0$ with $|\mathbf{x}(t_1)| \leq \gamma$, then for $t \geq t_1 + S$ we have*

$$V(t, \mathbf{x}(\cdot)) \leq W_2(\gamma) + m \sup_{t_1 \leq s \leq t} W_1(|\mathbf{x}(s)|).$$

Then $\mathbf{x} = 0$ is U.A.S.

Proof. (a) and (b) suffice for uniform stability. In fact, given $\varepsilon > 0$, determine $\delta > 0$ satisfying $W_2(\delta) + W_3(\delta) < W_1(\varepsilon)$. Then $t_0 \geq 0$ and $|\phi(t)| < \delta$ on $[\alpha, t_0]$, with $\mathbf{x}(t) = \mathbf{x}(t, t_0, \phi)$ and $t \geq t_0$, yields

$$W_1(|\mathbf{x}(t)|) \leq V(t, \mathbf{x}(\cdot)) \leq V(t_0, \phi)$$
$$\leq W_2(|\phi(t_0)|) + W_3(\|\phi\|^{[\alpha, t_0]})$$
$$\leq W_2(\delta) + W_3(\delta) < W_1(\varepsilon),$$

so that $|\mathbf{x}(t)| < \varepsilon$.

Now we show the U.A.S. Let $\varepsilon = 1$, find the δ of uniform stability, and take $\eta = \delta$.

Now, let $\varepsilon > 0$ be given. We must find $T > 0$ such that

$$t_0 \geq 0, \quad \|\phi\|^{[\alpha, t_0]} < \eta, \quad t \geq t_0 + T$$

imply $|\mathbf{x}(t, t_0, \phi)| < \varepsilon$.

For this $\varepsilon > 0$, find $\gamma > 0$ with $W_1(\varepsilon) > W_2(\gamma) + mW_1(\gamma)$. With this γ refer to part (c) and find $S = S(\gamma)$. Notice that if $|\mathbf{x}(t)| < \gamma$ on any interval $t_1 \leq t \leq t_1 + S$, then, for $t \geq t_1 + S$, we have

$$W_1(|\mathbf{x}(t)|) \le V(t, \mathbf{x}(\cdot)) \le V(t_1 + S, \mathbf{x}(\cdot))$$
$$\le W_2(|\mathbf{x}(t_1 + S)|) + m \sup_{t_1 \le s \le t_1 + S} W_1(|\mathbf{x}(s)|)$$
$$\le W_2(\gamma) + mW_1(\gamma) < W_1(\varepsilon),$$

so that $|\mathbf{x}(t)| < \varepsilon$.

Next, find $L > 1$ with $W_2(\gamma/L) \sum_{i=0}^{\infty} m^i < W_2(\gamma)$. Again refer to (c) with γ replaced by γ/L and find $S - S(\gamma/L)$. In the remainder of the proof S will denote $S(\gamma/L)$.

Now we note that $|\mathbf{x}(t)|$ will satisfy $|\mathbf{x}(t)| < \gamma/L$ at some point on every interval of sufficiently large length, say, P. This follows from

$$0 \le V(t, \mathbf{x}(\cdot)) \le V(t_0, \boldsymbol{\phi}) - \int_{t_0}^{t} W_4(|\mathbf{x}(s)|)\, ds.$$

Let $Q = S + P$ and select a sequence $\{t_i\}$ with $t_n + S < t_{n+1} \le T_n + Q$ and $|\mathbf{x}(t_i)| < \gamma/L$. Then form intervals $I_1 = [t_1, t_2]$, $I_2 = [t_2, t_3]$, ...

Now $t \in I_2$ implies

$$W_1(|\mathbf{x}(t)|) \le V(t, \mathbf{x}(\cdot)) \le V(t_2, \mathbf{x}(\cdot))$$
$$\le W_2(\gamma/L) + m \sup_{s \in I_1} W_1(|\mathbf{x}(s)|).$$

Similarly, $t \in I_3$ implies

$$W_1(|\mathbf{x}(t)| \le V(t, \mathbf{x}(\cdot))$$
$$\le V(t_3, \mathbf{x}(\cdot))$$
$$\le W_2(\gamma/L) + m \sup_{s \in I_2} W_1(|\mathbf{x}(s)|)$$
$$\le W_2(\gamma/L) + m\left[W_2(\gamma/L) + m \sup_{s \in I_1} W_1(|\mathbf{x}(s)|) \right]$$
$$\le W_2(\gamma/L)[1 + m] + m^2 \sup_{s \in I_1} W_1(|\mathbf{x}(s)|).$$

The pattern is clear. If $t \in I_n$, then

$$W_1(|\mathbf{x}(t)|) \le W_2(\gamma/L) \sum_{i=0}^{n-2} m^i + m^{n-1} \sup_{s \in I_1} W_1(|\mathbf{x}(s)|).$$

If n is large enough, say $n \ge N$, then $t \ge t_N$ yields $W_1(|\mathbf{x}(t)|) \le W_2(\gamma) + mW_1(\gamma) < W_1(\varepsilon)$, so that $|\mathbf{x}(t)| < \varepsilon$. We then select $T = NQ$ and conclude that $\mathbf{x} = 0$ is U.A.S.

An example can be obtained from Example 8.4.6 by choosing $\mathbf{P}(t, \mathbf{x}) \equiv 0$ or by letting $|\mathbf{P}(t, \mathbf{x})|/|\mathbf{x}| \to 0$ as $|\mathbf{x}| \to 0$ uniformly for $0 \le t < \infty$.

8.5   Limit Sets

Frequently the derivative of V is not negative definite in applications and we still wish to conclude asymptotic stability. Thus, we are interested in extending the results of Barbashin and Krasovskii discussed in Section 6.1 (see Theorems 6.1.4 and 6.1.5). It is clear from Krasovskii (1963, pp. 157-160, p. 174) that Krasovskii knew precisely how to extend those results, not only to autonomous functional differential equations of the type treated here, but also to periodic systems. Similar results for functional differential equations were obtained by Hale (1965) using Liapunov functionals and by Haddock and Terjéki (1983) using Razumikhin techniques.

We again adopt the notation of Section 8.3 and consider the autonomous system of functional differential equations

$$\mathbf{x}'(t) = \mathbf{f}(\mathbf{x}_t)\,, \quad t \geq 0\,, \tag{8.5.1}$$

where $\mathbf{f} : C_H \to R^n$ and $\mathbf{x}'(t)$ denotes the right-hand derivative. It is supposed that $\mathbf{f}$ satisfies a local Lipschitz condition.

In this notation we say $\mathbf{x}(\phi)$ is a solution of (8.5.1) with initial condition $\phi \in C_H$ at $t = 0$ if there is an $A > 0$ such that $\mathbf{x}(\phi)$ is a function from $[-h, A) \to R^n$ with $\mathbf{x}_t(\phi) \in C_H$ for $0 \leq t < A$, $\mathbf{x}_0(\phi) = \phi$, and $\mathbf{x}(t, \phi)$ satisfies (8.5.1) for $0 \leq t < A$.

Because (8.5.1) is autonomous t_0 may always be selected to be zero, so that some of the notation in the preceding paragraph is a simplification of that in Section 8.3.

We follow Hale's (1965) paper.

Definition 8.5.1.   *A motion through $\phi \in C_H$ is the set of functions in C given by*

$$\bigcup_{0 \leq t < A} \mathbf{x}_t(\phi)\,,$$

where $[-h, A)$ is the interval of definition of $\mathbf{x}(\phi)$.

Definition 8.5.2.   *An element $\psi \in C$ is the ω-limit set of ϕ, say $\Omega(\phi)$, if $\mathbf{x}(\phi)$ is defined on $[-h, \infty)$ and there is a sequence $\{t_n\} \to \infty$ with $\|\mathbf{x}_{t_n}(\phi) - \psi\| \to 0$ as $n \to \infty$.*

Definition 8.5.3.   *A set $M \subset C$ is an invariant set if for any $\phi \in M$, there exists a function $\mathbf{x}$, depending on ϕ, defined on $(-\infty, \infty)$, $\mathbf{x}_t \in M$ for $t \in (-\infty, \infty)$, $\mathbf{x}_0 = \phi$, such that if $\mathbf{x}^*(\sigma, \mathbf{x}_\sigma)$ is the solution of (8.5.1) with initial value $\mathbf{x}_\sigma$ at σ, then $\mathbf{x}^*(\sigma, \mathbf{x}_\sigma) = \mathbf{x}_t$ for all $t \geq \sigma$.*

Lemma 8.5.1. Hale (1965) *Let $\mathbf{x}(\phi)$ be a solution of (8.5.1) with $\mathbf{x}_0(\phi) = \phi$ defined on $[-h, \infty)$ and $\|\mathbf{x}_t(\phi)\| \leq H_1 < H$ for all $t \in [0, \infty)$. Then the family of functions $\{\mathbf{x}_t(\phi),\ t \geq 0\}$ belongs to a compact subset of C; that is, the motion through ϕ belongs to a compact subset of C.*

One proves the lemma by noting that for any $H_1 < H$, there is a constant L with $|\mathbf{f}(\phi)| \leq L$ for all $\phi \in C_{H_1}$.

Lemma 8.5.2. Hale (1965) *If $\phi \in C_H$ is such that the solution $\mathbf{x}(\phi)$ of (8.5.1) with $\mathbf{x}_0(\phi) = \phi$ is defined on $[-h, \infty)$ and $\|\mathbf{x}_t(\phi)\| \leq H_1 < H$ for $t \in [0, \infty)$, then $\Omega(\phi)$ is a nonempty, compact, connected, invariant set and $\mathrm{dist}\,(\mathbf{x}_t(\phi), \Omega(\phi)) \to 0$ as $t \to \infty$.*

Proof. By Lemma 8.5.1 the family $\mathbf{x}_t(\phi)$, $t \geq h$, belongs to a compact subset $S \subset C$ and S can be chosen as the set of $\psi \in C$ with $\|\psi\| \leq H_1$, $\|\psi'\| \leq K$ for some constant K. Thus, $\Omega(\phi)$ is nonempty and bounded.

Next, we show $\Omega(\phi)$ is invariant. If $\psi \in \Omega(\phi)$, then there is a $\{t_n\} \to \infty$ as $n \to \infty$ with $\|\mathbf{x}_{t_n}(\phi) - \psi\| \to 0$ as $n \to \infty$. For any integer N, there exists a subsequence of the t_n, say $\{t_n\}$ again, and a function $\mathbf{g}_\tau(\phi)$ defined for $-N \leq \tau \leq N$ such that $\|\mathbf{x}_{t_n+\tau}(\phi) - \mathbf{g}_\tau(\phi)\| \to 0$ as $n \to \infty$ uniformly for τ in $[-N, N]$. By the diagonalization process, one can choose the t_n so that $\|\mathbf{x}_{t_n+\tau}(\phi) - \mathbf{g}_\tau(\phi)\| \to 0$ as $n \to \infty$ uniformly on all compact subsets of $(-\infty, \infty)$. In particular, $\{\mathbf{x}_{t_n+\tau}(\phi)\}$ defines a function $\mathbf{g}_\tau(\phi)$ for $-\infty < \tau < \infty$. Certainly, $\mathbf{g}_\tau(\phi)$ satisfies (8.5.1). Because $\mathbf{g}_0(\phi) = \psi$, the solution $\mathbf{x}_t(\psi)$ of (8.5.1) with $\mathbf{x}_0(\psi) = \psi$ is defined for all $t \in (-\infty, \infty)$, and furthermore, it is in $\Omega(\phi)$, because $\|\mathbf{x}_{t_n+\tau}(\phi) - \mathbf{x}_t(\psi)\| \to 0$ as $n \to \infty$ for fixed t. Thus $\Omega(\phi)$ is invariant.

It is clear that $\Omega(\phi)$ is connected.

To see that $\Omega(\phi)$ is closed, suppose $\{\psi_n\}$ is a sequence in $\psi(\phi)$ and $\psi_n \to \psi$ as $n \to \infty$. There exists an increasing sequence of $t_n = t_n(\psi_n) \to \infty$ as $n \to \infty$ with $\|\mathbf{x}_{t_n}(\phi) - \psi_n\| \to 0$ as $n \to \infty$. Given $\varepsilon > 0$, choose n so large that $\|\psi_n - \psi\| < \varepsilon/2$ and $\|\mathbf{x}_{t_n}(\phi) - \psi_n\| < \varepsilon/2$. Then $\|\mathbf{x}_{t_n}(\phi) - \psi\| < \varepsilon$ for n large. Thus $\psi \in \Omega(\phi)$ and so $\Omega(\phi)$ is closed. But $\Omega(\phi) \subset S$ and S is compact so $\Omega(\phi)$ is compact.

Finally, suppose there is an increasing sequence $\{t_n\} \to \infty$ and an $\alpha > 0$ with $\|\mathbf{x}_{t_n}(\phi) - \psi\| \geq \alpha$ for all ψ in $\Omega(\phi)$. Because $\mathbf{x}_t(\phi)$ belongs to a compact subset of C there is a subsequence which converges to some $\psi \in C$, so $\psi \in \Omega(\phi)$. This is a contradiction to the inequality, and the proof is complete.

Theorem 8.5.1. Hale (1965) *Let $V : C_H \to (-\infty, \infty)$ be a continuous function satisfying a local Lipschitz condition. Let U_r denote the region where $V(\phi) < r$ and suppose there exists a constant $K \geq 0$ with $|\phi(0)| \leq$*

K, $V(\phi) \geq 0$, and $V'_{(8.5.1)}(\phi) \leq 0$ for all $\phi \in U_r$. If R is the set of points in U_r, where $V'_{(8.5.1)}(\phi) = 0$, and M is the largest invariant set in R, then every solution of $(8.5.1)$ with initial values in U_r approaches M as $t \to \infty$.

Proof. These conditions imply that $V(\mathbf{x}_t(\phi))$ is a nonincreasing function of t and that $V(\mathbf{x}_t(\phi))$ is bounded from below within U_r. Hence, $\phi \in U_r$ implies $\mathbf{x}_t(\phi) \in U_r$ and $|\mathbf{x}(t, \phi)| \leq K$ for all $t \geq 0$, implying $\|\mathbf{x}_t(\phi)\| \leq K$ for all $t \geq 0$; that is, $\mathbf{x}_t(\phi)$ is bounded, and by Lemma 8.5.2, $\mathbf{\Omega}(\phi)$ is an invariant set. Now $V(\mathbf{x}_t(\phi))$ has a limit $r_0 < r$ as $t \to \infty$ and $V = r_0$ on $\mathbf{\Omega}(\phi)$. Hence, $\mathbf{\Omega}(\phi)$ is in U_r and $V'_{(8.5.1)} = 0$ on $\mathbf{\Omega}(\phi)$. Thus $\mathbf{\Omega}(\phi)$ invariant implies $\mathbf{\Omega}(\phi)$ in M, and Lemma 8.5.2 yields $\mathbf{x}_t(\phi) \to M$ as $t \to \infty$. This completes the proof.

Corollary. Hale (1965)   *If the conditions of Theorem 8.5.1 are satisfied and $V'_{(8.5.1)}(\phi) < 0$ for all $\phi \neq \mathbf{0}$ in U_r, then every solution of $(8.5.1)$ with initial value in U_r approaches zero as $t \to \infty$.*

Theorem 8.5.2.   Hale (1965)   *Suppose that $\mathbf{f}(\mathbf{0}) = \mathbf{0}$ and there is a locally Lipschitz function $V : C_H \to [0, \infty)$ with $V(\mathbf{0}) = \mathbf{0}$, $W(|\phi(0)|) \leq V(\phi)$ for some wedge W, and $V'_{(8.5.1)}(\phi) \leq 0$ for all $\phi \in C_H$.*

  (a) *Then the zero solution of $(8.5.1)$ is stable.*

  (b) *If, in addition, the only invariant set in $V'_{(8.5.1)}(\phi) = 0$ is $\mathbf{0}$, then $\mathbf{x} = \mathbf{0}$ is asymptotically stable.*

  (c) *Every solution $\mathbf{x}(t, \phi)$ of $(8.5.1)$ tends to zero if $V(\phi) < r_0 = W(H)$.*

Proof. We have $V(\mathbf{0}) = \mathbf{0}$ and V is locally Lipschitz, so there is a wedge W_1 with $V(\phi) \leq W_1(\|\phi\|)$ for $\|\phi\|$ sufficiently small.

   Given $\varepsilon > 0$, $\varepsilon < H$, choose $\delta > 0$ with $W_1(\delta) < W(\varepsilon)$. If $\phi \in C_\delta$, then $V(\mathbf{x}_t(\phi))$ nondecreasing implies

$$W(|\mathbf{x}(t, \phi)|) \leq V(\mathbf{x}_t(\phi)) \leq V(\phi) \leq W(\delta) < W(\varepsilon)$$

for $t \geq 0$. Thus, $\mathbf{x} = 0$ is stable.

   Parts (a) and (b) are consequences of the corollary to Theorem 8.5.1.

Corollary.   *Suppose $\mathbf{f}(\mathbf{0}) = \mathbf{0}$, $W(|\phi(0)|) \leq V(\phi) \leq W_1(\|\phi\|)$, and*

$$V'_{(8.5.1)}(\phi) \leq -W_2(|\phi(0)|)$$

for wedges W_1, W_2, and W_3. Then the zero solution of $(8.5.1)$ is asymptotically stable.

Theorem 8.5.3. Hale (1965)   Let $V : C_H \to (-\infty, \infty)$ be continuous and locally Lipschitz. Let $V(\phi) \geq 0$, $V'_{(8.5.1)}(\phi) \leq 0$ for all $\phi \in C_H$, and let R be the set of all $\phi \in C_H$ with $V'_{(8.5.1)}(\phi) = 0$. If M is the largest invariant set in R, then all solutions of (8.5.1) that are bounded for $t \geq 0$ approach M as $t \to \infty$.

The proof is essentially the same as that of Theorem 8.5.1 when we assume solutions bounded.

See Hale (1965) and Krasovskii (1963; pp. 157-160, p. 173) for examples, as well as partial extensions to equations with unbounded delays.

The following example is by Krasovskii (1963; p. 173).

Example 8.5.1. Consider the system

$$x'(t) = y(t),$$
$$y'(t) = -\psi(x(t), y(t))y(t) - g(x(t))$$
$$+ \int_{-h}^{0} g^*(x(t+s))y(t+s)\, ds, \tag{8.5.2}$$

in which $\psi : R^2 \to (0, \infty)$, $g : (-\infty, \infty) \to (-\infty, \infty)$, ψ is locally Lipschitz, g has a continuous derivative with

$$g^*(x) = dg(x)/dx,$$

and $xg(x) > 0$ if $x \neq 0$.

Define $G(x) = \int_0^\infty g(s)\, ds$ and

$$V(x_t, y_t) = 2G(x) + y^2 + b \int_{-h}^{0} \left(\int_{s}^{0} y^2(t+u)\, du \right) ds$$

and obtain

$$V'_{(8.5.2)}(x_t, y_t) = -2\psi(x, y)y^2 + 2y \int_{-h}^{0} g^*(x(t+s))y(t+s)\, ds$$
$$+ b \int_{-h}^{0} \left[y^2(t) - y^2(t+s) \right] ds.$$

We now ask that

$$\psi(x, y) \geq bh \quad \text{and} \quad |g^*(x)| \leq L. \tag{8.5.3}$$

Thus,

$$V'_{(8.5.2)}(x_t, y_t)$$

$$\leq \int_{-h}^{0} \left\{ -2by^2(t) + 2L|y(t)|\,|y(t+s)| + by^2(t) - by^2(t+s) \right\} ds$$

$$\leq (-b+L) \int_{-h}^{0} \left(y^2(t) + y^2(t+s) \right) ds \leq 0$$

if

$$b > L. \tag{8.5.4}$$

In Theorem 8.5.3 we then have $R = \left\{ (x_t, y_t) : y_t = 0 \right\}$. If $y_t = 0$ in (8.5.2), then

$$y'(t) = -g(x(t)),$$

which is nonzero unless $x = 0$. Hence, $M = \{(0,0)\}$ and all bounded solutions tend to $(0,0)$. If $\int_0^x g(s)\,ds \to \infty$ as $|x| \to \infty$, then all solutions are bounded.

Haddock and Terjéki (1983) provided a counterpart to the work of Hale using Liapunov functions and Razumikhin techniques instead of functionals.

Here, we are looking at the whole space $C = C\big([-h, 0] \to R^n\big)$ and we consider

$$\mathbf{x}' = \mathbf{f}(\mathbf{x}_t), \quad t \geq 0, \tag{8.5.1}$$

again with $\mathbf{f} : C \to R^n$ and $\mathbf{f}$ satisfying a local Lipschitz condition.

Definition 8.5.4. *A set $M \subset C$ is positively invariant if, for each $\phi \in M$, then $\mathbf{x}_t(\phi) \in M$ for all $t \geq 0$.*

Haddock and Terjéki call any function $V : R^n \to (-\infty, \infty)$ having continuous first partial derivatives a *Liapunov* or *Razumikhin function*.

Thus, if V is a Liapunov function then

$$V'_{(8.5.1)}(\phi) = \sum_{i=1}^{n} \frac{\partial V}{\partial x_i}\,(\phi(0)) f_i(\phi)$$

is a functional (which Driver stressed in our Section 8.1), even though V is a function.

Notation. If V is a Liapunov function and $G \subset C$, define

$$E_V(G) = \left\{ \phi \in G : \max_{-h \leq s \leq 0} V[\mathbf{x}_t(\phi)(s)] \right.$$

$$\left. = \max_{-h \leq s \leq 0} V[\phi(s)] \quad \text{for all} \quad t \geq 0 \right\}$$

and let $M_V(G)$ denote the largest subset of $E_V(G)$ that is invariant with respect to (8.5.1).

One may note that $M_V(G)$ is the set of functions $\phi \in G$ for every element of which there exists a solution $\mathbf{x}_t(\phi)$ through ϕ such that

$$\max_{-h \leq s \leq 0} V[\phi(s)] = \max_{-h \leq s \leq 0} V[\mathbf{x}_t(\phi)(s)]$$

for all $t \in (-\infty, \infty)$.

Note also that, for a Liapunov function V and for any $\phi \in E_V(G)$, $V'(\mathbf{x}_t(\phi)) = 0$ for any $t > 0$ such that $\max_{-h \leq s \leq 0} V[\mathbf{x}_t(\phi)(s)] = V[\mathbf{x}_t(\phi)(0)]$. This is a consequence of the definition of $E_V(G)$ and the fact that V must attain a relative maximum for such t.

Moreover, if $\phi \in M_V(G)$ with $\max_{-h \leq s \leq 0} V(\phi(s)) = V(\phi(0))$, then $V'(\phi) = 0$.

Remark 8.5.1. In the corresponding theory of ordinary differential equations, Barbashin (1968) points out that if the surface $V'(\mathbf{x}) = 0$ is given by the equation $\phi(x_1, \ldots, x_n) = 0$, then

$$\frac{d\phi}{dt} = \sum_{i=1}^{n} \frac{\partial \phi}{\partial x_i} x_i' \neq 0$$

is a necessary condition for the absence of integral trajectories lying on the surface. Similar observations in an intelligible form do not seem to have been made for functional differential equations using either functionals or functions. And it seems that such additions would be most welcome.

Theorem 8.5.4. Haddock-Terjéki (1983)   *Suppose that V is a Liapunov function and that $G \subset C$ is a closed and positively invariant set with respect to (8.5.1) with the property that*

$$V'(\phi) \leq 0 \quad \text{for all} \quad \phi \in G \quad \text{such that}$$
$$V(\phi(0)) = \max_{-h \leq s \leq 0} V(\phi(s)) . \tag{8.5.5}$$

Then for any $\phi \in G$ such that $\mathbf{x}(\phi)$ is bounded on $[-h, \infty)$, we have $\Omega(\phi) \subset M_V(G) \subset E_V(G)$. Hence, $\mathbf{x}_t(\phi) \to M_V(G)$ as $t \to \infty$.

Proof. Let $\phi \in G$ and $\mathbf{x}(\phi)(\cdot)$ be bounded on $[-h, \infty)$. Then $\mathbf{x}_t(\phi) \in G$ for all $t \geq 0$ and $\Omega(\phi)$ is nonempty. By (8.5.2) one may argue that the function $\max_{-h \leq s \leq 0} V[\mathbf{x}_t(\phi)(s)]$ is a nonincreasing function of t on $[0, \infty)$. Because V is bounded from below along this solution,

$$\lim_{t \to \infty} \left\{ \max_{-h \leq s \leq 0} V[\mathbf{x}_t(\phi)(s)] \right\} = c \quad \text{exists.}$$

We now show that $\psi \in \Omega(\phi)$ implies $\psi \in E_V(G)$. There is a sequence $\{t_n\} \to \infty$ with $x_{t_n} \to \psi$ as $n \to \infty$. Thus,

$$\lim_{t \to \infty} \left\{ \max_{-h \leq s \leq 0} V[\mathbf{x}_{t_n}(\phi)(s)] \right\} = \max_{-h \leq s \leq 0} V(\psi(s)) = c.$$

Because G is closed, $\psi \in G$ and $\mathbf{x}_t(\psi) \in \Omega(\phi) \cap G$ for all $t \geq 0$. Thus,

$$\max_{-h \leq s \leq 0} V[\mathbf{x}_t(\psi)(s)] = c = \max_{-h \leq s \leq 0} V(\psi(s))$$

for all $t \geq 0$. Hence, $\psi \in E_V(G)$ and we have $\Omega(\phi) \subset E_V(G)$. It now follows that

$$x_t(\phi) \to M_V(G) \supset \Omega(\phi)$$

as $t \to \infty$. This completes the proof.

Corollary. *Let* $\mathbf{f}(0) = \mathbf{0}$ *and suppose there is a Liapunov function* V *and a constant* $\alpha > 0$ *with:*

(a) $V(\mathbf{0}) = 0$, $V(\mathbf{x}) > 0$ *for* $0 < |\mathbf{x}| < \alpha$,

(b) $V'_{(8.5.1)}(\mathbf{0}) = 0$,

and

(c) $V'_{(8.5.1)}(\phi) < 0$ *for* $0 < \|\phi\| < \alpha$ *when* $\max_{-h \leq s \leq 0} V(\phi(s)) = V(\phi(0))$.

Then the zero solution of (8.5.1) *is asymptotically stable.*

Proof. Stability follows just as in the proof of Theorem 8.1.7. Note that $\mathbf{0} \in M_V(S(\mathbf{0}, \alpha)) \subset E_V(S(\mathbf{0}, \alpha))$, where $S(\mathbf{0}, \alpha) = \{\phi \in C : \|\phi\| < \alpha\}$. Also, one may argue that

$$\max_{-h \leq s \leq 0} V(\phi(s)) > \max_{-h \leq s \leq 0} V[\mathbf{x}_t(\phi)(s)]$$

for any $\phi \neq 0$ and $t > h$. Thus, for each ϕ satisfying $0 < \|\phi\| < \alpha$, then $\phi \notin E_V(S(\mathbf{0}, \alpha))$, and it follows that $\{\mathbf{0}\} = M_V(S(\mathbf{0}, \alpha)) = E_V(S(\mathbf{0}, \alpha))$. Hence, for each solution $\mathbf{x}(\phi)(\cdot)$ with $\|\mathbf{x}_t(\phi)\| < \alpha$ for $t \geq 0$, we conclude that $\mathbf{x}_t(\phi) \to \mathbf{0} \in C$ as $t \to \infty$. This completes the proof.

In the following scalar example we use

$$V(x) = x^2/2\,,$$

so that the condition

$$V(\phi(0)) = \max_{-h \le s \le 0} V(\phi(s))$$

is equivalent to the condition

$$|\phi(0)| = \|\phi\|\,.$$

Example 8.5.2. Consider the scalar equation

$$x'(t) = -ax(t) + \int_{-h}^{0} p(s)x(t+s)\,ds \ - x(t)\int_{-h}^{0} q(s)x^2(t+s)\,ds\,,$$

where $p, q : [-h, 0] \to (-\infty, \infty)$ are continuous, $a > 0$, $\int_{-h}^{0} |p(s)|\,ds \le a$, and $q(s) \ge 0$ for $s \in [-h, 0]$. Define

$$V(x) = x^2/2$$

and obtain

$$V'_{(8.5.1)}(\phi) = -a\phi^2(0) + \phi(0)\int_{-h}^{0} p(s)\phi(s)\,ds$$

$$- \phi^2(0)\int_{-h}^{0} q(s)\phi^2(s)\,ds$$

$$\le -a\phi^2(0) + |\phi(0)|\,\|\phi\|\int_{-h}^{0} |p(s)|\,ds$$

$$- \phi^2(0)\int_{-h}^{0} q(s)\phi^2(s)\,ds\,.$$

It is clear that $V'_{(8.5.1)}(\phi) \le 0$ for all ϕ with $|\phi(0)| = \|\phi\|$. If either

$$\left\{ \int_{-h}^{0} |p(s)|\,ds \le a \quad \text{and} \quad |p(s)| > 0\,, \right.$$

$$\left. q(s) > 0 \quad \text{for some} \quad s \in [-h, 0] \right\}$$

or

$$\left\{ \int_{-h}^{0} |p(s)|\,ds = a \quad \text{and} \quad q(0) > 0 \right\}\,,$$

then $V'(\phi) < 0$ for any $\phi \not\equiv 0$ with $|\phi(0)| = \|\phi\|$. Thus, (global) asymptotic stability results.

Numerous other examples and results along these lines are also presented by Haddock and Terjéki (1983).

The reader should beware of attempting to extend the asymptotic stability results to nonautonomous equations. Early results of that sort were false. Corrections were made by Krasovskii and Driver by introducing the function $f(r) > r$ as shown in Section 8.1.

8.6   Periodic Solutions

Probably no topic in the theory of ordinary differential equations has attracted more interest than that of the existence of a periodic solution. About half of the monograph [Burton (1985)] is devoted to the subject. We also saw some work on periodic solutions in Chapter 2 using the resolvent and Perron's theorem. Frequently, such results depend on fixed point theorems of analysis and topology. The most elementary such result is as follows for one dimension.

Theorem 8.6.1.   *If $f : [0,1] \to [0,1]$ is continuous, then f has a fixed point; that is, there is an $x_0 \in [0,1]$ with $f(x_0) = x_0$.*

The result is obvious if we think about it as follows. Draw the line $f(x) = x$ from $(0,0)$ to $(1,1)$. Now, consider the two points $A = (0, f(0))$ and $B = (1, f(1))$. The graph $(x, f(x))$ for $0 \le x \le 1$ is a connected one from A to B and, hence, must intersect the line $f(x) = x$ at some point $(x_0, f(x_0)) = (x_0, x_0)$.

Applying this to differential equations, we consider the scalar equation

$$x' = -x^3 + \cos t, \quad t \ge 0. \tag{8.6.1}$$

Notice that when $x = +1$, then $x' \le 0$, whereas $x' \ge 0$ when $x = -1$. Because solutions of (8.6.1) depend continuously on initial conditions, we may use (8.6.1) to define a continuous map f from $[-1,1]$ to $[-1,1]$ as follows. Denote the solution of (8.6.1) through $(0, x_0)$ by $x(t, x_0)$. Then define $f(x_0) = x(2\pi, x_0)$. We have f continuous and $f : [-1,1] \to [-1,1]$. Certainly, Theorem 8.6.1 is true with $[0,1]$ replaced by $[-1,1]$, so there is a fixed point. That is, there is an x_0 with $x(2\pi, x_0) = x_0$. But the initial-value problem

$$x' = -x^3 + \cos t, \quad x(2\pi) = x_0$$

is indistinguishable from that of

$$x' = -x^3 + \cos t, \quad x(0) = x_0.$$

Hence, the solution $x(t, x_0)$ on $[0, 2\pi]$ repeats itself on $[2\pi, 4\pi]$ and, in fact, on $[2k\pi, 2(k+1)\pi]$.

It is often of great interest to know that there is only one fixed point; but Theorem 8.6.1 is of no help there. Every point on $[0, 1]$ is a fixed point of the map $f(x) = x$.

Theorem 8.6.1 is, of course, a case of the Brouwer fixed point theorem. Its n-dimensional statement is as follows.

Theorem 8.6.2. Brouwer   *Suppose that $A \subset R^n$ is homeomorphic to the closed unit ball in R^n and that $f : A \to A$ is continuous. Then f has a fixed point.*

From 1940 to the present scores, if not hundreds, of papers have been published applying Theorem 8.6.2 to the scalar equation

$$x'' + h(x, x')x' + g(x) = e(t) , \qquad (8.6.2)$$

in which h and g are locally Lipschitz, $xg(x) > 0$ if $x \neq 0$, $h(x, y) > 0$ for large $x^2 + y^2$, $e(t)$ is continuous, and $e(t + T) = e(t)$ for all t and some $T > 0$. The paper by Massera (1950) gives a fundamental background for such studies.

One may write (8.6.2) as a system

$$x' = y ,$$
$$y' = -h(x, y)y - g(x) + e(t) . \qquad (8.6.3)$$

The object, then, is to construct a simple closed curve in the xy plane bounding all solutions that start inside or on the curve for $t \geq 0$. If that curve, together with its interior, is denoted by A, then a mapping of A into itself may be defined by the solutions of (8.6.3) from $t = 0$ to $t = T$.

The books by Sansone and Conti (1964) and Reissig *et al.* (1964) are substantially devoted to this problem. More recent bibliographies may be found in Graef (1972) and in Burton and Townsend (1968, 1971).

Much investigation concerning the existence of a unique periodic solution of (8.6.3) was carried out in the early 1940s concerning the proper operation of communication equipment used by the military. Early papers of M. L. Cartwright and J. E. Littlewood provided motivation for such research, which continues to this day.

Autonomous two-dimensional systems of the form

$$x' = P(x, y) ,$$
$$y' = Q(x, y)$$

are periodic in t of any period and may have periodic solutions of any (and every) period. When P and Q are locally Lipschitz, then the Brouwer fixed point theorem gives way to the Poincaré-Bendixson theorem [see Lefschetz (1957)] which states that if a solution of (8.6.4) is bounded for $t \geq 0$ then it is

(a) periodic, or
(b) it approaches a periodic solution spirally, or
(c) it has an equilibrium point in its ω-limit set.

A companion to the Poincaré-Bendixson theorem is the Bendixson-du Lac criterion [see Clark (1976), p. 202], which states that if P and Q are C^1 functions, if there is a C^1 function $B(x, y)$, and if there is a simply connected region L such that $\partial(PB)/\partial x + \partial(QB)/\partial y$ does not change sign or vanish identically on any open subset of L, then there is no periodic solution of $(x' = P, y' = Q)$ lying entirely in L.

The question of the uniqueness of periodic solutions is answered by using the du Lac criterion, which states that if P and Q are C^1, if there is a C^1 function B, and if there is an annular ring L such that $\partial(PB)/\partial x + \partial(QB)\partial y$ does not change sign or vanish identically on any open subset of L, then there is at most one periodic solution lying entirely in L.

In dimensions higher than two the geometry becomes difficult, but Liapunov's direct method can be used very effectively. Consider a system of ordinary differential equations

$$\mathbf{x}' = \mathbf{f}(t, \mathbf{x}) \tag{8.6.4}$$

with $\mathbf{f} : (-\infty, \infty) \times R^n \to R^n$ continuous and locally Lipschitz in $\mathbf{x}$. Suppose also that $\mathbf{f}(t + T, \mathbf{x}) = \mathbf{f}(t, \mathbf{x})$ for all $t \geq 0$ and some $T > 0$. To prove the existence of a periodic solution we construct a function $V : (-\infty, \infty) \times R^n \to [0, \infty)$ that is continuous in t and locally Lipschitz in $\mathbf{x}$, together with wedges W_1, W_2, and W_3 with

$$W_1(|\mathbf{x}|) \leq V(t, \mathbf{x}) \leq W_2(|\mathbf{x}|) \,,$$

$W_1(r) \to \infty$ as $r \to \infty$, and

$$V'_{(8.6.4)}(t, \mathbf{x}) \leq -W_3(|\mathbf{x}|) \quad \text{if} \quad |\mathbf{x}| \geq M \,,$$

for some $M > 0$. (See Theorem 6.1.2 and the remarks surrounding it.) One may prove that solutions of (8.6.4) are uniformly bounded and uniformly ultimately bounded. It may then be argued that solutions of (8.6.4) define a continuous mapping of an n-ball into itself by the solution from $\mathbf{x}(t_0, t_0, \mathbf{x}_0)$ to $\mathbf{x}(t_0 + kT, t_0, \mathbf{x}_0)$ for some sufficiently large integer k. We conclude that there is a fixed point and a periodic solution of period not greater than kT.

Exercise 8.6.1. Gather the material from Eq. (8.6.4) into a theorem and prove it. Apply the result to

$$\mathbf{x}' = A\mathbf{x} + \mathbf{e}(t),$$

in which e is periodic and the characteristic roots of A have negative real parts. Use

$$V(\mathbf{x}) = \mathbf{x}^T B \mathbf{x}$$

where $A^T B + BA = -I$. Show also that when $V = V(\mathbf{x})$ and when $\{\mathbf{x} \in R^n : V(\mathbf{x}) \leq$ constant$\}$ is homeomorphic to a closed n-ball, then $k = 1$.

The books by Fink (1974), Hale (1963), and Yoshizawa (1975) contain extensive results on periodic and almost periodic solutions of ordinary differential equations.

We now return to (8.3.1) which we write as

$$\mathbf{x}'(t) = \mathbf{F}(t, \mathbf{x}_t) \tag{8.6.5}$$

with the same notation as in Section 8.3: C is the space of continuous functions ϕ mapping $[-h, 0]$ into R^n with the supremum norm, C_H is the H-ball in C, $F : [0, \infty) \times C \to R^n$ is continuous and locally Lipschitz in x_t, and a solution is denoted by $x(t_0, \phi)$ with value at t being $x(t, t_0, \phi)$, and with $x_{t_0}(t_0, \phi) = \phi$. Refer to Definitions 8.2.1 and 8.2.2 for boundedness.

We are going to require a certain periodicity property for (8.6.5) which is somewhat unusual. We will ask that $F(t, \phi)$ be periodic with a fixed period $T > 0$ whenever ϕ is also T-periodic. The condition is an outgrowth of the method of proof which will be used here. Throughout the book we have used fixed point theorems and we have always formulated the fixed point mapping directly from the differential equation, usually by a simple integration. Our fixed point method here is very different. The mapping will be a translate of the solution, called the Poincaré map. For a given continuous initial function ϕ we will need to know that the solution $x(t, 0, \phi)$ can be defined at least on $[0, T]$. We will then define a mapping of ϕ by $(P\phi)(t) = x(t + T, 0, \phi)$ for $-h \leq t \leq 0$. If the mapping has a fixed point ϕ, then we want to be able to say that this is a periodic solution. Since ϕ is a fixed point, we see that $x(t, 0, \phi)$ and $x(t + T, 0, \phi)$ are both solutions with the same initial function and by uniqueness, they are the same. Hence, the solution is T-periodic. For this to work, we need to know the following facts:

(i) If $x(t)$ is a solution, so is $x(t + T)$. This is our periodicity condition on (8.6.5) and we will illustrate it below.

(ii) For a given continuous initial function ϕ, the solution $x(t, 0, \phi)$ is uniquely determined by ϕ and it exists at least on $[0, T]$.

The following result is proved in Burton(1985, p. 249), but it is also a consequence of a result which we later prove in this section.

Theorem 8.6.3.   *Suppose that $x(t + T)$ is a solution of (8.6.5) whenever $x(t)$ is a solution of (8.6.5). If solutions of (8.6.5) are uniform bounded and uniform ultimate bounded for bound D, then (8.6.5) has a periodic solution of period T and is bounded by D.*

Examples of uniform ultimate boundedness can be found throughout Burton (1985), while Burton (1990) is devoted entirely to the question. We will give a particularly simple example next in which we use a Liapunov functional and have a differential inequality.

Example 8.6.1.   Consider the scalar equation

$$x' = -x + \int_{t-h}^{t} C(t, s) x(s)\, ds\ + f(t) \tag{8.6.6}$$

with C continuous for $-\infty < s \leq t < \infty$, f continuous on $(-\infty, \infty)$, $f(t+T) = f(t)$, $C(t+T, s+T) = C(t, s)$, $|f(t)| \leq M$ for some $M > 0$. We are going to derive conditions under which solutions are uniform bounded and uniform ultimate bounded and then give two exercises so that the interested reader can extend the result.

Suppose that $x(t)$ is a solution of (8.6.6). Then

$$x'(t + T) = -x(t + T) + \int_{t+T-h}^{t+T} C(t + T, s) x(s)\, ds\ + f(t + T)$$

$$= -x(t + T) + \int_{t-h}^{t} C(t + T, s + T) x(s + T)\, ds + f(t)$$

so $x(t + T)$ is also a solution.

Define the Liapunov functional

$$V(t, x_t) = |x| + K \int_{t-h}^{t} \int_{t}^{\infty} |C(u, s)|\, du\, |x(s)|\, ds\,, \tag{8.6.7}$$

so that the derivative of V along solutions of (8.6.6) satisfies

$$V'(t, x_t) \leq -|x| + \int_{t-h}^{t} |C(t,s)|\,|x(s)|\,ds + M$$

$$+ K \int_{t}^{\infty} |C(u,t)|\,du\,|x| - K \int_{t-h}^{t} |C(t,s)|\,|x(s)|\,ds$$

$$\leq \left[-1 + K \int_{t}^{\infty} |C(u,t)|\,du \right] |x|$$

$$+ (-K+1) \int_{t-h}^{t} |C(t,s)|\,|x(s)|\,ds + M\,.$$

We now suppose that there are positive numbers $K > 1$, α, and P with

$$-1 + K \int_{t}^{\infty} |C(u,t)|\,du \leq -\alpha$$

$$\int_{t-h}^{t} \int_{t}^{\infty} |C(u,s)|\,du\,ds \leq P\,.$$

Next, let

$$C(t,s) = C(t-s)\,, \quad C(t) > 0\,, \quad \text{and let} \quad C(t) \text{ be decreasing.}$$

We can then argue that

$$V(t, x_t) \leq |x| + \int_{0}^{\infty} C(u)\,du \int_{t-h}^{t} |x(s)|\,ds$$

and

$$V'(t, x_t) \leq -\alpha|x| + (-K+1)C(h) \int_{t-h}^{t} |x(s)|\,ds + M\,.$$

It then follows that there is a positive constant β with

$$V'(t, x_t) \leq -\beta V(t, x_t) + M$$

so that

$$V(t, x_t) \leq e^{-\beta t} V(t_0, x_{t_0}) + \int_{t_0}^{t} e^{-\beta(t-u)} M\,du\,.$$

This yields uniform boundedness and uniform ultimate boundedness. By Theorem 8.6.3 there is a T periodic solution of (8.6.6).

Exercise 8.6.2. Carry out the details when C is not of convolution type, but $C(t,s) > 0$. Carry out the details when the equation is

$$x' = -g(x) + \int_{t-h}^{t} C(t,s)h(x(s))\,ds \qquad (8.6.8)$$

with $xg(x) > 0$ if $x \neq 0$, g is strictly increasing, and $|h(x)| \leq |g(x)|$. Use

$$V(t, x_t) = |x| + K \int_{t-h}^{t} \int_{t}^{\infty} |C(u,s)|\,du|h(x(s))|\,ds\,.$$

Investigators can sometimes obtain differential inequalities for pointwise delay equations. Moreover, the technique introduced in Section 2.5 with arc length can also be very effective. The following result is proved in Burton (1983, pp. 255-256).

Theorem 8.6.4. *Suppose there are a continuous functional $V : [0,\infty] \times C_H \to [0,\infty)$ that is locally Lipschitz in ϕ, positive constants c, U, β, and μ, and a wedge W_2 with*

(a) $V'_{(8.6.5)}(t, x_t) \leq -\mu|F(t, x_t)| + c$ *if* $|x(t)| \geq U$,
(b) $V'_{(8.6.5)}(t, x_t) \leq \beta$ *if* $|x(t)| < U$,
(c) $0 \leq V(t, x_t) \leq W_2(\|x_t\|)$.

Then solutions of (8.6.5) are uniform bounded and uniform ultimate bounded.

Exercise 8.6.3. Consider the scalar equation

$$x' = -a(t)x + b(t)x(t-h) + f(t) \qquad (8.6.9)$$

with a, b, f continuous and T-periodic functions. Suppose there are positive constants c, M, $k > 1$ with

$$-a(t) + k|b(t+h)| \leq -c < 0\,, \quad |f(t)| \leq M\,.$$

Let

$$V(t, x_t) = |x| + \int_{t-h}^{t} k|b(s+h)x(s)|\,ds \qquad (8.6.10)$$

so that

$$
\begin{aligned}
V'(t, x_t) &\leq -a(t)|x| + |b(t)x(t-h)| + k|b(t+h)x(t)| \\
&\quad - k|b(t)x(t-h)| + |f(t)| \\
&\leq -c\,|x(t)| + (1-k)|b(t)x(t-h)| + M \\
&\leq -\delta\big(|x(t)| + |x'(t)|\big) + (1+\delta)M\,.
\end{aligned}
$$

The conditions of Theorems 8.6.3 and 8.6.4 are satisfied so there is a T-periodic solution of (8.6.9).

We turn now to a functional differential equation with infinite delay expressed as

$$\mathbf{x}'(t) = \mathbf{f}(t, \mathbf{x}_t) \tag{8.6.11}$$

where $\mathbf{x}_t(s) = \mathbf{x}(t+s)$ for $-\infty < s \leq 0$. It will be assumed that if $\mathbf{x}(t)$ is a solution, so is $\mathbf{x}(t+T)$ for a fixed positive number T. The goal is to show that if solutions are uniform ultimate bounded in the supremum norm, then there is a periodic solution. Kato (1980) has shown that for such systems uniform ultimate boundedness does not imply uniform boundedness. Thus, the result actually goes beyond our stated Theorem 8.6.3 even when restricted to the case of finite delay. Seifert's work in Example 8.2.2 shows that some kind of fading memory is required for ultimate boundedness and a fading memory will be central here. This work may be found in Burton-Zhang (1990).

Let $(X, \| \cdot \|)$ be the Banach space of bounded continuous functions $\phi : (-\infty, 0] \to R^n$ with the supremum norm. In addition to the already stated periodicity assumption, let

$$f : (-\infty, \infty) \times X \to R^n \tag{8.6.12}$$

and suppose that

> for each $\phi \in X$ there is a unique solution $\mathbf{x}(t, 0, \phi)$
> satisfying (8.6.11) on $0 \leq t < \infty$ with $x_0(\cdot, 0, \phi) = \phi$. $\tag{8.6.13}$

In the way of fulfillment of (8.6.13), Sawano (1982) asks that

(H_1) if $\mathbf{x} : (-\infty, A) \to R^n$ is bounded and continuous, then $\mathbf{f}(t, \mathbf{x}_t)$ is measurable in $t \in [0, A)$,

(H_2) for any bounded set $V \subset X$ there exists a function $m(t) = m_V(t)$, locally integrable on R^+, such that $|\mathbf{f}(t, \phi)| \leq m(t)$ for any $\phi \in V$,

and

(H_3) $\mathbf{f}(t, \phi)$ is continuous in ϕ for each $t \in R^+$.

He then shows that (8.6.11) has a solution on some interval $0 \leq t \leq \alpha$. Moreover,

(H_4) if there is a locally integrable function $\eta(t) = \eta_V(t)$ such that $|\mathbf{f}(t, \phi) - \mathbf{f}(t, \psi)| \leq \eta(t)\|\phi - \psi\|$ on $R^+ \times V$, then the solution is unique. Finally, if the solution is defined on $[0, \alpha)$ and is noncontinuable beyond α, then $\limsup_{t \to \alpha^-} |\mathbf{x}(t, 0, \phi| = \infty$. Since our result asks that solutions be U.U.B., they will be continuable to $+\infty$.

The following notation will be adopted.

R^n denotes n-dimensional Euclidean space, R^-, R^+, R mean the intervals, $(-\infty, 0]$, $[0, \infty)$, and $(-\infty, \infty)$ respectively.

For every $t \geq 0$, let $\mathbf{P}_t : X \to X$ be defined by

$$(\mathbf{P}_t\phi)(s) = \mathbf{x}(t + s, 0, \phi) \quad \text{for} \quad -\infty < s \leq 0\,.$$

G denotes the set of continuous non-increasing functions $g : (-\infty, 0] \to [1, \infty)$ such that $g(r) \to \infty$ as $r \to -\infty$ and $g(0) = 1$.

For a given $g \in G$, then $(X_g, |\cdot|_g)$ denotes the Banach space of continuous functions $\phi : R^- \to R^n$ for which

$$|\phi|_g = \sup_{-\infty < t \leq 0} |\phi(t)/g(t)|$$

exists.

Let $\mathbf{x} : [a, b] \to R^n$ and define

$$\|\mathbf{x}\|^{[a,b]} = \sup\left\{|\mathbf{x}(t)| : a \leq t \leq b\right\}.$$

Definition 8.6.1. *Solutions of* (8.6.11) *are* uniformly bounded *at* $t = 0$ (U.B.) *if for each* $B_1 > 0$ *there exists* $B_2 > 0$ *such that* $\left[\|\phi\| \leq B_1,\ t \geq 0\right]$ *imply that* $|\mathbf{x}(t, 0, \phi)| < B_2$. *Solutions of* (8.6.11) *are* uniformly ultimately bounded *for bound* B *at* $t = 0$ (U.U.B.) *if for each* $B_3 > 0$ *there exists* $K > 0$ *such that* $\left[\|\phi\| \leq B_3,\ t \geq K\right]$ *imply that* $|\mathbf{x}(t, 0, \phi)| \leq B$.

Definition 8.6.2. *Let* $\mathbf{\Omega} \subset X$. *We say that* $\mathbf{P}_t$ *is* continuous *in* $(\mathbf{\Omega}, G)$ *if there is a* $g \in G$ *and for every* $\phi_1 \in \mathbf{\Omega}$, $J > 0$, *and* $\mu > 0$ *there exists a* $\delta > 0$ *such that* $\left[\phi_2 \in \mathbf{\Omega},\ |\phi_1 - \phi_2|_g < \delta\right]$ *imply that* $|\mathbf{P}_J\phi_1 - \mathbf{P}_J\phi_2|_g < \mu$.

Definition 8.6.3. *Equation* (8.6.11) *is said to have a* weakly fading memory *in* $\mathbf{\Omega} \subset X$ *if for any* $J > 0$, $D > 0$, *and* $\mu > 0$ *there exists a* $K > 0$ *such that*

$$\begin{gathered}\phi, \phi_1 \in \mathbf{\Omega}\,, \quad \|\phi\| < D\,, \quad \|\phi_1\| < D\,,\\ \phi(s) = \phi_1(s) \quad on \quad [-K, 0]\,, \quad 0 \leq t \leq J\end{gathered}$$

imply that $|\mathbf{f}(t, \phi) - \mathbf{f}(t, \phi_1)| < \mu$.

The following result was proved in Burton-Feng (1991).

Proposition. *Suppose that (8.6.13) holds and that*

(i) *equation (8.6.11) has a weakly fading memory,*

(ii) $\mathbf{f}(t, \boldsymbol{\phi})$ *is continuous at every* $(t, \boldsymbol{\phi})$ *of* $[0, T] \times X$ *with respect to the supremum norm,*

(iii) *for each* $M > 0$ *and* $\alpha > 0$ *there exists* $L > 0$ *such that* $\big[\|\boldsymbol{\phi}\| \leq M$, $0 \leq t \leq \alpha\big]$ *imply that* $|\mathbf{f}(t, \boldsymbol{\phi})| < L$.

Then for every $M > 0$ *and* $\boldsymbol{\Omega} = \big\{\boldsymbol{\phi} \in X : \|\boldsymbol{\phi}\| \leq M\big\}$, $\mathbf{P}_t$ *is continuous in* $(\boldsymbol{\Omega}, G)$.

Horn's theorem will now be stated for reference.

Theorem. Horn (1970) *Let* $S_0 \subset S_1 \subset S_2$ *be convex subsets of a Banach space* X *with* S_0 *and* S_2 *compact and* S_1 *open relative to* S_2. *Let* $P : S_2 \to X$ *be a continuous function such that for some integer* $m > 0$,

(a) $P^j S_1 \subset S_2$, $\quad 1 \leq j \leq m - 1$

and

(b) $P^j S_1 \subset S_0$, $\quad m \leq j \leq 2m - 1$.

Then P *has a fixed point in* S_0.

The proof of the existence of periodic solutions of dissipative systems utilizes a translation map $\mathbf{P}_t$ which

(a)   must be continuous at least in the supremum norm. In order for solutions of (8.6.11) to be defined even locally it is necessary that

(b) f take bounded sets into bounded sets. As f is T-periodic, this takes the form of (ii) in the next theorem. In order for solutions to be U.U.B., by Seifert's example,

(c)   some type of fading memory is required. In view of the referenced proposition of the last section, conditions (ii) and (iii) of the next result are in some sense necessary.

The following theorem yields a T-periodic solution of (8.6.11) without asking that solutions be U.B.

Theorem 8.6.5. *Suppose that* (8.6.12), (8.6.13), *and the following hold:*

(i) *Solutions of* (8.6.11) *are U.U.B.*

(ii) *For each* $M > 0$ *there exists* $L > 0$ *such that* $\big[\|\boldsymbol{\phi}\| \leq M$, $t \geq 0\big]$ *imply that* $|\mathbf{f}(t, \boldsymbol{\phi})| < L$.

(iii) *For every bounded (in the supremum norm) set* $\boldsymbol{\Omega} \subset X$, $\mathbf{P}_t$ *is continuous in* $(\boldsymbol{\Omega}, G)$.

Then (8.6.11) *has a T-periodic solution.*

Proof. We first prepare the sets for Horn's theorem. Since solutions of (8.6.11) are U.U.B., there is an $N > 0$ such that

$$\left[\phi \in X,\ \|\phi\| \le 2B,\ t \ge N\right] \quad \text{imply that} \quad |\mathbf{x}(t, 0, \phi)| \le B\,.$$

By (ii), there is an $L_B > 0$ such that

$$\|\phi\| \le 2B \quad \text{implies that} \quad |\mathbf{f}(t, \phi)| < L_B \quad \text{for all}\quad t \ge 0\,. \qquad (8.6.14)$$

Let

$$S_B = \left\{\phi \in X : \|\phi\| \le 2B,\ |\phi(u) - \phi(v)| \le L_B|u - v|\right.$$
$$\left.\text{for } u, v \in R^-\right\}. \qquad (8.6.15)$$

By (iii) there exists a $g^* \in G$ such that P_N is continuous in the g^*-norm on S_B. Also, S_B is compact in the g^*-norm; hence $P_N(S_B)$ is bounded in the g^*-norm and, being bounded for $t \le -N$, is bounded in the supremum norm: there exists $B^* > 0$ such that

$$\phi \in S_B \quad \text{implies that} \quad \|\mathbf{P}_N(\phi)\| \le B^* \quad \text{and} \quad |\mathbf{P}_N(\phi)|_{g^*} \le B^*\,. \quad (8.6.16)$$

In particular, there exists a $B_1 > B$ such that

$$\phi \in S_B \quad \text{implies that} \quad \|\mathbf{x}(\cdot, 0, \phi)\|^{[0,N]} \le B_1\,. \qquad (8.6.17)$$

Let

$$B_2 = B_1 + B \quad \text{and find} \quad L > L_B \quad \text{with} \quad |\mathbf{f}(t, \phi)| < L$$
$$\text{if} \quad t \ge 0 \quad \text{and} \quad \|\phi\| \le B_2\,. \qquad (8.6.18)$$

By (iii), there is a $g \in G$ such that P_t is continuous in the g-norm on

$$\mathbf{\Omega} = \left\{\phi \in X : \|\phi\| \le B_2,\ |\phi(u) - \phi(v)| \le L|u - v|\right.$$
$$\left.\text{for } u, v \in R^-\right\} \qquad (8.6.19)$$

where B_2 and L are defined in (8.6.18).

By the U.U.B., for the B_2 of (8.6.18) there exists $K > 0$ such that

$$\left\{\phi \in X,\ \|\phi\| \le B_2,\ t \ge K\right] \quad \text{imply that} \quad |\mathbf{x}(t, 0, \phi)| \le B\,. \qquad (8.6.20)$$

As $\mathbf{P}_K$ is continuous on the compact set $\mathbf{\Omega}$ in the g-norm, it is uniformly continuous; thus, there exists a $\delta > 0$ such that

$$\left[\phi_1, \phi \in \mathbf{\Omega},\ |\phi - \phi_1|_g < 2\delta\right] \quad \text{imply that}$$
$$\sup_{0 \le t \le K}\ |\mathbf{x}(t, 0, \phi) - \mathbf{x}(t, 0, \phi_1)| < B/2\,. \qquad (8.6.21)$$

In view of (8.6.17) and (8.6.21),

$$\left[\boldsymbol{\phi} \in S_B\,,\ \boldsymbol{\phi}_1 \in \boldsymbol{\Omega}\,,\ |\boldsymbol{\phi} - \boldsymbol{\phi}_1|_g < 2\delta\right] \quad \text{imply that}$$

$$\sup_{0 \le t \le K} |\mathbf{x}(t, 0, \boldsymbol{\phi}_1)| \le \sup_{0 \le t \le K} |\mathbf{x}(t, 0, \boldsymbol{\phi})| + \sup_{0 \le t \le K} |\mathbf{x}(t, 0, \boldsymbol{\phi}) - \mathbf{x}(t, 0, \boldsymbol{\phi}_1)|$$

$$\le B_1 + (B/2) < B_2\,.$$

Now define

$$S_2 = \left\{\boldsymbol{\phi} \in X : \|\boldsymbol{\phi}\| \le B_2\,,\ |\boldsymbol{\phi}(u) - \boldsymbol{\phi}(v)| \le L|u - v|\,,\ u, v \in R^-\right\},$$

$$Q_1 = \bigcup_{\boldsymbol{\phi}_1 \in S_B} \left\{\boldsymbol{\phi} \in X_g : |\boldsymbol{\phi} - \boldsymbol{\phi}_1|_g < 2\delta\right\},$$

$$Q_0 = \bigcup_{\boldsymbol{\phi}_1 \in S_B} \left\{\boldsymbol{\phi} \in X_g : |\boldsymbol{\phi} - \boldsymbol{\phi}_1|_g \le \delta\right\},$$

$$S_1 = Q_1 \cap S_2\,,$$

$$S_0 = Q_0 \cap S_2\,,$$

Now S_2 is compact in $(X_g, |\cdot|_g)$, while Q_1 is open in $(X_g, |\cdot|_g)$. We will show that Q_0 is closed in $(X_g, |\cdot|_g)$ since S_B is a compact set. Thus, S_1 is open relative to S_2 and S_0 is compact. Moreover, it can be verified that $S_0 \subset S_1 \subset S_2$ are all convex.

To see that Q_0 is closed in $(X_g, |\cdot|_g)$, let $\{\boldsymbol{\psi}_n\} \subset Q_0$ and $|\boldsymbol{\psi}_n - \boldsymbol{\psi}|_g \to 0$ as $n \to \infty$ for some $\boldsymbol{\psi} \in X_g$. For each $\boldsymbol{\psi}_n$, there exists a $\boldsymbol{\phi}_n \in S_B$ such that $|\boldsymbol{\psi}_n - \boldsymbol{\phi}_n|_g \le \delta$. Since S_B is compact in $(X_g, |\cdot|_g)$, there exists a subsequence $\{\boldsymbol{\phi}_{n_k}\}$ of $\{\boldsymbol{\phi}_n\}$ and a $\boldsymbol{\phi} \in S_B$ such that $|\boldsymbol{\phi}_{n_k} - \boldsymbol{\phi}|_g \to 0$ as $k \to \infty$. Now

$$|\boldsymbol{\psi} - \boldsymbol{\phi}|_g \le |\boldsymbol{\psi} - \boldsymbol{\psi}_{n_k}|_g + |\boldsymbol{\psi}_{n_k} - \boldsymbol{\phi}_{n_k}|_g + |\boldsymbol{\phi}_{n_k} - \boldsymbol{\phi}|_g$$

$$\le \delta + |\boldsymbol{\psi} - \boldsymbol{\psi}_{n_k}|_g + |\boldsymbol{\phi} - \boldsymbol{\phi}_{n_k}|_g\,.$$

Letting $k \to \infty$ yields $|\boldsymbol{\psi} - \boldsymbol{\phi}|_g \le \delta$. This implies that $\boldsymbol{\psi} \in Q_0$ and Q_0 is closed in $(X_g, |\cdot|_g)$.

Define $P : S_2 \to X_g$ by

$$P(\boldsymbol{\phi}) = \mathbf{x}_T(\cdot, 0, \boldsymbol{\phi}) \quad \text{for} \quad \boldsymbol{\phi} \in S_2\,. \tag{8.6.22}$$

That is $\mathbf{P} = \mathbf{P}_T$ in terms of the notation above. In preparation for part (a) of Horn's theorem, we now show that $P^j S_1 \subset S_2$ for $j = 1, 2, \ldots$.

For every $\phi \in S_1$ there is a $\phi_1 \in S_B$ such that $|\phi - \phi_1|_g < 2\delta$. Thus, by (8.6.17), (8.6.21), and the fact that $B_1 > B$ we have

$$\sup_{0 \le t \le K} |\mathbf{x}(t, 0, \phi)| \le \sup_{0 \le t \le K} |\mathbf{x}(t, 0, \phi_1)| + \sup_{0 \le t \le K} |\mathbf{x}(t, 0, \phi) - \mathbf{x}(t, 0, \phi_1)|$$

$$\le B_1 + (B/2) < B_2 \,.$$

Also, $\phi \in S_1$ implies that $\|\phi\| \le B_2$ which, together with (8.6.20), yields $|\mathbf{x}(t, 0, \phi)| \le B$ for $t \ge K$. Moreover, $|\mathbf{f}(t, \mathbf{P}_t(\phi))| \le L$ for $t \ge 0$ by (8.6.18). As $\mathbf{P}^j(\phi) = \mathbf{P}_{jT}(\phi) = \mathbf{x}_{jT}(\cdot, 0, \phi)$, it is clear that $\mathbf{P}^j(\phi) \in S_2$ for $j = 1, 2, \dots$.

Next, we find an m and J with $\mathbf{P}^j(S_1) \subset S_0$ for $m+J \le j \le 2(m+J)-1$. First, there is a $J > 0$ such that $4B_2 < \delta g(-JT)$ where δ is defined just before (8.6.21). Use the fact that $\mathbf{P}_{JT}$ is continuous on the compact set $\mathbf{\Omega}$ [see (8.6.19)] to find a $\mu > 0$ such that

$$\left[\phi, \phi_1 \in \mathbf{\Omega}, \ |\phi - \phi_1|_g < \mu, \ 0 \le t \le JT \right] \text{ imply that}$$
$$|\mathbf{x}(t, 0, \phi) - \mathbf{x}(t, 0, \phi_1)| \le \min\{\delta, B\}/2 \,. \tag{8.6.23}$$

Find $H > 0$ such that $4B_2 < \mu g(-HT)$. By (8.6.20) we have

$$\left[\phi \in \mathbf{\Omega}, \ \mathbf{P}_{kT}(\phi) \in \mathbf{\Omega} \ \text{ for } \ k = 0, 1, 2, \dots, \ mT > K + HT \,, \right.$$
$$\left. -HT \le \theta \le 0 \right] \text{ imply that } |\mathbf{x}(mT + \theta, 0, \phi| \le B \,. \tag{8.6.24}$$

Define

$$\bar{\mathbf{x}}(\theta) = \begin{cases} \mathbf{x}(mT + \theta, 0, \phi) & \text{if } -HT \le \theta \le 0 \\ \mathbf{x}(mT - HT, 0, \phi) & \text{if } -\infty < \theta \le -HT \,. \end{cases}$$

Then

$$|\bar{\mathbf{x}} - \mathbf{P}_{mT}(\phi)|_g = \sup_{\theta \le -HT} |\bar{\mathbf{x}}(\theta) - \mathbf{P}_{mT}(\phi)|/g(\theta)$$

$$\le 2B_2/g(-HT) < \mu/2 \tag{8.6.25}$$

by choice of H. This implies that

$$|\mathbf{x}(t, 0, \bar{\mathbf{x}}) - \mathbf{x}(t, 0, \mathbf{P}_{mT}(\phi))| < \min\{\delta, B\}/2 \quad \text{on} \quad [0, JT] \tag{8.6.26}$$

by (8.6.23) since (8.6.25) holds, $\|\bar{\mathbf{x}}\| \le B$, $B_2 > 2B$, and so $\mathbf{P}_{mT}(\phi)$ and $\bar{\mathbf{x}}$ are both in $\mathbf{\Omega}$. This yields

$$|\mathbf{x}(t, 0, \bar{\mathbf{x}}| \le (B/2) + |\mathbf{x}(t, 0, \mathbf{P}_{mT}(\phi))| < 2B \quad \text{on} \quad [0, JT] \tag{8.6.27}$$

since $\mathbf{x}(t, 0, \mathbf{P}_{mT}(\phi)) = \mathbf{x}(t + mT, 0, \phi)$. Hence,

$$|\mathbf{x}'(t, 0, \bar{\mathbf{x}}| = |\mathbf{f}(t, \mathbf{P}_t(\bar{\mathbf{x}}))| \le L_B \quad \text{on} \quad [0, JT] \tag{8.6.28}$$

by (8.6.14).

Let

$$\mathbf{y}(t) = \begin{cases} \mathbf{x}(\bar{0}) & \text{for } t \leq 0 \\ \mathbf{x}(t, 0, \bar{\mathbf{x}}) & \text{for } 0 \leq t \leq JT. \end{cases} \tag{8.6.29}$$

It follows that $\mathbf{y}_{JT} \in S_B$ by (8.6.27) [see (8.6.15)] and that for $\phi \in S_2$ then

$$\begin{aligned}
&\left| \mathbf{P}_{(m+J)T}(\phi) - \mathbf{y}_{JT} \right|_g \\
&= \left| \mathbf{P}_{JT}(\mathbf{P}_{mT}(\phi)) - \mathbf{y}_{JT} \right|_g \\
&= \sup_{\theta \leq 0} \left| \mathbf{x}(JT + \theta, 0, \mathbf{P}_{mT}(\phi)) - \mathbf{y}(JT + \theta) \right| / g(\theta) \\
&\leq \sup_{\theta \leq -JT} \left| \mathbf{x}(JT + \theta, 0, \mathbf{P}_{mT}(\phi)) - \mathbf{y}(JT + \theta) \right| / g(\theta) \\
&\quad + \sup_{-JT \leq \theta \leq 0} \left| \mathbf{x}(JT + \theta, 0, \mathbf{P}_{mT}(\phi)) - \mathbf{y}(JT + \theta) \right| / g(\theta) \\
&\leq 2B_2/g(-JT) + \sup_{-JT \leq \theta \leq 0} \left| \mathbf{x}(JT + \theta, 0, \mathbf{P}_{mT}(\phi)) - \mathbf{x}(JT + \theta, 0, \bar{\mathbf{x}}) \right| \\
&\leq (\delta/2) + \sup_{0 \leq t \leq JT} \left| \mathbf{x}(t, 0, \mathbf{P}_{mT}(\phi)) - \mathbf{x}(t, 0, \bar{\mathbf{x}}) \right| \leq \delta
\end{aligned}$$

by choice of J [see the material just before (8.6.23)] and by (8.6.26). This proves that if $\phi \in \Omega$ and $\mathbf{x}_{kT}(\cdot, 0, \phi) = \mathbf{P}_{kT}(\phi) \in \Omega$ for $k = 1, 2, \ldots$, then

$$\mathbf{x}_{(m+J)T}(\cdot, 0, \phi) = \mathbf{P}_{(m+J)T}(\phi) \in S_0 \tag{8.6.30}$$

by definition of Q_0.

In particular, if $\phi \in S_1$, then $\mathbf{P}_{(m+J)T}(\phi) \in S_0$. Now consider $\mathbf{P}_{(m+J+1)T}(\phi)$ for $\phi \in S_1$. It follows that $\mathbf{P}_T(\phi) \in S_2$ and $\mathbf{P}_{kT}(\mathbf{P}_T(\phi)) \in S_2$ for $k = 1, 2, \ldots$. By (8.6.30) we have $\mathbf{P}_{(m+J)T}(\mathbf{P}_T(\phi)) \in S_0$. But $\mathbf{P}_{(m+J+1)T}(\phi) = \mathbf{P}_{(m+J)T}(\mathbf{P}_T(\phi))$. Thus, $\mathbf{P}_{(m+J+1)T}(\phi) \in S_0$. In this way we argue that

$$P^j S_1 \subset S_2 \quad \text{for} \quad 1 \leq j \leq m + J - 1,$$

$$P^j S_1 \subset S_0 \quad \text{for} \quad m + J \leq j \leq 2(m + J) - 1.$$

Also, P is continuous in the g-norm by (iii). By Horn's theorem, there is a $\phi \in S_0$ with $\mathbf{P}\phi = \phi$. Since $\mathbf{x}(t, 0, \phi)$ and $\mathbf{x}(t + T, 0, \phi)$ are both solutions of (8.6.11) with the same initial function, by uniqueness, they are equal. This completes the proof.

Many examples of U.U.B. are to found in Arino-Burton-Haddock (1985), Burton (1985), and Burton-Dwiggins-Feng (1989).

8.7   Limit Sets and Unbounded Delays

Once more consider the system (8.1.1), which we denote by

$$\mathbf{x}' = \mathbf{F}(t, \mathbf{x}(s); \alpha \leq s \leq t), \quad t \geq 0, \tag{8.7.1}$$

with $\alpha \geq -\infty$ and $\mathbf{F}$ continuous when $t \geq 0$ and $\mathbf{x}$ is a continuous function in R^n.

The results here concern the behavior of solutions of (8.7.1) when there is a scalar functional $V(t, \mathbf{x}(\cdot))$ continuous for $t \geq 0$ and locally Lipschitz in $\mathbf{x}(\cdot)$ satisfying the relation $V'_{(8.7.1)}(t, \mathbf{x}(\cdot)) \leq 0$. The discussion closely follows Burton (1979b).

In earlier chapters we developed alternatives to the Marachkov requirement of $\mathbf{x}'(t)$ being bounded for $\mathbf{x}$ bounded. Some of those alternatives consisted of asking that $V' \leq -\delta|\mathbf{F}|$ or that V satisfied a one-sided Lipschitz condition.

The Marchkov hypothesis was acceptable in 1940 for ordinary differential equations because it was the only idea available, but it is entirely unacceptable for functional differential equations today.

For reference we state the Yoshizawa (1966, p. 60) generalization of the Marachkov result.

Theorem 8.7.1.   *Let* $\mathbf{P} : [0, \infty) \times R^n \rightarrow R^n$ *be continuous,* $V : [0, \infty) \times R^n \rightarrow [0, \infty)$ *be continuous in* t *and locally Lipschitz in* $\mathbf{x}$, $W : R^n \rightarrow [0, \infty)$ *be continuous and let* $V'(t, \mathbf{x}) \leq -W(\mathbf{x})$ *along any solution of*

$$\mathbf{x}' = \mathbf{P}(t, \mathbf{x}). \tag{8.7.2}$$

Then every bounded solution of (8.7.2) *approaches* $E = \big\{\mathbf{x} : W(\mathbf{x}) = 0\big\}$.

In the first result here, we suppose that there is a differentiable function

$$H : R^n \rightarrow (-\infty, \infty) \tag{8.7.3}$$

and a continuous function $K : [0, \infty) \rightarrow [0, \infty)$ with

$$\operatorname{grad} H(\mathbf{x}) \cdot \mathbf{F}\big(t, \mathbf{x}(s); \ \alpha \leq s \leq t\big) \leq K(M) \tag{8.7.4}$$

whenever $\mathbf{x}$ is a continuous function satisfying $|\mathbf{x}(s)| \leq M$.

Theorem 8.7.2.   *Let* V *be a nonnegative functional and* W_1 *a wedge with* $V'(t, \mathbf{x}(\cdot)) \leq -W_1(|\mathbf{x}(t)|)$ *and suppose there is a function* H *satisfying* (8.7.3) *and* (8.7.4). *Then each solution* $\mathbf{x}(t)$ *of* (8.7.1) *bounded on* $[t_0, \infty)$ *satisfies* $H(\mathbf{x}(t)) \rightarrow H(\mathbf{0})$ *as* $t \rightarrow \infty$.

Proof. First notice that the new function $H(\mathbf{x}) - H(\mathbf{0})$ also satisfies (8.7.3) and (8.7.4). Thus, we suppose $H(\mathbf{0}) = 0$. If the theorem is false, then there is a solution $\mathbf{x}(t)$ of (8.7.1) on $[t_0, \infty)$ with $|\mathbf{x}(t)| \leq M$ for some $M > 0$, an $\varepsilon > 0$, and a sequence $\{T_n\}$ tending to infinity with $|H(\mathbf{x}(T_n)) - H(\mathbf{0})| \geq \varepsilon$. Because we have supposed $H(\mathbf{0}) = 0$, we have $|H(\mathbf{x}(T_n))| \geq \varepsilon$. First suppose that there is a subsequence, say $\{T_n\}$ again, with $H(\mathbf{x}(T_n)) \geq \varepsilon$.

Now $V \geq 0$ and $V'_{(8.7.1)}(t, \mathbf{x}) \leq -W_1(|\mathbf{x}(t)|)$ so

$$0 \leq V(t, \mathbf{x}) \leq V(t_0, \boldsymbol{\phi}) - \int_{t_0}^{t} W_1(|\mathbf{x}(s)|)\, ds\,.$$

Thus, $\mathbf{x}(t)$ is not bounded strictly away from zero. There is a sequence $\{t_n\}$ tending to infinity such that $\mathbf{x}(t_n) \to \mathbf{0}$. Hence, $H(\mathbf{x}(t_n)) \to 0$. Because H is continuous, there exists $\delta > 0$ with $|H(\mathbf{x})| < \varepsilon/2$ if $|\mathbf{x}| < \delta$.

We can find sequences $\{s_n\}$ and $\{S_n\}$ increasing to infinity with $H(\mathbf{x}(s_n)) = \varepsilon/2$, $H(\mathbf{x}(S_n)) = 3\varepsilon/4$, and $|\mathbf{x}(t)| \geq \delta$ if $s_n \leq t \leq S_n$. That is, $H(\mathbf{x}(t))$ is continuous, and if $t_k < T_j$, then $H(\mathbf{x}(t))$ moves from near zero to near ε as t goes from t_k to T_j when k is large.

As $H'_{(8.7.1)}(\mathbf{x}) \leq K(M)$ on each interval $[s_n, S_n]$ (as can be seen from (8.7.4) and the chain rule), we obtain

$$\varepsilon/4 = H(\mathbf{x}(S_n)) - H(\mathbf{x}(s_n)) \leq K(M)(S_n - s_n)$$

or

$$S_n - s_n \geq \varepsilon/4K(M) \overset{\text{def}}{=} T\,.$$

If $H(\mathbf{x}(T_n)) \leq -\varepsilon$, then choose $\{S_n\}$ and $\{s_n\}$ with $H(\mathbf{x}(s_n)) = -3\varepsilon/4$, $H(\mathbf{x}(S_n)) = -\varepsilon/2$, and $s_n < S_n$. Proceed as before to $S_n - s_n \geq T$.

Because $V'_{(8.7.1)}(t, \mathbf{x}) \leq -W_1(|\mathbf{x}(t)|) \leq -W_1(\delta)$ for $s_n \leq t \leq S_n$, if we integrate from t_0 to $t > S_n$ we obtain

$$V(t, \mathbf{x}) \leq V(t_0, \boldsymbol{\phi}) - \int_{t_0}^{t} W_1(|\mathbf{x}(s)|)\, ds$$

$$\leq V(t_0, \boldsymbol{\phi}) - \sum_{i=1}^{n} \int_{s_i}^{S_i} W_1(\delta)\, ds$$

$$\leq V(t_0, \boldsymbol{\phi}) - nTW_1(\delta)\,,$$

which tends to $-\infty$ as $n \to \infty$. This contradiction completes the proof.

Example 8.7.1. In the scalar delay equation

$$x'(t) = -(t^2 + 4)x(t) + x(t - t/2)$$

we take

$$V(t,x) = x^2(t) + \int_{t/2}^t x^2(s)\,ds$$

and obtain

$$V'(t,x) = -2(t^2+4)x^2(t) + 2x(t)x(t/2) + x^2(t) - \tfrac{1}{2}x^2(t/2)$$
$$\leq -5x^2(t) \stackrel{\text{def}}{=} -W_4(|x(t)|).$$

The last inequality was obtained by completing the square. Then take $H(x) = x^2$ and obtain $H' \leq +2x(t)x(t/2)$, which is bounded above for $x(\cdot)$ bounded. Because $V \geq x^2$ and $V' \leq 0$, all solutions are bounded. Thus, all tend to zero.

Example 8.7.2. Suppose that J is a Liapunov function for the differential equation (8.7.2) such that $J : [0,\infty) \times R^n \to [0,\infty)$, $J'_{(8.7.2)}(t,\mathbf{x}) \leq -W_2(|\mathbf{x}(t)|)$, $J(t,\mathbf{0}) = 0$, and $J(t,\mathbf{x}) \geq W_1(|\mathbf{x}|)$. Let $\mathbf{P}(t,\mathbf{x}) = (P_1, \ldots, P_n)$ and suppose that some P_i is bounded for $\mathbf{x}$ bounded. Now $\mathbf{x} = \mathbf{0}$ is Liapunov stable. Let $\mathbf{x}(t)$ be a bounded solution of (8.7.2) and suppose $\mathbf{x}(t) = (x_1, \ldots, x_n)$. Then $x_i(t) \to 0$.

To see this, take $H(x) = x_i^2$ so that $H'_{(8.7.2)}(\mathbf{x}) = 2x_i P_i(t,\mathbf{x})$, which is bounded for $\mathbf{x}$ bounded. Hence, $x_i^2(t) \to 0$ as $t \to \infty$.

Corollary. Marachkov   *If $J(t,\mathbf{x})$ is positive definite, if $J'_{(8.7.2)}(t,\mathbf{x})$ is negative definite, and if $\mathbf{P}(t,\mathbf{x})$ is bounded for $\mathbf{x}$ bounded, then $\mathbf{x} = 0$ is asymptotically stable.*

Proof. Take $H(\mathbf{x}) = x_1^2 + \cdots + x_n^2$ and obtain $H'_{(8.7.2)}(x)$ bounded for $\mathbf{x}$ bounded. Thus, $H(\mathbf{x}(t)) \to H(\mathbf{0}) = 0$, and so bounded solutions approach zero. Because J is positive definite and $J'_{(8.7.2)} \leq 0$, $\mathbf{x} = 0$ is Liapunov stable. This completes the proof.

The effect of this corollary is that if the conditions of Marachkov's result hold, then the conditions of Theorem 8.7.2 hold. The reverse is, of course, false.

Marachkov's result yielded asymptotic stability when J was not decrescent. Results of the class of Theorem 8.7.1 were direct extensions of Marchkov's result. These extensions introduced the set E in place of $\{0\}$ as the location where J' could tend to zero. They also replaced J positive definite by the (tacit) assumption that some solutions were bounded.

This result is an extension of Marchkov's in an entirely different direction. It replaces $\mathbf{F}$ (or $\mathbf{P}$) bounded with the requirements (8.7.3) and

(8.7.4); the conclusion still involves zero instead of a general set E. The next result generalized Marachkov's result by again allowing $\mathbf{F}$ (or $\mathbf{P}$) to be unbounded, but also allowing E to be a general closed set in R^n. However, we now need restrictions on H.

The formulation here is patterned after that given by Yoshizawa (1966, pp. 116–117) for stability of a compact set E.

In the following, $N(a, E)$ denotes the a-neighborhood of a set E, $N^c(a, E)$ denotes its complement, and $d(\mathbf{x}, E)$ is the distance from $\mathbf{x}$ to E.

Definition 8.7.1. *Let $E \subset R^n$ be a closed set, U an open neighborhood of E, and let $H : [0, \infty) \times U \to [0, \infty)$ be a differentiable function. H is a pseudo-Liapunov function for (8.7.1) and E if for each compact subset K of U:*

(a) *for any $\varepsilon > 0$, there exists $\lambda > 0$ such that $H(t, \mathbf{x}) > \lambda$ for $\mathbf{x} \in K \cap N^c(\varepsilon, E)$,*

(b) *for any $\lambda > 0$, there exists $\eta > 0$ such that $H(t, \mathbf{x}) < \lambda/2$ for $\mathbf{x} \in K \cap N(\eta, E)$, and*

(c) *if $\mathbf{x}(t)$ is a bounded solution of (8.7.1) on $[t_0, \infty)$ then for $\mathbf{x}(t) \in K$ we have $\operatorname{grad} H \cdot \mathbf{F} + \partial H/\partial t$ bounded above.*

Example 8.7.3. In the system

$$x_1' = x_2$$
$$x_2' = -\psi\bigl(t, x_1(t), x_2(t)\bigr)x_2(t) - g\bigl(x_1(t - r(t))\bigr) \qquad (8.7.5)$$

let $\psi \geq 0$, ψ, g, and r be continuous, and $t \geq r(t) \geq 0$. If $E = \bigl\{(x_1, x_2) \mid x_2 = 0\bigr\}$ and $H(t, x_1, x_2) = x_2^2$, then (a) and (b) hold, and $H_{(8.7.5)}' = -2\psi(\cdot)x_2^2 - 2x_2 g\bigl(x_1(t - r(t))\bigr) \leq -2x_2 g\bigl(x_1(t - r(t))\bigr)$, so that if $x_1(t)$ is bounded on $[t_0, \infty)$, then H' is bounded above. Thus, H is a pseudo-Liapunov function for (8.7.5).

Definition. *A function $L : R^n \to [0, \infty)$ is positive definite relative to a closed set $E \subset R^n$ if, for any compact set $K \subset R^n$ and any $\varepsilon > 0$, there exists $L_0 > 0$ such that $L(\mathbf{x}) > L_0$ for $\mathbf{x} \in K \cap N^c(\varepsilon, E)$.*

Theorem 8.7.3. *Let V be a nonnegative functional with $V_{(8.7.1)}'(t, \mathbf{x}) \leq -L(\mathbf{x}(t))$, where L is positive definite relative to a closed set $E \subset R^n$. If U is an open neighborhood of E and $H : [0, \infty) \times U \to [0, \infty)$ is a pseudo-Liapunov function for (8.7.1) and E, then each bounded solution of (8.7.1) approaches E as $t \to \infty$.*

Proof. If the theorem is false, then there is a bounded solution $\mathbf{x}(t)$ of (8.7.1) that does not approach E. Then there is a compact set $\bar{K} \subset R^n$ with $\mathbf{x}(t) \in \bar{K}$ for $t_0 \leq t < \infty$. Also, there is an $\varepsilon > 0$ with $N(2\varepsilon, E) \cap \bar{K} \subset U$ and a sequence $\{\bar{t}_n\}$ tending to infinity with $d(\mathbf{x}(\bar{t}_n), E) = \varepsilon$. (That is, we first say that $d(\mathbf{x}(t_n), E) \geq \varepsilon$; but because $V' \leq -L(\mathbf{x}(t))$ we argue that $\mathbf{x}(t)$ approaches E along a sequence. Thus, we can select $\bar{t}_n$ so that equality holds.) Let $K = \bar{K} \cap \bar{N}(\varepsilon, E)$ where $\bar{N}$ is the closure of N. For this compact subset K of U and this $\varepsilon > 0$, find λ and η of parts (a) and (b) of the definition of a pseudo-Liapunov function with $0 < 2\eta < \varepsilon$.

Because $V'_{(8.7.1)}(t, \mathbf{x}) \leq -L(\mathbf{x}(t))$, there is a sequence $\{\bar{T}_n\}$ tending to infinity with $d(\mathbf{x}(\bar{T}_n), E) \to 0$ as $n \to \infty$. It is then possible to find sequences $\{t_n\}$ and $\{T_n\}$ tending to infinity with $d(\mathbf{x}(t_n), E) = \varepsilon$, $d(\mathbf{x}(T_n), E) = \eta/2$, and $\eta/2 \leq d(\mathbf{x}(t), E) \leq \varepsilon$ if $T_n \leq t \leq t_n$.

If we compute the derivative of $H(t, \mathbf{x})$ along solutions of (8.7.1) we find (using (c) in the definition of the pseudo Liapunov function) $H'_{(8.7.1)}(t, \mathbf{x}) = \text{grad } H \cdot \mathbf{F} + \partial H/\partial t$. In particular, if $T_n \leq t \leq t_n$, then $H'_{(8.7.1)}(t, \mathbf{x}) \leq P$. Integrating $H'_{(8.7.1)}(t, \mathbf{x})$ from T_n to t_n we obtain $H(t_n, \mathbf{x}(t_n)) - H(T_n, \mathbf{x}(T_n)) \leq P(t_n - T_n)$. Because $H(t_n, \mathbf{x}(t_n)) > \lambda$ and $H(T_n, \mathbf{x}(T_n)) < \lambda/2$ we have

$$\lambda - \lambda/2 \leq H(t_n, \mathbf{x}(t_n)) - H(T_n, \mathbf{x}(T_n))$$
$$\leq P(t_n - T_n)$$

or

$$t_n - T_n \geq \lambda/2P \overset{\text{def}}{=} T.$$

Because

$$V'_{(8.7.1)}(t, \mathbf{x}(\cdot)) \leq -L(\mathbf{x}(t)),$$

for $t > t_n$, we have

$$V(t, \mathbf{x}(\cdot)) \leq V(t_0, \boldsymbol{\phi}) - \int_{t_0}^{t} L(\mathbf{x}(s))\, ds$$

$$\leq V(t_0, \boldsymbol{\phi}) - \sum_{i=1}^{n} \int_{T_i}^{t_i} L(\mathbf{x}(s))\, ds$$

$$\leq V(t_0, \boldsymbol{\phi}) - \sum_{i=1}^{n} \int_{T_i}^{t_i} L_0\, ds$$

$$\leq V(t_0, \boldsymbol{\phi}) - nL_0 T$$

for some $L_0 > 0$ by definition of L being positive definite relative to E. Here, $\boldsymbol{\phi}$ is the initial function for $\mathbf{x}$. As $n \to \infty$, we contradict $V \geq 0$. This completes the proof.

Example 8.7.4. In Example 8.7.3, we can take $r(t) = 0$, $V(t, x_1, x_2) = x_2^2 + 2G(x_1)$ where $G(x_1) = \int_0^{x_1} g(s) \, ds$, $V'_{(8.7.5)} = -2\psi(t, x_1, x_2)x_2^2$, $H(t, x) = x_2^2$, and $H'_{(8.7.5)}(t, x) = -2\psi(t, x_1, x_2)x_2^2 - 2x_2g(x_1)$. If for each $M > 0$, there exists $\alpha > 0$ with $\psi(t, x_1, x_2) \geq \alpha$ when $|x_2| \leq M$ then the conditions of the theorem are satisfied and all bounded solutions approach the x_1 axis. If $G(x_1) \to \infty$ as $|x_1| \to \infty$, then all solutions are bounded. If $x_1 g(x_1) > 0$ for $x_1 \neq 0$, then an argument similar to Burton (1965) yields all solutions bounded.

Example 8.7.5. In Example 8.7.3, let $r(t) > 0$, $r'(t) \leq \alpha < b/2$, $|g'(x)| \leq L$, and $(x_1, x_2) = (x, y)$. Suppose $2\psi(t, x, y)/r(t) \geq b + a/r(t)$ for some positive constants a and b with $4L^2 \leq b^2 - 2\alpha b$.

Write the system as

$$x'(t) = y(t)$$

$$y'(t) = -\psi(t, x(t), y(t))y(t) - g(x(t))$$

$$+ \int_{-r(t)}^{0} g'(x(t+s))y(t+s) \, ds \tag{8.7.6}$$

and take

$$V(t, x(\cdot), y(\cdot)) = 2G(x(t)) + y^2(t)$$

$$+ (b/2) \int_{-r(t)}^{0} \left(\int_{s}^{0} y^2(t+u) \, du \right) ds$$

with $G(x) = \int_0^x g(s) \, ds$. We obtain

$$V'_{(8.7.6)} = -2\psi(t, x, y)y^2 + 2y \int_{-r(t)}^{0} g'(x(t+s))y(t+s) \, ds$$

$$+ (b/2) \left(\int_{-r(t)}^{0} \{y^2(t) - y^2(t+s)\} \, ds \right.$$

$$\left. + r'(t) \int_{-r(t)}^{0} y^2(t+s) \, ds \right).$$

If we use the assumption on ψ we find $V'_{(8.7.6)} \leq -ay^2(t)$.

Then $L(x, y) = -ay^2$ is negative definite relative to $E = \{(x, y) \mid y = 0\}$. Here $U = R \times R$.

Example 8.7.3 showed that $H = y^2$ is a pseudo-Liapunov function. By our result we conclude that all bounded solutions approach the x axis. If $G(x) \to \infty$ as $|x| \to \infty$, all solutions are bounded.

Theorem 8.7.3 is related to a result of Hale (1977, pp. 118–126) for autonomous, functional differential equations $x'(t) = R(x_t)$. In that reference, Hale and Somolinos apply the result to our system (8.7.6) when $r(t)$ is constant, $\psi(t, x, y) = a/r$ and $g(x) = (b/r)\sin x$. They conclude that bounded solutions approach invariant sets located on the x axis.

Corollary. *Let V be a functional with $V'_{(8.7.1)}(t, \mathbf{x}) \leq -W_1(|\mathbf{x}(t)|)$. Suppose there is a differentiable function $H : [0, \infty) \times R^n \to [0, \infty)$ with $W_2(|\mathbf{x}|) \leq H(t, \mathbf{x}) \leq W_3(|\mathbf{x}|)$ and $\operatorname{grad} H \cdot \mathbf{F} + \partial H / \partial t$ bounded above for any continuous function $\mathbf{x}(t)$ that is defined and bounded on $[0, \infty)$. Then every bounded solution of (8.7.1) tends to zero as $t \to \infty$. If, in addition, $W_4(|\mathbf{x}(t)|) \leq V(t, \mathbf{x})$ and $V(t, \mathbf{0}) = 0$, then $\mathbf{x} = \mathbf{0}$ is asymptotically stable.*

Proof. H is a pseudo-Liapunov function when $E = \{\mathbf{0}\}$ and $U = R^n$. Thus, by Theorem 8.7.3 all bounded solutions approach zero. With the additional hypotheses, the zero solution is Liapunov stable. Bounded solutions tend to zero, so we have asymptotic stability.

Example 8.7.1 is also an example of the corollary. The corollary complements Theorem 8.7.2 by allowing $H = H(t, \mathbf{x})$, but restricts the sign of H.

Remark 8.7.1. If the conditions of Theorem 8.7.1 hold for (8.7.2), then the conditions of Theorem 8.7.3 also hold. To see this, let the conditions of Theorem 8.7.1 hold and let ρ be the distance function from a point $\mathbf{x} \in R^n$ to C. Langenhop (1973) points out that $|\rho(\mathbf{x}) - \rho(\mathbf{y})| \leq |\mathbf{x} - \mathbf{y}|$, a global Lipschitz condition, and if E is convex, then ρ^2 is actually differentiable. Let $H(t, \mathbf{x}) = \rho(\mathbf{x})$ and compute $\rho'_{(8.7.2)}$. We have

$$\left| \lim_{h \to 0^+} (1/h)\{\rho(\mathbf{x} + h\mathbf{P}(t, \mathbf{x})) - \rho(\mathbf{x})\} \right|$$
$$\leq \lim_{h \to 0^+} (1/h)\left| \mathbf{x} + h\mathbf{P}(t, \mathbf{x}) - \mathbf{x} \right| = |\mathbf{P}(t, \mathbf{x})|,$$

which is bounded for $\mathbf{x}$ bounded. Thus, we easily see that ρ is a pseudo-Liapunov function for E and (8.7.2).

We turn now to the system

$$\mathbf{x}' = \mathbf{G}(t, \mathbf{x}_t) \qquad (8.7.7)$$

and obtain a generalization of the classical result of Yoshizawa (1966, p. 191), retaining the notation of Section 8.3.

We ask that $G : [0, \infty) \times C_M \to R^n$ be continuous and take bounded sets into bounded sets.

It has become standard to ask that $V'_{(8.7.7)}(t, \mathbf{x}_t) \leq -W_4(|\mathbf{x}(t)|)$. It should be clear in the following proof that a similar result and proof hold if $V'_{(8.7.7)}(t, \mathbf{x}_t) \leq -W_4(\|\mathbf{x}_t\|_{(h_1,h_2)})$, where

$$\|\mathbf{x}_t\|_{(h_1,h_2)} = \sup_{-h_1 \leq \theta \leq -h_2} |\mathbf{x}_t(\theta)|$$

and $-h \leq -h_1 \leq -h_2 \leq 0$. The latter was Yoshizawa's form.

Theorem 8.7.4.   *Let* $V : [0, \infty) \times C_M \to [0, \infty)$ *be continuous, satisfy a Lipschitz condition in the second argument, and let* $W_1(|\mathbf{x}(t)|) \leq V(t, \mathbf{x}_t) \leq W_2(\|\mathbf{x}_t\|)$ *with* $V'_{(8.7.7)}(t, \mathbf{x}_t) \leq -W_4(|\mathbf{x}(t)|)$. *Let* $E = \{\mathbf{0}\}$, U *be a neighborhood of* E *in* R^n, $H : [0, \infty) \times U \to [0, \infty)$ *with* $W_5(|\mathbf{x}|) \leq H(t, \mathbf{x}) \leq W_6(|\mathbf{x}|)$. *If* $H'_{(8.7.7)}(t, \mathbf{x})$ *is bounded above for* $\mathbf{x} \in U$ *and* $\mathbf{x}_t \in C_M$, *then* $\mathbf{x} = \mathbf{0}$ *is uniformly asymptotically stable for* (8.7.7).

Remark 8.7.2.   When $\mathbf{G}(t, \mathbf{x}_t)$ is bounded for $\mathbf{x}_t \in C_M$, then Yoshizawa's hypotheses for U.A.S. are identical to those in the first sentence of this theorem. The present result replaces $\mathbf{G}$ bounded by the existence of H with H' bounded above. To see that Yoshizawa's result is a corollary of Theorem 8.7.4, if $\mathbf{G}$ is bounded for $\mathbf{x}_t \in C_M$ take $H(t, \mathbf{x}) = x_1^2 + \cdots + x_n^2$ and obtain $H'_{(8.7.7)}(t, \mathbf{x}) \leq |G(t, \mathbf{x}_t)|(2nM)$ for $|\mathbf{x}| \leq M$ and $\mathbf{x}_t \in C_M$.

Proof of Theorem 8.7.4. Let $\varepsilon > 0$ be given with $\varepsilon < M$. There exists $\delta > 0$ with $W_2(\delta) < W_1(\varepsilon)$. If $t \geq t_0$ and if $\|\phi\| < \delta$, then for $\mathbf{x}(t) = \mathbf{x}(t, t_0, \phi)$ we have $W_1(|\mathbf{x}(t)|) \leq V(t, \mathbf{x}_t) \leq V(t_0, \phi) \leq W_2(\|\phi\|) < W_2(\delta) < W_1(\varepsilon)$, so $|\mathbf{x}(t)| < \varepsilon$. This shows uniform stability.

By the uniform stability, $\eta > 0$ may be chosen so that $t_0 \geq 0$, $t \geq t_0$, and $\|\phi\| < \eta$ implies $|\mathbf{x}(t, t_0, \phi)| < M/2$.

Now let $\varepsilon > 0$ be given. To complete the proof we must find $T > 0$ such that $t_0 \geq 0$, $\|\phi\| < \eta$, and $t \geq t_0 + T$ imply $|\mathbf{x}(t, t_0, \phi)| < \varepsilon$. For this $\varepsilon > 0$, find $\delta > 0$ by uniform stability so that $t_0 \geq 0$, $\|\phi\| < \delta$, and $t \geq t_0$ imply $|\mathbf{x}(t, t_0, \phi)| < \varepsilon$.

Let $t_0 \geq 0$ be arbitrary and consider intervals

$$I_1 = [t_0, t_0 + h],$$
$$I_2 = [t_0 + h, t_0 + 2h],$$
$$\vdots$$
$$I_{k+1} = [t_0 + kh, t_0 + (k+1)h], \ldots.$$

If $\|\phi\| < \eta$ and $\mathbf{x}(t) = \mathbf{x}(t, t_0, \phi)$, then for each i either $|\mathbf{x}(t)| < \delta$ on I_1 (and hence, $|\mathbf{x}(t)| < \varepsilon$ for $t \geq t_0 + ih$) or there exists $t_i \in I_i$ with $|\mathbf{x}(t_i)| \geq \delta$.

Now there exists $\gamma > 0$ such that $W_6(\gamma) < W_5(\delta)$. Also, there exists $\bar{T} > 0$ such that if $|\mathbf{x}(t_i)| \geq \delta$ then $|\mathbf{x}(t)| \geq \gamma$ for $t_i - \bar{T} \leq t \leq t_i$. To see this, if for some $\bar{t}_i < t_i$ we have $|\mathbf{x}(\bar{t}_i)| = \gamma$, $|\mathbf{x}(t_i)| \geq \delta$, and $\gamma \leq |\mathbf{x}(t)|$ for $\bar{t}_i \leq t \leq t_i$, then we have $H'_{(8.7.7)}(t, \mathbf{x}(t)) \leq P$ for some $P > 0$, so

$$W_5(|\mathbf{x}(t_i)|) \leq H(t_i, \mathbf{x}(t_i))$$

$$\leq H(\bar{t}_i, \mathbf{x}(\bar{t}_i)) + P(t_i - \bar{t}_i)$$

$$\leq W_6(|\mathbf{x}(\bar{t}_i)|) + P(t_i - \bar{t}_i)$$

or

$$W_5(\delta) \leq W_6(\gamma) + P(t_i - \bar{t}_i) \,,$$

so that

$$\bar{T} \stackrel{\text{def}}{=} [W_5(\delta) - W_6(\gamma)]/P \leq t_i - \bar{t}_i \,.$$

Let $Q = \min[\bar{T}, h]$, pick $N > W_2(\eta)/W_4(\gamma)Q$, and choose $T = 2Nh$. We now show that $|\mathbf{x}(t)| < \delta$ for some t in $[t_0, t_0 + T]$, so that $|\mathbf{x}(t)| < \varepsilon$ for $t \geq t_0 + T$. If such a t does not exist, then in each I_i there is a t_i with $|\mathbf{x}(t_i)| \geq \delta$, and hence, $|\mathbf{x}(t)| \geq \gamma$ for $t_i - Q \leq t \leq t_i$. Thus $V'_{(8.7.7)}(t, \mathbf{x}_t) \leq -W_4(\gamma)$ for $t_i - Q \leq t \leq t_i$. Let $t \geq t_{2N}$ so that

$$V(t, \mathbf{x}_t) \leq V(t_0, \boldsymbol{\phi}) - \sum_{i=1}^{N} \int_{t_{2i}-Q}^{t_{2i}} W_4(\gamma)\, dt$$

$$= V(t_0, \boldsymbol{\phi}) - NQW_4(\gamma)$$

$$\leq W_2(\eta) - NQW_4(\gamma) < 0 \,,$$

a contradiction. In our integration it was necessary to skip the intervals $[t_i - Q, t_i]$ with i odd to avoid possible overlap.

Example 8.7.6. Consider the scalar equation

$$x'(t) = -a(t)x(t) + b(t)x(t - r(t))$$

with $a(t) \geq a_0 > 0$, $r(t) \geq 0$, $r'(t) \leq \alpha < 1$, $r(t) \leq h$ for some $h > 0$, and $|b(t)| \leq a_1\sqrt{1-\alpha} < a_0\sqrt{1-\alpha}$ for some $a_1 > 0$.

Take

$$V(t, x(\cdot)) = x^2(t) + a_0 \int_{t-r(t)}^{t} x^2(s)\, ds$$

so that

$$
\begin{aligned}
V' &= -2a(t)x^2(t) + 2b(t)x(t)x(t - r(t)) \\
&\quad + a_0 \big[x^2(t) - x^2(t - r(t))(1 - r'(t))\big] \\
&\leq [-2a(t) + a_0]x^2(t) + 2\big|b(t)\big|\,\big|x(t)x(t - r(t))\big| \\
&\quad + a_0(\alpha - 1)x^2(t - r(t)) \\
&\leq -a_0 x^2(t) + 2\big|b(t)\big|\,\big|x(t)x(t - r(t))\big| \\
&\quad + a_0(\alpha - 1)x^2(t - r(t)) \\
&\leq -a_0\big\{1 - \big[b^2(t)/a_0^2\,(1 - \alpha)\big]\big\}x^2(t)\,,
\end{aligned}
$$

which can be seen by completing the square. This yields $V'(t,x) \leq -\gamma x^2$ for some $\gamma > 0$. We take $H = x^2$ and have $H' \leq 2a_1\sqrt{1 - \alpha}\,\big|xx(t - r(t))\big|$. The conditions of Theorem 8.7.4 are satisfied and $x = 0$ is U.A.S.

8.8   Liapunov Theory for Integral Equations

Prior to 1992 Liapunov theory for integral equations was limited to cases in which the integral equation could be differentiated. This is discussed in some detail in Miller (1971, p. 337), Lakshmikantham and Leela (1959), and Gripenberg *et al.* (1990, p. 426). But two things go wrong in this process. First, it may be that the functions involved in the equation are not differentiable. But even if a function is differentiable, the process can produce a far more complicated equation since differentiation tends to destroy smoothness. The proper place to treat an integral equation is in its given form.

In 1992 we constructed a number of examples of Liapunov functions for integral equations [see Burton (1996)]. Kato (1994) significantly improved the process by showing that Liapunov functions for integral equations could be obtained by constructing a Liapunov function for an associated differential equation, about which much is known. That differential equation was not the derivative of the integral equation. There then appeared a long series of papers establishing a theory of Liapunov's direct method for integral equations of both ordinary and partial type. Examples and bibliographies may be found in Burton-Furumochi (1994) and in Burton-Furumochi-Huang (1995).

References

Antosiewicz, H. A. (1958). A survey of Lyapunov's second method. *Ann. Math. Studies* **41**, 141–166.

Appleby, J. A. D. and Reynolds, D. W. (2002). Subexponential solutions of linear Volterra integro-differential equations and transient renewal equations. *Roy. Soc. Edinburgh Proc. A* **132A**, 521–543.

Appleby, J. A. D. and Reynolds, D. W. (2003). Corrigendum: Subexponential solutions of linear Volterra integro-differential equations and transient renewal equations. *Roy. Soc. Edinburgh Proc. A* **133A**, 1421.

Appleby, J. A. D. and Reynolds, D. W. (2004). On the non-exponential decay to equilibrium of solutions of nonlinear scalar Volterra integro-differential equations. *E. J. Qualitative Theory of Diff. Equ.(http://www.math.u-szeged.hu/ejqtde), Proc. 7th Coll. Qualitative Theory of Diff. Equ.* **3**, 1-14.

Arino, O. A., Burton, T. A., and Haddock, J. R. (1985). Periodic solutions of functional differential equations. *Roy. Soc. Edinburgh Proc. A* **101A**, 253–271.

Banach, S. (1932). "Théorie des Opérations Linéairs" (reprint of the 1932 ed.). Chelsea, New York.

Barbashin, E. A. (1968). The construction of Lyapunov functions. *Differential Equations* **4**, 1097–1112.

Becker, L. C. (1979). Stability considerations for Volterra integro-differential equations. Ph.D. dissertation. Southern Illinois University, Carbondale, Illinois.

Becker, L. C., Burton, T. A., and Krisztin, T. (1988). Floquet theory for a Volterra equation. *J. London Math. Soc.* **37**, 141–147.

Becker, L. C., Burton, T. A., and Zhang, S. (1989). Functional differential equations and Jensen's inequality. *J. Math. Anal. Appl.* **138**, 137–156.

Bellman, R. and Cooke, K. L. (1963). "Differential-Difference Equations." Academic Press, New York.

Belousov, B. P. (1959). A periodic reaction and its mechanism. *Ref. Radiat. Med. Medgiz*, p. 145.

Bloom, F. (1981). "Ill-posed Problems for Integrodifferential Equations in Mechanics and Electromagnetic Theory." SIAM, Philadelphia.

Brauer, F. (1978). Asymptotic stability of a class of integrodifferential equations. *J. Differential Equations* **28**, 180–188.

Burton, T. A. (1965). The generalized Liénard equation. *SIAM J. Control* **3**, 223–230.

Burton, T. A. (1977). Differential inequalities for Liapunov equations. *Nonlinear Analysis: Theory, Methods, and Appl.* **1**, 331–338.

Burton, T. A. (1978). Uniform asymptotic stability in functional differential equations. *Proc. Amer. Math. Soc.* **68**, 195–199.

Burton, T. A. (1979a). Stability theory for delay equations. *Funkcial. Ekvac.* **22**, 67–76.

Burton, T. A. (1979b). Stability theory for functional differential equations. *Trans. Amer. Math. Soc.* **255**, 263–275.

Burton, T. A. (1979c). Stability theory for Volterra equations. *J. Differential Equations* **32**, 101–118.

Burton, T. A. (1980a). Uniform stabilities for Volterra equations. *J. Differential Equations* **36**, 40–53.

Burton, T. A. (1980b). ed. "Modeling and Differential Equations in Biology." Dekker, New York.

Burton, T. A. (1980c). An integrodifferential equation. *Proc. Amer. Math. Soc.* **79**, 393–399.

Burton, T. A. (1982a). Construction of Liapunov functionals for Volterra equations. *J. Math. Anal. Appl.* **85**, 90–105.

Burton, T. A. (1982b). Perturbed Volterra equations. *J. Differential Equations* **43**, 168–183.

Burton, T. A. (1982c). Boundedness in functional differential equations. *Funkcial. Ekvac.* **25**, 51–77.

Burton, T. A. (1983a). Volterra equations with small kernels. *J. Integral Equations* **5**, 271–285.

Burton, T. A. (1983b). "Volterra Integral and Differential Equations." Academic Press, Orlando

Burton, T. A. (1985). "Stability and Periodic Solutions of Ordinary and Functional Differential Equations." Academic Press, Orlando.

Burton, T. A. (1987). Liapunov's direct method for delay equations. Proc. Eleventh Conf. Nonlin. Oscilations, Budapest. M. Farkas, V. Kertesz, and G. Stepan, eds. pp. 26–33.

Burton, T. A. (1990). Uniform boundedness for delay equations. *Acta Math. Hung.* **56**, 259–268.

Burton, T. A. (1996). Examples of Lyapunov functionals for non-differentiated equations. Proc. First World Congress Nonlinear Analysts, 1992. V. Lakshmikantham, ed. Walter de Gruyter, New York. pp. 1203–1214.

Burton, T. A. (2003a). Stability by fixed point theory or Liapunov theory: a comparison. *Fixed Point Theory* **4**, 15–32.

Burton, T. A. (2003b). Perron-type stability theorems for neutral equations. *Nonlinear Analysis* **55**, 285–297.

Burton, T. A. (2005). Fixed points, Volterra equations, and Becker's resolvent. *Acta Math. Hungar.* **109**, to appear.

Burton, T. A., Casal, A., and Somolinos, A. (1987). Krasovskii's stability theory. *in* Nonlinear Analysis and Applications. V. Lakshmikantham, ed. Marcel Dekker, New York. pp. 99–104.

Burton, T. A., Casal, A., and Somolinos, A. (1989). Upper and lower bounds for Liapunov functionals. *Funkcial. Ekvac.* **32**, 23–55.

Burton, T. A., Dwiggins, D. P., and Feng, Y. (1989). Periodic solutions of functional differential equations. *J. London Math. Soc.* **40**, 81–88.

Burton, T. A., and Feng, Y. (1991). Continuity in functional differential equations. *Acta Mathematica Applicatae Sinica* **7**, 229–244.

Burton, T. A., and Furumochi, Tetsuo (1994). A stability theory for integral equations. *J. Integral Equations and Applications* **6**, 445–477.

Burton, T. A., and Furumochi, Tetsuo (2001). Fixed points and problems in stability theory for ordinary and functional differential equations. *Dynamical Systems and Appl.* **10**, 89–116.

Burton, T. A., Furumochi, Tetsuo, and Huang, Qichang (1995). Stability in several measures and a differential inequality for a partial integral equation. *Nonlinear Analysis* **25**, 885–898.

Burton, T. A., and Grimmer, R. (1971). On continuability of solutions of second order differential equations. *Proc. Amer. Math. Soc.* **29**, 277–283.

Burton, T. A., and Grimmer, R. (1972). On the asymptotic behavior of solutions of $x'' + a(t)f(x) = e(t)$. *Pacific J. Math.* **41**, 43–55.

Burton, T. A., and Grimmer, R. (1973). Oscillation, continuation, and uniqueness of solutions of retarded differential equations. *Trans. Amer. Math. Soc.* **179**, 193–209.

Burton, T. A., and Grimmer, R. (1974). Erratum to Oscillation, continuation, and uniqueness of solutions of retarded differential equations. *Trans. Amer. Math. Soc.* **187**, 429.

Burton, T. A., and Hatvani, L. (1989). Stability theorems for nonautonomous functional differential equations by Liapunov functionals. *Tohoku Math. J.* **41**, 65–104.

Burton, T. A., and Hatvani, L. (1990). On nonuniform asymptotic stability for nonautonomous functional differential equations. *Differential and Integral Equations* **3**, 285–293.

Burton, T. A., and Hering, R. H. (1994). Liapunov theory for functional differential equations. *Rocky Mountain J. Math.* **24**, 3–17.

Burton, T. A., and Mahfoud, W. E. (1983) Stability criteria for Volterra equations. *Trans. Amer. Math. Soc.* **279**, 143–174.

Burton, T. A., and Mahfoud, W. E. (1985) Stability by decompositions for Volterra equations. *Tohoku Math. J.* **37**, 489–511.

Burton, T. A., and Makay, G. (1994). Asymptotic stability for functional differential equations. *Acta Math. Hungar.* **65**, 243–251.

Burton, T. A., and Townsend, C. G. (1968). On the generalized Liénard equation with forcing function. *J. Differential Equations* **4**, 620–633.

Burton, T. A., and Townsend, C. G. (1971). Stability regions of the forced Liénard equation. *J. London Math. Soc.* **3**, 393–402.

Burton, T. A., and Zhang, B. (1990). Uniform ultimate boundedness and periodicity in functional differential equations. *Tohoku Math. J.* **42**, 93–100.

Churchill, R. V. (1958). "Operational Mathematics." McGraw-Hill, New York.

Clark, C. W. (1976). "Mathematical Bioeconomics." Wiley, New York.

Corduneanu, C. (1971). "Principles of Differential and Integral Equations." Allyn and Bacon, Rockledge, New Jersey.

Corduneanu, C. (1991). "Integral Equations and Applications." Cambridge Univ. Press, Cambridge, U.K.

Cushing, J. M. (1976). Forced asymptotically periodic solutions of predator-prey systems with or without hereditary effects. *SIAM J. Appl. Math.* **30**, 665–674.

Davis, H. T. (1962). "Introduction to Nonlinear Differential and Integral Equations." Dover, New York.

Davis, P. L. (1975). Hyperbolic integrodifferential equations arising in the electromagnetic theory of dielectrics. *J. Differential Equations* **18**, 170–178.

Driver, R. D. (1962). Existence and stability of solutions of a delay-differential system. *Arch. Rational Mech. Anal.* **10**, 401–426.

Driver, R. D. (1963). Existence theory for a delay-differential system. *Contrib. Differential Equations* **1**, 317–335.

El'sgol'ts, L. E. (1966). "Introduction to the Theory of Differential Equations with Deviating Arguments." Holden-Day, San Francisco.

Erhart, J. (1973). Lyapunov theory and perturbations of differential equations. *SIAM J. Math. Anal.* **4**, 417–432.

Feller, W. (1941). On the integral equation of renewal theory. *Ann. Math. Statist.* **12**, 243–267.

Fink, A. M. (1974). "Almost Periodic Differential Equations." Springer Publ., New York.

Gantmacher, F. R. (1960). "The Theory of Matrices," Vol. II. Chelsea, Bronx, New York.

Goel, N. S., Maitra, S. C., and Montroll, E. W. (1971). "On the Volterra and other Nonlinear Models of Interacting Populations." Academic Press, New York.

Graef, J. R. (1972). On the generalized Liénard equation with negative damping. *J. Differential Equations* **12**, 34–62.

Grimmer, R. (1979). Existence of periodic solutions of functional differential equations. *J. Math. Anal. Appl.* **72**, 666–673.

Grimmer, R., and Seifert, G. (1975). Stability properties of Volterra integrodifferential equations. *J. Differential Equations* **19**, 142–166.

Gripenberg, G., Londen, S. O., and Staffans, O. (1990). "Volterra Integral and Functional Equations." Cambridge Univ. Press, Cambridge, U.K.

Grossman, S. I., and Miller, R. K. (1970). Perturbation theory for Volterra integrodifferential systems. *J. Differential Equations* **8**, 457–474.

Haddock, J. R. (1972a). A remark on a stability theorem of Marachkoff. *Proc. Amer. Math. Soc.* **31**, 209–212.

Haddock, J. R. (1972b). Liapunov functions and boundedness and global existence of solutions. *Applicable Analysis* **1**, 321–330.

Haddock, J. R. (1974). On Liapunov functions for nonautonomous systems. *J. Math. Anal. Appl.* **47**, 599–603.

Haddock, J. R., and Terjéki, J. (1983). Liapunov-Razumikhin functions and an invariance principle for functional differential equations. *J. Differential Equations* **48**, 95–122.

Hahn, W. (1963). "Theory and Application of Liapunov's Direct Method." Prentice-Hall, Englewood Cliffs, New Jersey.

Halanay, A., and Yorke, J. (1971). Some new results and problems in the theory of differential delay equations. *SIAM Rev.* **13**, 55–80.

Hale, J. K. (1963). "Oscillations in Nonlinear Systems." McGraw-Hill, New York.

Hale, J. K. (1965) Sufficient conditions for stability and instability of autonomous functional-differential equations. *J. Differential Equations* **1**, 452–482.

Hale, J. K. (1969). "Ordinary Differential Equations." Wiley, New York.

Hale, J. K. (1971). "Functional Differential Equations." Springer Publ., New York.

Hale, J. K. (1977). "Theory of Functional Differential Equations." Springer Publ., New York.

Hartman, P. (1964). "Ordinary Differential Equations." Wiley, New York.

Hatvani, L. (1978). Attractivity theorems for non-autonomous systems of differential equations. *Acta Sci. Math.* (Szeged) **40**, 271–283.

Hatvani, L. (1990). On the asymptotic stability of the solutions of functional differential equations. *in* Differential Equations: Qualitative Theory, *Colloq. Math. Soc. Janos Bolyai* **53**, North-Holland, Amsterdam. pp. 227–238.

Hatvani, L. (1996). On the asymptotic stability in differential systems by Lyapunov direct method. *in* Proc. World Congress of Nonlinear Analysts. V. Lakshmikantham, ed. pp. 1341–1348.

Hatvani, L. (1997a). On the asymptotic stability by Lyapunov functionals with semidefinite derivatives. *Nonlinear Analysis* **30**, 4713–4721.

Hatvani, L. (1997b). Annulus arguments in the stability theory for functional differential equations. *Differential and Integral Equations* **10**, 975–1002.

Hatvani, L. (2000). On the asymptotic stability for functional differential equations by Lyapunov functionals. *Nonlinear Analysis* **40**, 251–263.

Hatvani, L. (2002). On the asymptotic stability for nonautonomous functional differential equations by Lyapunov functionals. *Trans. Amer. Math. Soc.* **354**, 3555–3571.

Hatvani, L., and Krisztin, T. (1995). Asymptotic stability for a differential-difference equation containing terms with and without a delay. *Acta Sci. Math.* (Szeged) **60**, 371–384.

Herdman, T. L. (1980). A note on noncontinuable solutions of a delay differential equation. *in* "Differential Equations" (S. Ahmad, M. Keener, and A. C. Lazer eds.). Academic Press, New York.

Hino, Y., and Murakami, S. (1991a). Stability properties of linear Volterra equations. *J. Differential Equations* **89**, 121–137.

Hino, Y., and Murakami, S. (1991b). Total stability and uniform asymptotic stability for linear Volterra equations. *J. London Math. Soc.* **43**, 305–312.

Hino, Y., and Murakami, S. (1996). Stabilities in linear integrodifferential equations. *Lecture Notes in Numerical and Applied Analysis* **15**, 31–46.

Hopkinson, J. (1877). The residual charge of the Leyden jar, *Philos. Trans. Roy. Soc. London* **167**, 599–626.

Horn, W. A. (1970). Some fixed point theorems for compact maps and flows in Banach spaces. *Trans. Amer. Math. Soc.* **149**, 391–404.

Kamke, E. (1930). "Differentialgleichungen reeller Funktionen." Akademie-Verlag, Berlin.

Kato, J. (1980). An autonomous system whose solutions are uniformly ultimately bounded but not uniformly bounded. *Tohoku Math. J.* **32**, 499–504.

Kato, J. (1994). Stability criteria for difference equations related with Liapunov functions for delay-differential equations. *Dynamic Systems and Applications* **3**, 75–84.

Kato, J. (1996). A conjecture in Lyapunov method for functional differential equations. Proc. First World Congress Nonlinear Analysts, 1992. V. Lakshmikantham, ed. Walter de Gruyter, New York. pp. 1240–1246.

Kato, J., and Strauss, A. (1967). On the global existence of solutions and Liapunov functions. *Ann. Math. Pura. Appl.* **77**, 303–316.

Krasovskii, N. N. (1963). "Stability of Motion." Stanford Univ. Press.

Lakshmikantham, V., and Leela, S. (1969). "Differential and Integral Inequalities," Vol. I. Academic Press, New York.

Lakshmikantham, V., and Rao, M. Rama Mohana. (1995) "Theory of Integrodifferential Equations." Gordon and Breach, Lausanne, Switzerland.

Langenhop, C. E. (1973). Differentiability of the distance to a set in Hilbert space. *J. Math. Anal. Appl.* **44**, 620–624.

Lebedev, A. A. (1957). On a method of constructing Liapunov functions. *P. M. M.* **21**, 121–124.

Lefschtetz, S. (1957). "Differential Equations: Geometric Theory." Wiley, New York.

Lefschetz, S. (1965). "Stability of Nonlinear Control Systems." Academic Press, New York.

Leitman, M. J., and Mizel, V. J. (1974). On fading memory spaces and hereditary integral equations. *Arch. Rational Mech. Anal.* **55**, 18–51.

Leitman, M. J., and Mizel, V. J. (1978). Asymptotic stability and the periodic solutions of $x(t) + \int_{-\infty}^{t} a(t-s)g(s,x(s))\,ds = f(t)$. *J. Math. Anal. Appl.* **66**, 606–625.

Leslie, P. H. (1948). Some further notes on the use of matrices in population mathematics. *Biometrika* **35**, 213–245.

Levin, J. J. (1963). The asymptotic behavior of a Volterra equation. *Proc. Amer. Math. Soc.* **14**, 434–451.

Levin, J. J. (1968). A nonlinear Volterra equation not of convolution type. *J. Differential Equations* **4**, 176–186.

Levin, J. J., and Nohel, J. A. (1960). On a system of integrodifferential equations occuring in reactor dynamics. *J. Math. Mechanics* **9**, 347–368.

Liouville, J. (1838). (title unavailable). *J. Math. Pures Appl.* (1)**2**, 19.

Lotka, A. J. (1924). "Elements of Physical Biology." Reprint as "Mathematical Biology." Dover, New York, 1956.

Lotka, A. J. (1939). A contribution to the theory of self-renewing aggregates with special reference to industrial replacement. *Ann. Math. Statist.* **10**, 1–25.

MacCamy, R. C. (1977a). An integro-differential equation with application in heat flow. *Quart. Appl. Math.* **35**, 1–19.

MacCamy, R. C. (1977b). A model for one-dimensional, nonlinear viscoelasticity. *Quart. Appl. Math.* **35**, 21–33.

Makay, G. (1991). On the asymptotic stability in terms of two measures for functional differential equations. *Nonlinear Anal.* **16**, 721–727.

Makay, G. (1994). An example on the asymptotic stability for functional differential equations. *Nonlinear Anal.* **23**, 365–368.

Massera, J. L. (1949). On Liapunov's condition of stability. *Ann. of Math.* **50**, 705–721.

Massera, J. L. (1950). The existence of periodic solutions of a system of differential equations. *Duke Math. J.* **17**, 457–475.

Maynard Smith, J. (1974). "Models in Ecology." Cambridge Univ. Press, London and New York.

Miller, R. K. (1968). On the linearization of Volterra integral equations. *J. Math. Anal. Appl.* **23**, 198–208.

Miller, R. K. (1971a). "Nonlinear Volterra Integral Equations." Benjamin, New York.

Miller, R. K. (1971b). Asymptotic stability properties of linear Volterra integrodifferential equations. *J. Differential Equations* **10**, 485–506.

Myshkis, A. D. (1951). General theory of differential equations with retarded argument. *Amer. Math. Soc. Transl.* No. 55.

Natanson, I. P. (1960). "Theory of Functions of a Real Variable," Vol. II. Ungar, New York.

Nohel, J. A. (1973). Asymptotic equivalence of Volterra equations. *Ann. Mat. Pura. Appl.* **96**, 340–347.

Peirce, Charles S. (1957). "Essays in the Philosophy of Science." Bobbs-Merrill, New York.

Perron, O. (1930). Die stabilitatsfrage bei differential-gleichungssysteme. *Math. Z.* **32**, 703–728.

Picard, E. (1890). Mémorie sur la théorie des équations aux dérivées partielles et la méthode des approximations successive. *J. Math. Pures Appl.* (4)**6**, 145–210 (esp. pp. 197–210).

Picard, E. (1907). La mécanique classique et ses approximations successive. *Révista de Scienza* **1**, 4–15.

Pielou, E. C. (1969). "An Introduction to Mathematical Ecology." Wiley, New York.

Prigogine, I. (1967). "Introduction to Thermodynamics of Irreversible Processes," 3rd ed. Wiley, New York.

Reissig, R., Sansone, G., and Conti, R. (1963). "Qualitative Theorie Nichtlinear Differentialgleichungen." Edizioni Cremonese, Roma.

Rosenzweig, M. L. (1969). Why the prey curve has a hump. *Amer. Natur.* **103**, 81–87.

Rosenzweig, M. L., and MacArthur, R. H. (1963). Graphical representation and stability conditions of predator-prey interactions. *Amer. Natur.* **97**, 209–223.

Rudin, W. (1966). "Real and Complex Analysis." McGraw-Hill, New York.

Rutherford, D. E. (1960). "Classical Mechanics," 2nd ed. Oliver and Boyd, Edinburgh.

Sansone, G., and Conti, R. (1964). "Nonlinear Differential Equations." Macmillan, New York.

Sawano, K. (1982). Some considerations on the fundamental theorems for functional differential equations with infinite delay. *Funkcial. Ekvac.* **25**, 97–104.

Seifert, G. (1973). Liapunov-Razumikhin conditions for stability and boundedness of functional differential equations of Volterra type. *J. Differential Equations* **14**, 424–430.

Seifert, G. (1974). Liapunov-Razumikhin conditions for asymptotic stability in functional differential equations of Volterra type. *J. Differential Equations* **16**, 289–297.

Seifert, G. (1981). Almost periodic solutions for delay-differential equations with infinite delays. *J. Differential Equations* **41**, 416–425.

Seiji, Saito (1989). Global stability of solutions for quasilinear ordinary differential systems. *Jath. Japonica* **34**, 821–829.

Serban, M.-A. (2001). Global asymptotic stability for some difference equations via fixed point technique. Seminar on Fixed Point Theory, Cluj-Napoca **2**, 87–96.

Snyder, H. H. (1975). On simple models for field theories of certain electron-beam devices. *Internat. J. Electron.* **38**, 97–115.

Somolinos, A. (1977). Stability of Lurie-type functional equations. *J. Differential Equations* **26**, 191–199.

Troy, W. C. (1980). The existence of traveling wave front solutions of a model of the Belousov-Zhabotinskii chemical reaction. *J. Differential Equations* **36**, 89–98.

Volterra, V. (1913). "Lecons sur les équations intégrales et les équations intégro-differentielles." Collection Borel, Paris.

Volterra, V. (1928). Sur la théorie mathématique des phénomès héréditaires. *J. Math. Pur. Appl.* **7**, 249–298.

Volterra, V. (1931). "Lecons sur la théorie mathématique de la lutte pour la vie." Paris.

Volterra, V. (1959). "Theory of Functionals and of Integral and Integro-differential Equations." Dover, New York.

Wang, Tingxiu (1992). Equivalent conditions on stability of functional differential equations. *J. Math. Anal. Appl.* **170**, 138–157.

Wang, Tingxiu (1994a). Asymptotic stability and the derivatives of solutions of functional differential equations. *Rocky Mountain J. Math.* **24**, 403–427.

Wang, Tingxiu (1994b). Upper bounds for Liapunov functionals, integral Lipschitz condition and asymptotic stability. *Differential and Integral Equations* **7**, 441–452.

Wangersky, P. J., and Cunningham, W. J. (1957). Time lag in prey-predator population models. *Ecology* **38**, 136–139.

Yoshizawa, T. (1963). Asymptotic behavior of solutions of a system of differential equations. *Contrib. Differential Equations* **1**, 371–387.

Yoshizawa, T. (1966). "Stability Theory by Liapunov's Second Method." Math. Soc. Japan, Tokyo.

Yoshizawa, T. (1975). "Stability Theory and the Existence of Periodic Solutions and Almost Periodic Solutions." Springer Publ., New York.

Zhang, B. (1995). Asymptotic stability in functional-differential equations by Liapunov functionals. *Trans. Amer. Math. Soc.* **347**, 1375–1382.

Zhang, B. (1997). Asymptotic stability criteria and integrability properties of the resolvent of Volterra and functional equations. *Funkcial. Ekvac.* **40**, 335–351.

Zhang, B. (2001). Formulas of Liapunov functions for systems of linear ordinary and delay differential equations. *Funkcial. Ekvac.* **44**, 253–278.

Author Index

349

Subject Index

Mathematics in Science and Engineering
Edited by C.K. Chui, Stanford University

Recent titles:
I. Podlubny, *Fractional Differential Equations*
E. Castillo, A. Iglesias, R. Ruíz-Cobo, *Functional Equations in Applied Sciences*
V. Hutson, J.S. Pym, M.J. Cloud, *Applications of Functional Analysis and Operator Theory (Second Edition)*
V. Lakshmikantham and S.K. Sen, *Computational Error and Complexity in Science and Engineering*

Lightning Source UK Ltd.
Milton Keynes UK
UKOW05n0029090317

296199UK00014B/98/P